Human Perspectives in Health Sciences and Technology

Volume 1

Series Editor
Marta Bertolaso, Campus Bio-Medico University of Rome, Rome, Italy

The Human Perspectives in Health Sciences and Technology series publishes volumes that delve into the coevolution between technology, life sciences, and medicine. The distinctive mark of the series is a focus on the human, as a subject and object of research. The series provides an editorial forum to present both scientists' cutting-edge proposals in biomedical sciences that are able to deeply impact our human biological, emotional and social lives, and thought-provoking theoretical reflections by philosophers and scientists alike on how those scientific achievements affect not only our lives, but also the way we understand and conceptualize how we produce knowledge and advance science, so contributing to refine the image of ourselves as human knowing subjects. The series addresses ethical issues in a unique way, i.e. an ethics seen not as an external limitation on science, but as internal to scientific practice itself; as well as an ethics characterized by a positive attitude towards science, trusting the history of science and the resources that, in science, may be promoted in order to orient science itself towards the common good for the future. This is a unique series suitable for an interdisciplinary audience, ranging from philosophers to ethicists, from bio-technologists to epidemiologists as well to public health policy makers.

More information about this series at http://www.springer.com/series/16128

Marta Bertolaso • Fabio Sterpetti
Editors

A Critical Reflection on Automated Science

Will Science Remain Human?

Springer

Editors
Marta Bertolaso
Campus Bio-Medico University of Rome
Rome, Italy

Fabio Sterpetti
Department of Philosophy
Sapienza University of Rome
Rome, Italy

ISSN 2661-8915 ISSN 2661-8923 (electronic)
Human Perspectives in Health Sciences and Technology
ISBN 978-3-030-25000-3 ISBN 978-3-030-25001-0 (eBook)
https://doi.org/10.1007/978-3-030-25001-0

This Springer imprint is published by the registered company Springer Nature Switzerland AG.
The registered company address is: Gewerbestrasse 11, 6330 Cham, Switzerland

Foreword: The Social Trends Institute

The Social Trends Institute (STI) is a nonprofit international research center dedicated to fostering understanding of globally significant social trends. The individuals and institutions that support STI share a conception of society and the individual that commands a deep respect for the equal dignity of human beings, and for freedom of thought, as well as a strong desire to contribute to the social progress and the common good. To this end, STI organizes experts meetings around specific topics in its areas of priority study and brings together the world's leading thinkers, taking an interdisciplinary and international approach. Currently, these priority research areas are Family, Bioethics, Culture and Lifestyles, Governance, and Civil Society. The findings are disseminated to the media and through scholarly publications. Founded in New York City, STI, currently headed by Carlos Cavallé, Ph.D., also has a delegation in Barcelona, Spain.

This volume, *A Critical Reflection on Automated Science: Will Science Remain Human?*, is the result of one such expert meeting held in Rome in March 2018, under the academic leadership of Marta Bertolaso to explore the question "Will Science Remain Human?" This query is particularly suited to STI's multidisciplinary approach. To fully explore the challenges that arise from ever-increasing automation of science, STI gathered epistemologists, philosophers of science, moral philosophers, and logicians.

The results of their research are presented in this book. Without endorsing any particular viewpoint, STI hopes that as a whole, these contributions will deepen the readers' understanding of this very important question.

Secretary General
Social Trends Institute
Barcelona, Spain
2019

Tracey O'Donnell
TODonnell_STI@socialtrendsinstitute.org

Acknowledgments

We want to thank the Social Trends Institute (www.socialtrendsinstitute.org) for its generous financial and organizational support, which made possible the Experts Meeting "Will Science Remain Human? Frontiers of the Incorporation of Technological Innovations in the Bio-Medical Sciences" that gave rise to the structure and contents of *this volume*.

Marta Bertolaso
Fabio Sterpetti

Contents

Introduction. Human Perspectives on the Quest for Knowledge

Marta Bertolaso and Fabio Sterpetti

Introducing the New Series

This volume is the very first one to appear in the new Springer series *Human Perspectives in Health Sciences and Technology* (HPHST). It is first of all worth clarifying that the term 'human' in the series' title has not to be understood as synonymous with the term 'humanities' as it is usually understood in the extant literature. Indeed, the HPHST series is not another series devoted to what is usually referred to by the expression 'medical humanities', which nowadays is a quite precisely delimited disciplinary field.

The HPHST series aims to provide an editorial forum to present both scientists' cutting-edge proposals in biomedical sciences which are able to deeply impact our human biological, emotional and social lives, and thought-provoking reflections by scientists and philosophers alike on how those scientific achievements affect not only our lives, but also the way we understand and conceptualize how we produce knowledge and advance science, so contributing to refine the image of ourselves as human knowing subjects. The main idea that led to the creation of the series is indeed that those are two sides of the same 'being-human' coin: scientific achievements can affect both our lives and ways of thinking, and, on the other hand, a critical scrutiny of those achievements may suggest new directions to scientific inquiry. So, although epistemological, social, and ethical issues are certainly all central for the series, what distinguishes it from other already existing series dealing with

M. Bertolaso (✉)
Faculty of Engineering & Institute of Philosophy of Scientific and Technological Practice, Campus Bio-Medico University of Rome, Rome, Italy
e-mail: m.bertolaso@unicampus.it

F. Sterpetti
Department of Philosophy, Sapienza University of Rome, Rome, Italy
e-mail: fabio.sterpetti@uniroma1.it

M. Bertolaso, F. Sterpetti (eds.), *A Critical Reflection on Automated Science*, Human Perspectives in Health Sciences and Technology 1,
https://doi.org/10.1007/978-3-030-25001-0_1

similar issues, is that HPHST aims to address those issues from a less rigidly predefined disciplinary perspective and be especially attractive for non-mainstream views and innovative ideas in the biomedical as well as in the philosophical field.

The HPHST series aims to deal both with general and theoretical issues that spread across disciplines, such as the one we deal with in *this volume*, and with more specific issues related to scientific practice in specific domains. The series focuses especially, although not exclusively, on health sciences and technological disciplines, since technology, bio-medicine, and health sciences more in general are coevolving in unprecedented ways, and much philosophical work needs to be done to understand the implications of this process.

Technological development has always offered new opportunities for scientific and social advancement. In the last decades, technology has entered in intimate partnership with the life sciences, providing tools to isolate and modify experimental systems in vitro, offering computational power to expand our cognitive capacities and grasp features of complex living systems, then introducing digital models and simulations, providing methods to modify and 'rewrite' certain life processes and organismal traits, and allowing more daring and smooth hybridizations between artificial design and natural systems.

Incredible improvements to human life have come from these techno-scientific developments. Complex ethical questions have also arisen regarding the impact and the use of these developments. The life sciences have been shaking their paradigm and transforming the ways we conceptualize and deal with organisms and living systems, including humans (see e.g. Soto et al. 2016). Bio-medicine is also experiencing ever increasingly difficult challenges (see e.g. Ioannidis 2016). In fact, we live in a time of widespread worries and fears about science, and also of scepticism and resignation regarding the extreme complexity of living and social systems. Great expectations are placed on technology, and much of science is said to be technology driven, but human choices and responsibility remain ineliminable ingredients in the task of elaborating on the acquired knowledge and thus improving our scientific understanding about the natural world. A driving persuasion of HPHST series is that a new trust in science is possible, but it must be based on a sound and up-to-date epistemology, and the recognition of the inherent ethical dimension of science as a human endeavor: the only way to understand and govern science and communities for the better is a view of science rich in all its human aspects, requiring the contribution of philosophy, as well as natural and social sciences. Our hope is that the HPHST series can contribute to such a renewed trust in science and to a real improvement of it.

The Theme of the Volume

This volume originated from the conference "Will Science Remain Human? Frontiers of the Incorporation of Technological Innovations in the Bio-Medical Sciences," which took place in Rome, at the University Campus Bio-Medico, in

March 2018, thanks to the support of the Social Trends Institute. When working on *this book*, we decided to deal with the very same challenging question, namely whether science will remain human, but to focus a little bit more sharply, although not exclusively, on the issue of whether science will remain human notwithstanding the increasing automation of science. So, we thought, together with Floor Oosting, Springer Executive Editor of Applied Ethics, Social Sciences, and Humanities, whom we wish to thank for all her support and advice, that *A Critical Reflection on Automated Science – Will Science Remain Human?* would be an appropriate title for *this book*.

According to some philosophers and scientists, humans are becoming more and more dispensable in the pursuing of knowledge, since scientific research can be automated (Sparkes et al. 2010; King et al. 2009; Anderson 2008; Allen 2001). More precisely, according to some philosophers and scientists, the very aim of Artificial Intelligence today is not merely to mimic human intelligence under every respect, rather it is automated scientific discovery (Sweeney 2017; Colton 2002). Those claims are very appealing and ever more shared by many scientists, philosophers and lay people. Yet, they raise both epistemological and ethical concerns, and rely on assumptions that are disputable, and indeed have been disputed. From an epistemological point of view, consider, for instance, that assuming that scientific discovery can be automated means to assume that ampliative reasoning can be mechanized, i.e. that it is algorithmic in character. But proving that this is the case is not an easy task (see e.g. Sterpetti and Bertolaso 2018). From an ethical standpoint, consider, for instance, issues of responsibility on data management processes (which entail consistency of data organization and transmission with the original scientific question, contextualization of data, etc.) or on possible (and often unforeseen) consequences of completely automated researches. As an analogy, think of the difficulties we have in acknowledging limits of the majority of target therapies, as well as of the debate on who is responsible for deaths provoked by self-driving cars. So, although there is an increasing enthusiasm for the idea that machines can substitute scientists in crucial aspects of scientific practice, the current explosion of technological tools for scientific research seems to call also for a renewed understanding of the human character of science, which is going to be sometime less central but not for this less fundamental.

The topic *this volume* deals with, namely whether the computational tools we developed in last decades are changing the way we humans do science, is a very hot topic in the philosophy of science (see e.g. Gillies 1996; Humphreys 2004; Nickles 2018). The question that many are trying to answer is the following: Can machines *really* replace scientists in crucial aspects of *scientific advancement*? Despite its interest, it is a topic on which, to the best of our knowledge, one can find very few consistent works in the literature. This is why *this volume* brings together philosophers and scientists with different opinions to address from different perspectives the issue of whether machines can replace scientists. The book's aim is to contribute to the debate with valuable insights and critical suggestions which might be able to further it toward more reasoned and aware stances on the problem. Another feature of *this volume* that might be interesting for readers, is that it does not only deal with the issue of automated science in general, but it also focuses on biology and medicine, which are often ignored when abstract and general issues such as whether scientific research can be automated are discussed.

Most of the times, works that are devoted to the topic at stake try to support or deny the hypothesis that science can be automated from an 'engaged' perspective. On the contrary, *this volume* tries to scrutinize that hypothesis without prejudices or any previous theoretical commitment. There are indeed different opinions among the contributors to *this volume* on the automation of science, but overall the book integrates reasons for thinking that current computational tools are changing our way to do science, while they also ask for a deeper philosophical reflection about science itself as a rational human endeavor. This opens to innovative thoughts about the peculiar way humans know and understand the world and gives us reasons for thinking that the role of the human knowing subject in the process of scientific discovery is not deniable nor dispensable.

We decided that *this volume* would be the first one of the HPHST series for several reasons. First of all, *this book* shares the focus of the series on the interplay between the human subject and technology. As we said, between the great promises of technology and the great fears that technology may replace the human in driving important aspects of our life, to get the right attitude for future scientific practice we need to develop an adequate epistemology. To this end, the series invokes contributions from philosophy as well as natural and social sciences, and this is exactly what *this book* aims to provide. Also, *this volume* deals with the human subject as an object and a subject of inquiry, an approach that perfectly fits with the series' approach. Indeed, in order to claim that scientific discovery can be automated, scientific discovery needs to be clearly understood. And in order to understand scientific discovery, it is usually believed that human reasoning needs to be clearly understood. This means that we need to improve our understanding of a qualifying feature of human nature, i.e. reasoning, and of what do humans really do when they do science. Moreover, *this volume* deals with epistemological, ethical, and technological issues, and focuses specifically on the bio-medical sciences and technology. Finally, in accordance with the series' aims, *this volume* tries to provide both scientists' and philosophers' reflections on practical and theoretical aspects of science, and so to further our understanding of how we produce knowledge and advance science. For all those reasons, it was natural for us to think about this volume as the most adequate one to inaugurate the HPHST series.

Overview of the Volume

The book is divided into three parts. The first part, *Can Discovery Be automated?*, addresses the question of whether scientific discovery can be automated from a general and theoretical perspective. This part consists of five chapters.

The first chapter of this part is Paul Humphreys' *Why Automated Science Should Be Cautiously Welcomed*, which focuses on the notion of 'representational opacity', a notion Humphreys develops in analogy with the notion of 'epistemic opacity' that he introduced in previous works, in order to clarify in what sense the introduction of automated methods in scientific practice is epistemologically relevant. Humphreys argues for a moderately optimistic view of the role that automated methods can play in the advancement of scientific knowledge. He also draws some interesting parallels between the problem of scientific realism and the problem of internal representations in deep neural nets.

In the second chapter, *Instrumental Perspectivism: Is AI Machine Learning Technology like NMR Spectroscopy?*, Sandra Mitchell addresses the issue of whether something crucial is lost if deep learning algorithms replace scientists in making decisions, by considering whether the ways in which new learning technologies extend beyond human cognitive aspects of science can be treated *instrumentally*, i.e. in analogy with the ways in which telescopes and microscopes extended beyond human sensory perception. To illustrate her proposal, Mitchell compares machine learning technology with nuclear magnetic resonance technology in protein structure prediction.

In chapter three, *How Scientists Are Brought Back into Science – The Error of Empiricism*, Mieke Boon argues that despite machine learning might be very useful for some specific epistemic tasks, such as classification and pattern recognition, for many other epistemic tasks, such as, for instance, searching for analogies that can help to interpret a problem differently, and so to find a solution to that problem, the production of *comprehensible* scientific knowledge is crucial. According to Boon, such kind of knowledge cannot be produced by machines, since machine learning technology is such that it does not provide understanding.

In the fourth chapter, *Information at the Threshold of Interpretation: Science as Human Construction of Sense*, Giuseppe Longo investigates the origin of an ambiguity that led to serious epistemological consequences, namely the ambiguity concerning the use of the concept of 'information' in artificial intelligence and biology. According to Longo, there is still a confusion between the process of knowledge-production and that of information-processing. Science is dehumanized because information is thought to be directly embedded in the world. In order to avoid this shortcoming, we must distinguish between information as formal elaboration of signs, and information as production of meaning.

The fifth chapter of this part is Fabio Sterpetti's *Mathematical Proofs and Scientific Discovery*. Sterpetti claims that the idea that science can be automated is deeply related to the idea that the method of mathematics is the axiomatic method. But, he argues, since the axiomatic view is inadequate as a view of the method of mathematics and we should prefer the analytic view, it cannot really be claimed that science can be automated. Indeed, if the method of mathematics is the analytic method, then the advancement of mathematical knowledge cannot be mechanized, since non-deductive reasoning plays a crucial role in the analytic method, and non-deductive reasoning cannot be mechanized.

The second part of the book, *Automated Science and Computer Modelling*, deals with an analysis of the consequences of using automated methods that is more focused on biology, medicine and health technologies. In particular, some epistemological issues related to the role that computer modelling, computer simulations and virtual reality play in scientific practice are discussed. This part consists of five chapters.

Fridolin Gross's *The Impact of Formal Reasoning in Computational Biology* investigates the role played by computational methods in molecular biology by focusing on the meaning of the concept of computation. According to Gross, computational methods do not necessarily represent an optimized version of informal reasoning, rather they are best understood as cognitive tools that can support, extend, and also transform human cognition. In this view, an analysis of computational methods as tools of formal reasoning allows for an analysis of the differences between human and machine-aided cognition and of how they interact in scientific practice.

Emanuele Ratti, in his *Phronesis and Automated Science: The Case of Machine Learning and Biology*, supports the thesis that, since Machine Learning is not independent from human beings, as it is often claimed, it cannot form the basis of automated science. Indeed, although usually computer scientists conceive of their work as being a case of Aristotle's *poiesis* perfected by *techne*, which can be reduced to a set of rules, Ratti argues that there are cases where at each level of computational analysis, more than just *poiesis* and *techne* is required for Machine Learning to work. In this view, biologists need to cultivate something analogous to *phronesis*, which cannot be automated.

A Protocol for Model Validation and Causal Inference from Computer Simulation, by Barbara Osimani and Roland Poellinger, aims to fill in a gap, namely to give a clear formal analysis of computational modelling in systems biology, which is still lacking. To this end, they present a theoretical scheme, which is able to visualize the development of a computer simulation, explicate the relation between different key concepts in the simulation, and trace the epistemological dynamics of model validation. To illustrate such conceptual scheme, they use as a case study the discovery of the functional properties of a protein, E-cadherin, which seems to have a key role in metastatic processes.

Eric Winsberg's *Can Models Have Skill?* aims to determine whether the idea that a model has skill is a step in the direction toward post-human science by focusing on climate science. Winsberg considers the paradigm of verification and validation which comes from engineering and shows that this paradigm is unsuitable for climate science. He argues that when one deals with models of complex non-linear systems, the best one can find to justify such models is the modeler's explanation to his peers of why it was rational to use a certain approximation technique to solve a particular problem for some specific and contextual reasons. And this shows that science will probably remain human.

In *Virtually Extending the Bodies with (Health) Technologies*, Francesco Bianchini suggests an analogy between the extended mind thesis and a so-called extended body thesis, with particular respect to new technologies connected to health care. According to Bianchini, if one accepts the three main principles which characterize what extends the mind to make it something cognitive, one might wonder whether similar principles are valid for a new vision of the body, which is extended by interactive health technologies. In this perspective, boundaries between what cognition is and what is not could change, as well as boundaries between mind and body could become more blurred.

Finally, the third part of the book, *Automated Science and Human Values*, addresses some relevant ethical issues related to the automation of science and the scientific endeavor more generally. This part consists of four chapters.

The first chapter of this part is Benjamin Hurlbut's *Behold the Man: Figuring the Human in the Development of Biotechnology*. Accounts of what the human is in debates about biotechnology have mostly focused on the human as an object of technological intervention and control. But, according to Hurlbut, the human being is not separated from the ways of being human together. So, Hurlbut explores the question of whether science will remain human by taking the political conditions of possibility for asking that question as an object of analysis, attending to the ways those political conditions have been transformed in conjunction with the development of biological sciences.

The chapter *The Dehumanization of Technoscience* by Alfredo Marcos addresses the problem of the so-called dehumanization of technoscience, both at the time of its production and at the time of its application. The causes of this twofold dehumanization are found in an oversimplified ontology and in an erratic anthropology, swinging between nihilism and radical naturalism. As an alternative to such perspective, Marcos proposes a pluralistic ontology and an anthropology of Aristotelian inspiration. In this perspective, technoscience becomes valuable and meaningful when it is part of a wider human horizon.

Christopher Tollefsen, in *What is 'Good Science'?*, approaches the question concerning the human character of science from a rather different perspective. In his view, it is not only automation which threatens the human character of science. Recent controversies over scientific endeavors led some commentators to assert the impropriety of imposing moral limits on scientific inquiry. According to Tollefsen, those claims aim to minimize the human character of science too, since there is instead a deep relationship between good science and morality, which he analyses along three axes, namely the external ethics of science, the social ethics of science, and the internal ethics of science.

In the final chapter, *Cultivating Humanity in Bio- and Artificial Sciences*, Mariachiara Tallacchini offers some reflections on Hurlbut's, Marcos', and Tollefsen's different approaches to the human characters of science presented in previous chapters of this part. She underlines how, although the accounts of the potential threats to the humanness of science proposed in those chapters follow different trajectories, similar or complementary arguments run across their narratives, revealing that critical voices toward a potential loss of humanity through technology come from multiple and pluralistic perspectives, which are often under-represented in the public debate.

Overall, the contributed chapters in one way or another underline that *something* is missing in the view that science can be made a completely human-independent endeavor, and that philosophical reflection is required nowadays in order to reinforce our understanding of science itself. As it often happens in history, apparent threats turn into constructive challenges both at the individual and collective level. The initial question "Will science remain human?" should thus be reframed by deepening the aspects that humanly characterize scientific praxis, knowledge and understanding. We think that opening up and framing the issue in a wider context of debate is one of the valuable contributions of *this volume* that offers important clue for further researches.

References

Allen, J.F. 2001. In Silico Veritas. Data-Mining and Automated Discovery: The Truth Is in There. *EMBO Reports* 2: 542–544.

Anderson, C. 2008. The End of Theory: The Data Deluge Makes the Scientific Method Obsolete. *Wired Magazine*, 23 June.

Colton, S. 2002. *Automated Theory Formation in Pure Mathematics*. London: Springer.

Gillies, D. 1996. *Artificial Intelligence and Scientific Method*. Oxford: Oxford University Press.

Humphreys, P. 2004. *Extending Ourselves: Computational Science, Empiricism, and Scientific Method*. Oxford: Oxford University Press.

Ioannidis, J.P.A. 2016. Why Most Clinical Research Is Not Useful. *PLOS Medicine* 13 (6): e1002049.

King, R.D., et al. 2009. The Automation of Science. *Science* 324: 85–89.

Nickles, T. 2018. Alien Reasoning: Is a Major Change in Scientific Research Underway? *Topoi. An International Review of Philosophy*. https://doi.org/10.1007/s11245-018-9557-1.

Soto, A.M., G. Longo, and D. Noble, eds. 2016. From the Century of the Genome to the Century of the Organism: New Theoretical Approaches (Special Issue). *Progress in Biophysics and Molecular Biology* 122 (1): 1–82.

Sparkes, A., et al. 2010. Towards Robot Scientists for Autonomous Scientific Discovery. *Automated Experimentation* 2: 1. https://doi.org/10.1186/1759-4499-2-1.

Sterpetti, F., and M. Bertolaso. 2018. The Pursuit of Knowledge and the Problem of the Unconceived Alternatives. *Topoi. An International Review of Philosophy*. https://doi.org/10.1007/s11245-018-9551-7.

Sweeney, P. 2017. Automated Science as a Vision for AI. *Medium*, available at: https://medium.com/inventing-intelligent-machines/ai-bloatware-why-the-popular-vision-of-ai-is-misleading-ca98c0680f4.

Part I
Can Discovery Be Automated?

Why Automated Science Should Be Cautiously Welcomed

Paul Humphreys

Introduction

Does the increasing use of technology within science and the use of automated scientific methods have the potential to epistemologically harm the scientific enterprise? Here, by 'automated scientific method' I mean any scientific method the operation of which can be implemented without the intervention of humans other than to initiate and, if necessary, train the process.[1] While there are legitimate concerns about the role of technology in science, I am going to argue here for conclusions that lie somewhat but not wholly counter to the stated purpose of the workshop in which this paper originated.[2] Few philosophers working outside the phenomenological tradition would argue that the addition of artificial devices to the arsenal of scientific research tools is something that should be discouraged, not the least because technologically based methods are now indispensable to a number of areas in science and crippling science is in nobody's interest. A more interesting angle is to obtain some sense of how the use of technologically based methods can shift the epistemology of science and whether those shifts are beneficial or detrimental.

Although there are important moral questions that stem from the use of technology in society, the primary concern within science is epistemological–will we harm

[1] The operation of the automated process may or may not include interpretative processes. The honorific 'scientific' in 'automated scientific method' does not entail that the method in question will be effective, accurate, reliable, or reproducible. Nor does the fact that a method is automated entail that it can be effectively used without significant human input before the process is initiated.

[2] *Will Science Remain Human?* University Campus Bio-Medico, Rome, March 5–6, 2018.

P. Humphreys (✉)
Corcoran Department of Philosophy, University of Virginia, Charlottesville, VA, USA
e-mail: pwh2a@virginia.edu

M. Bertolaso, F. Sterpetti (eds.), *A Critical Reflection on Automated Science*, Human Perspectives in Health Sciences and Technology 1,
https://doi.org/10.1007/978-3-030-25001-0_2

the knowledge seeking enterprise of science by moving too far in the direction of automated science? Lurking in the background to this question is an argument that is often implicit in discussions of this topic. The argument is that because science is an activity done by and for the benefit of humans, the ultimate epistemic authority in science should always be humans.[3]

We should not underestimate the attraction of anthropocentrism, one indication being that there is no generally accepted antonym for the term in English. But that attraction is not unjustified, for there are powerful philosophical and scientific reasons why one might want science to remain a primarily human activity. If human input is lessened, perhaps that will reduce the kind of thoughtful and creative inputs that are currently essential to good science, and so an unconstrained enthusiasm for these new technological tools is unacceptable. Any scientific development can be used in a formulaic and mindless way; skillful uses on the other hand bring great benefits, one example being in relieving scientists from drudge work using automated genomic analysis. There will continue to be thoughtless applications of automated tools, just as there are untutored applications of econometric methods and statistical packages for social sciences, as well as naïve applications of digital humanities tools within history and literature. All of these things are properly to be discouraged.

Some Advantages of Automated Science

In specific applications one can quantitatively demonstrate the advantages of automated methods over traditional methods. For example, as far back as 2001, many machine learning methods, including naive Bayes, neural networks, and k-nearest neighbor methods outperformed clinicians on classifying single cases of ischaemic heart disease in both positive and negative classifications, where the positive and the negative diagnoses are taken to be reliable if the predictive success of presence or absence of the disease, respectively, is greater than 0.90.[4] The accuracy of machine learning predictions of cancer susceptibility, recurrence, and mortality improved by 15%–20% over physician alone assessments in the early 2000s. (Kourou et al. 2015, p. 9). The discovery of the Higgs boson would not have been possible without automated event detectors and the design of the ATLAS detectors themselves was heavily dependent upon computer simulations of what kinds of data would be received in the real experiments.[5] A final example is the collection and processing of data in

[3]Although I shall not argue it here, retaining moral responsibility by humans over the pursuit of science and the use of its products is essential. The epistemological and moral aspects cannot be completely separated but enough of a separation can be maintained to sustain a reasonable set of arguments about the former domain.

[4]"Compared to physicians the naive Bayesian classifier improves the number of reliably classified positive cases by 17% and the number of reliably classified negative cases by 37%" (Kononenko 2001, p. 105).

[5]These uses are not new. Such uses of neural nets in high energy physics go back more than two decades. For more recent methods see Baldi et al. (2014).

contemporary astrophysics. The Square Kilometer Array radio telescope, assuming it is built, will produce around 10^{19} bytes of data per day, one hundred times greater than the LHC, requiring 100 petaflops of computation to process. That type of data analysis would be impossible to analyze if humans were in the loop.

Styles of Automated Representation

Styles of reasoning, a conceptual framework introduced by Ludwig Fleck (Fleck 1979) and elaborated by the historian of science A.C. Crombie (Crombie 1981) and the philosopher Ian Hacking (Hacking 1992), concerns modes of representation, reasoning, and evidence assessment that are characteristic of an intellectual domain. Canonical examples of styles of reasoning are statistical models and inference, proof theoretic methods in mathematics, and laboratory experimentation. That literature is largely about reasoning, but the approach can be generalized. One of the most remarkable features of the human mind is its ability to extend its ways of representing and thinking about various subject matters. Some historical examples of these extensions have been the introduction of non-Euclidean geometries, the use of computer-assisted proofs, the acceptance of non-representational art, our understanding of non-native cultures, and the development of new moral doctrines. Since automated scientific methods can involve new ways of representing the world, we have the question: are there distinctive styles of representation that are involved in automated science and if so, can humans adapt their cognitive resources to understand them?[6]

One possible approach to this question is to examine domain-specific expertise and to ask: in which areas are humans superior to artificial devices and in which areas are they inferior? There are some informative answers to those questions, but unless there are human epistemic activities that by their nature could never be reproduced or surpassed by artificial devices, these answers would inevitably be dependent on the state of technological development. Such eternally superior human abilities may exist; perhaps one example is the kind of craft knowledge that expert experimentalists have. One of the current problems with automated image classification is unpredictable failures. Having successfully separated images of humans from images of monkeys, if you give the monkey a guitar, the output of some programs will classify it as a human. Occlude part of an image of a horse and many machine learning programs will be unable to classify it, whereas humans are very good at identifying types of objects only parts of which are visible. So here we have an example of a domain in which humans are currently superior to extant technology.

Concerns about such unpredictable failures are addressed in various versions of the Precautionary Principle, which in its European Union form reads in part "When human activities may lead to morally unacceptable harm that is scientifically plau-

[6] For some earlier thoughts about styles of reasoning and automated science, see Humphreys (2011).

sible but uncertain, actions shall be taken to avoid or diminish that harm." That principle refers to moral harm. Given that I set aside such questions at the outset of this paper, without denying their importance, what is needed here is an epistemic principle, which might be formulated by replacing 'moral' with 'epistemic' and requiring that error rates be established for and permanently attached to releases of scientific software.

Nevertheless, if there is one thing we should have learned from the last 50 years of technology, it is that one should be very careful indeed when claiming that artificial devices cannot perform a given task, whether cognitive or operational. So, a useful alternative approach is to ask instead: What epistemic value do the modes of representation used in automated science contribute to the scientific enterprise and what are their epistemic dangers?

Two Views on Science

To address this question, we need to separate two views about science. The first, *the intellectual view*, is the view that science is an activity, the primary purpose of which is to provide knowledge and understanding of the world. The second, *the practical view*, is that science has as its primary purpose providing the means to predict and manipulate the world. Because providing knowledge of the world can in many cases lead to predictive and manipulative success, the difference between the two views can be a matter of methodological emphasis, but they are in tension with one another and, in some versions, the practical view denies that the intellectual view should or can be pursued.

We can now identify four central concerns about the negative aspects of automated science. One possible concern from the intellectual view is that easing humans away from the centre of science could result in a reduction or loss of understanding of the world if we are unable to understand the products or processes of automated scientific investigations. Call this *the worry about understanding*. There is a second worry emanating from the intellectual view, which is that the likelihood of errors, many undetected, will increase if humans do not understand the relevant science. Call this *the worry about error*. Third, also stemming from the intellectual view, is the concern that a reduction in human input could result in a loss of potential applications because if we cannot understand the science, the number of applications will be reduced. Call this *the worry about applications*. Finally, we have the *worry about creativity*–the concern that to effectively do science requires creativity and machines will never have that feature. I shall not discuss this last worry here, despite its importance.

These are all genuine worries but with respect to the first one we ought to allow for a kind of scientific knowledge, under an expansive conception of knowledge, that is beyond the current understanding of humans. The lack of understanding of carcinogenesis for many types of cancer, of the mechanism of action underlying aspirin, and of the etiology of cholera, did not prevent effective prediction of these phenomena nor the successful manipulation of systems involving them. More theoretically, chemistry advanced before the nature of the chemical bond was understood and quantum mechanics has been a predictively successful part of science for 90 years without being able to represent the measurement problem in a way that allows for a satisfactory explanation.[7] As a selective realist, I would argue that practical effectiveness generally improves when we discover previously unknown explanatory features but members of the practical school generally reject the need for such realist arguments and so our first and third types of worry are less of a problem under the practical view. The worry about error remains a concern under both views because sudden and unpredictable failures of automated methods can occur when the conditions for stability of the methods are not met.

Epistemic Opacity

In Humphreys (2004, 2011) I discussed the concept of epistemic opacity. The introduction of computer simulations seemed to have introduced a distinctively new style of reasoning that involved nonhuman agents and my original concern was with the move away from the explicit and transparent justifications of traditional mathematically formulated scientific theories and the human inability to follow the details of computer simulations. The first three worries about epistemic losses in science mentioned above are most noticeable when the scientific representations used are beyond human understanding. This is because explanation, upon which much understanding is based, is considered to be, at least in part, an epistemological activity and thus puts anthropocentric constraints on explanations. To go beyond these constraints is therefore to transgress the limits of explanation. One example of this occurs in neuroscience with the so-called 'explanatory gap between brain processes and phenomenological content, where it is difficult and perhaps impossible to match the representations for the qualitative contents of conscious experience with the representations used for neural processes.[8] Arguments that this situation is permanent have been given in Noë and Thompson (2004).

Returning to epistemic opacity, a process is epistemically opaque relative to a cognitive agent X at time t just in case X does not know at t all of the epistemically relevant elements of the process. Examples of epistemically relevant elements of a process are the justification for moving from one step in the process to another, the representational correctness of semantic elements of the process, and the defeasibility of the process. A process is essentially epistemically opaque to X if and only if it is impossible, given the nature of X, for X to know all of the epistemically relevant elements of the process.

[7] The Bohmian theory can avoid the measurement problem but it is a different theory than standard quantum mechanics.

[8] I am assuming here that such phenomenological content is representational, at least in part. One can reasonably disagree.

The original thrust of epistemic opacity was that due to the speed, model complexity, and richness of data found in computational methods, humans are unable to certify in detail that the computational processes being used are a warrant for trusting the outputs of computer simulations. Epistemic opacity is central to scientific concerns about predictions from climate models, concerns about it are up front in research into program verification techniques, and related worries accompanied the early and (in some areas) continuing resistance to computer assisted proofs in mathematics. Although those sciences are not completely automated, enough of their methods are to warrant the worries noted above.

Representational Opacity

There is a related concept that is often present when epistemic opacity occurs, the concept of *representational opacity*. Although this is a very general feature of many computationally based methods, it plays a central role in the contemporary field of explainable artificial intelligence. As machine learning methods have permeated many areas of artificial intelligence, their successes have often been accompanied by both a high degree of epistemic opacity and a degree of representational opacity.

A definition of representational opacity is given in this section but, to provide a focus, I shall use machine learning methods as a running example. As background history, one can roughly divide the history of artificial intelligence into three phases. The first somewhat arbitrarily dates from the 1956 Dartmouth AI workshop to the late 1980s.The second began with the resurgence of connectionism, stimulated by the publication of Rumelhart and McClelland's PDP books in 1987, and lasted for almost 30 years.[9] The third phase, which is ongoing, began around 2012 when improvements in computer hardware allowed the practical implementation of deep neural nets (DNN). I provide this brief history because some of the most prominent applications of machine learning are still in their infancy and novel technologies usually are far from error-free. So criticisms based on current problems may not extend to future versions of these methods.

In recent years deep neural networks have achieved significant levels of success in image recognition and image labeling.[10] While far from perfect, and unpredictable failures can occur in specific situations, success rates on classification recognition over controlled image sets is currently over 95%.[11] While predictively successful, how these types of neural nets achieve their predictive success is not fully understood. This has led to the push for what is called 'explainable artificial

[9] See Rumelhart and McClelland (1987).

[10] Convolutional neural nets, currently the most successful at image recognition, became effective in the 2010s because of increases in computational power.

[11] For a comparison with humans see: www.karpathy.github.io/2014/09/02/what-i-learned-from-competing-against-a-convnet-on-imagenet/, accessed July 5, 2018.

intelligence'. I note that we can be interested in whether artificial neural nets applied to image recognition have internal representations and if so, what type of representations are being used, without adjudicating the question of whether neural nets are good models of the human brain or of cognition. There is no sharp division between shallow and deep neural nets, although it is currently common to call a neural net with a single hidden layer 'shallow' and a neural net with more than one hidden layer 'deep'. Although in principle two-layer nets can replicate any task that a DNN can perform, limitations on the width of the layers makes the use of DNNs unavoidable in many cases.

At the most basic level, deep neural nets are function approximation devices, a fact that partly accounts for their flexibility and their predictive power. So one kind of internal representation is simply a numerical function between the input values and a probability distribution over the classification types. The Universal Approximation Theorem asserts that any measurable continuous function on the unit cube can be approximated arbitrarily closely by using a weighted sum of sigmoidal functions applied to filtered input values together with a bias function. (Cybenko 1989). Under the function approach, we have a compositional view of representation, where what is composed are functions between each layer in the DNN and the output of each function except the last, given a specific input, numerically represents the activation level at a node (or groups of nodes). But the functional approach is open to trivialization–there is always some function connecting inputs to the probability distribution over the outputs. Furthermore, most such functions are representationally opaque–the functions between layers are nonlinear and the composed functions are so complex that interpreting and understanding these functions in the form in which the DNN uses them is beyond the reach of humans. So, valuable as the functional approach is for the practical view, it does little for the intellectual view.

Problems with Automated Science

These questions about representational opacity take us to the second part of our discussion, the dangers of technologically augmented methods. We can begin by noting that an anthropocentric attitude towards epistemology was until recently embedded in many philosophical positions. It was sometimes overt in the commitment to empiricism, an epistemology squarely based upon human perceptual abilities. In other areas, such as computability, the influence was subtler. Gandy (1980) argued that the concept of Turing computability was defined in terms that were based on conformity to elementary human abilities in computation.[12] We are now in a post-empiricist age and once we move away from these older attitudes, the focus has changed from observations to data, resurrecting the empiricist hope that data from instruments is objective in ways that human observations are not. But that

[12] For an assessment of Gandy's claim see Shagrir (2016).

hope cannot be fulfilled fully in practice. The manipulation of data by humans is common in large data sets even when most processing is done computationally, and this undermines to a certain extent the objectivity of the data. Whereas what helped the demise of empiricism were arguments to show that observations were always interpreted within a theoretical framework, within data processing the problem is that human decisions can lessen the objectivity of the outputs.

The ways in which observations must usually be described in partly theoretical terms is a staple of undergraduate courses in the philosophy of science and there is no need to dwell on the arguments for that conclusion. The extent to which those theoretical dependencies frame and bias observations is a more complicated matter. Automated methods are sometimes thought to be freer of such influences, but even purely automated methods usually have some kind of human input and can thereby acquire similar biases. The need or lack thereof for human input in this area can be illustrated by the standard difference in machine learning between supervised and unsupervised learning. In supervised learning, the training examples are labeled by humans and the backpropagation learning algorithms are adjusted using success criteria related to correct classification of the inputs, where success means conformity to the usual human classifications. Any human biases that apply to the initial image classification are likely thereby to carry over to the training processes for the algorithms and the philosophical concerns about biases in observations thus carries over to this aspect of machine learning.

There is now a significant amount of evidence that biases can affect the outputs of image recognition algorithms because of the initial choice of training sets. In a much publicized case, Google Photo App misclassified photographs of two African Americans as gorillas in an image classification task because of the small number of African-American images in the training database.[13] The important thing is to recognize the biases so that they be corrected. However, there is an important difference between human and algorithmic biases. Human biases, even when subtle, can often be identified because other humans share that bias, whereas the various biases in algorithms and databases are usually statistical rather than psychological or sociological and, when subtle, are difficult to identify. The most dangerous biases are the assumptions that are so integral to a dominant research program that its practitioners are unable to recognize their existence or to identify what those assumptions are. Such a situation will carry over even to automated science when the methods are controlled by humans.

Objectivity for databases is different from objectivity with respect to traditional observations. With automated data processing there is no such thing as 'raw data' and it is common to clean and pre-process the data in order to improve predictive and classificatory accuracy. There is perhaps a small irony in the fact that altering the data often results in better science than does taking the data as it comes.[14] Because of the quantity of data in very large data sets, pre-selection of which data are collected is also common; most are just discarded or not collected. This is com-

---

[13] For other examples of algorithmic bias see Zarsky (2014).

[14] For examples in biological data sets, see Leonelli (2016).

mon with space-based telescopes where bandwidth limitations force selection of the data before it is transmitted to ground observatories. A greater concern is that such filtering of algorithmically collected data may leave science less open to genuinely novel and unanticipated discoveries than traditional, human supervised observations. Human choices are involved in all such filtering procedures. Dimensional reduction, often forced on us because of computational limitations in very high dimensional data sets, also often involves choices by humans as to what is salient in the data.

Types of Representation

There are many different types of representations, but I shall focus on the distinction between transparent and opaque representations. In the transparent type, we represent the states of a system in a way that is open to explicit scrutiny, analysis, interpretation, and understanding by humans, and transitions between those states are represented by rules that have similar transparent properties. A representation that is not transparent is an opaque representation. Whether or not a representation is opaque can be a temporary matter, for advances in knowledge can allow us to understand previously opaque representations. Discovering the key to an encrypted message is a simple example. We must also distinguish between opaque representations and nonreferring formalisms. An opaque representation refers to something, but we humans do not understand its representational content. A nonreferring formalism is an element of some interpreted formal system that picks out nothing in the intended domain of the formal system. As what I hope is only a mildly stipulative move, I require that all representations within DNN refer and so nonreferring formalisms are not representations of either kind.[15] Nonreferring formalisms can occur in scientific derivations and thus not all elements of the intermediate steps in a derivation need be representations. All that is required is that the transformations be mathematically legitimate. To take a simple example, if I use the third derivative of a function as a mathematical convenience in a scientific derivation, even though that third derivative does not refer to anything real, it is a legitimate and understandable element of the predictive process. Hence it is possible to understand a transformation without understanding what it is applied to. With this example, we can see what is at issue. Steps in the computational process are simply transformations of one state into another. So the question becomes: What transformations are permissible in arriving at an effective predictive process? A possible answer is: Any transformation that maps the referential content from one representational space to another representational space and that preserves the referential content of the initial representation is permissible. This is a sufficient condition only; other modes of permissible transformation are possible.

[15] This entails that misrepresentations are not representations. Although that entailment is false in general, in the base of image recognition methods, it is not unreasonable as a simplifying move.

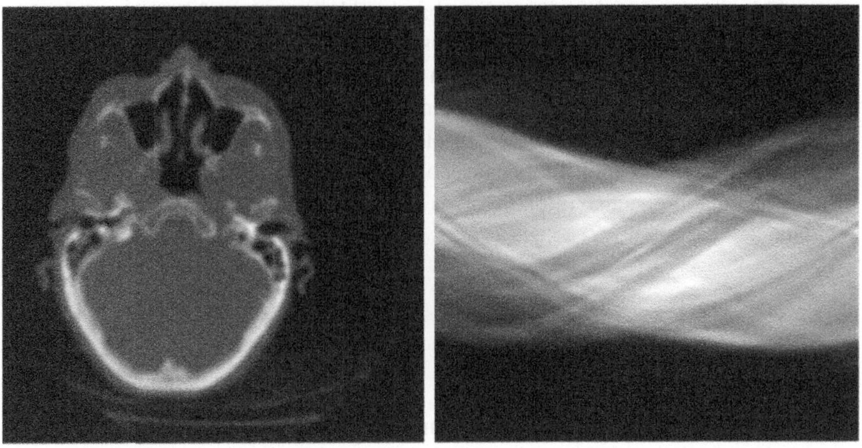

Fig. 1 Left: Computer Assisted Tomography image of a human skull. Right: Radon transformation of left hand image. (Images reproduced under Creative Commons License CC BY-NC-SA 3.0. Source: Nasser M. Abbasi, "Computed Tomography Simulation Using the Radon Transform", http://demonstrations.wolfram.com/ComputedTomographySimulationUsingTheRadonTransform/, on-line since: July 25, 2011, accessed July 5, 2018)

An example of such content preservation is the use of sinograms in computed tomography images. The representational content of the left hand image in Fig. 1 is familiar. The right hand image is the result of applying a Radon transformation to the left hand image. Because the original image can be recovered from the right hand image by applying an inverse Radon transformation, this is good evidence that the initial Radon transformation preserves the referential content.[16]

There is a sense here that the representational content of these two images is the same, it just happens that one is more amenable to machine interpretation and the other to human interpretation. But representational content is not always preserved under transformations. Rearranging the order of words in a sentence usually destroys the meaning and hence the representational content. Here is another, perhaps more interesting, example. The photographer Pavel Maria Smejkal has constructed a series of images called 'Fatescapes' that consists of historic photographs from which the iconic content has been removed. In one example, he has removed the images of the soldiers and the flag in the famous photograph of Marines raising the American flag at Iwo Jima, leaving only an image of a barren landscape.[17] The altered photograph can be arrived at by a straightforward set of transformations on a digital version of the original photograph. However, the content of the transformed photograph is essentially different from the content of the original – it is no longer an image of a flag raising.

[16] I note that in clinical applications of tomography the inverse transformation is applied not to the Radon transformation of an image but to the projections of densities in the specimen. For details of this example see Humphreys (2013).

[17] See: http://www.pavelmaria.com/fatescapes05.html, accessed July 5, 2018.

The familiar linguistic representations of the humanities and the formal representations of the sciences are usually transparent. One of the chief virtues of the axiomatic approach to theories is its explicit laying out of fundamental principles and the reduction of all knowledge in an area to those fundamental principles. In addition to theories, scientific models are also often transparent, as when a sequence of tosses of a coin is modeled by a Bernoulli process. Each part of the model - the independent tosses, the constancy of probability from toss to toss, and so on - is explicitly represented and understandable. Because some of the contemporary methods of machine learning, such as convolution neural nets and recurrent neural nets, use representations that are currently opaque and have features that do not correspond to familiar linguistic concepts, we are faced with the question of whether some of those representations are permanently unknowable by humans and whether some of the methods are representation-free. If so, the current quest for explainable AI is quixotic.

It is never wise, in the absence of an impossibility proof, to predict that something cannot be done and developments in machine learning are happening too rapidly to make definitive claims but perhaps the 'intuitions' developed by the AlphaZero chess playing software are an example of an essentially machine-like representation. AlphaZero plays the game at a very high level by starting with nothing but the rules of chess and playing millions of games against itself. The classification of positions within such programs is not transparent and unlike many other computationally based chess playing programs, AlphaZero seems to perform differently from both humans and more traditional chess playing software.

Representational opacity is an old phenomenon. A good example is Nicod's Axiom for propositional logic. Nicod discovered that he could replace the five axioms in Whitehead and Russell's Principia Mathematica with a single axiom from which all theorems of propositional logic follow. He did this by using as a primitive connective the Sheffer stroke | (which means 'not both … and …'):[18]

$$\big(A \mid (B \mid C)\big) \mid \Big(\big(D \mid (D \mid D)\big) \mid \Big(\big(E \mid B\big) \mid \big((A \mid E) \mid (A \mid E)\big)\Big)\Big).$$

In contrast to the original Principia axioms, which are all transparent, the truth to which Nicod's Axiom refers is very difficult for humans to interpret. Because Nicod's system is complete, there are transformations from Nicod's Axiom that will recover each of Principia's axioms, but those transformations do not help us understand the content of the axiom itself.

Concerns about representational opacity will differ from field to field. Consider the issue put in terms of inputs and outputs. The worry about representational opacity in pure mathematics is understandable. The inputs are the axioms (or previously proved theorems) and the output is the conclusion at hand. Note that we can identify each of the intermediate steps as valid, even though the individual representations are opaque. Within mathematics, nonreferring formalisms will not refer to anything,

[18] The single inference rule of Nicod's axiomatization is: from A|(B|C) and A infer C.

but they must be open to inspection in order for the proof to count as a proof. The intermediate steps could also be referentially opaque in the sense that humans were unable to interpret them, but a purely formal inspection of the proof could show it to be valid. Many proofs in Nicod's system would thus have steps that were both representationally opaque and contained nonreferring formalisms. The emphasis in the logical tradition has been on valid transformations, those that are necessarily truth preserving. In the computational realm, we need to examine which transformations are reference preserving.

There are imperfect parallels here with the problem of scientific realism. Scientific representations were needed to give us access to the realm beyond human observational capacities. Because those representations were constructed by humans, they were, for the most part, representationally transparent, with others turning out to be nonreferring formalisms. The interface between the representations and the 'unobservable' world was, again in many although not all cases, provided by scientific instruments. Those instruments were constructed by humans and so are, to a certain extent, understood. It was those instruments that told us, via much interpretation, whether the theoretical representations were referring or nonreferring. In other cases, where instruments were unavailable or unsuccessful, the decision about referentiality had to be made on the basis of inferences from data.

How much like this scenario is the problem of representational opacity in deep neural nets? One positive dimension of the analogy is that humans designed and constructed the computational hardware, together with the general algorithmic structure of the software. Another is that in some cases, the activation levels of the network can be directly probed and measured. But the most significant negative dimension is that the details of the internal representations are constructed by the computer, not by humans. This immediately opens the prospect of representational opacity and as a consequence, inferences from the output data about the internal structure of those representations is often profoundly difficult. These difficulties include two distinct problems. One is whether those internal representations refer to aspects of reality, but aspects of reality to which humans are not accustomed, and perhaps not even able to interpret. The other is whether parts of the internal processes, or even the whole, are nonreferring and are artifacts of the computational processes. Similar questions have long dogged machine learning methods used in topic modeling for textual analysis.[19]

For traditional scientific predictions, the inputs and outputs are statements involving observables and we require the intermediate steps, which usually involve theoretical statements, to be interpretable and understandable in order for the prediction to be justified, hence the traditional problem of scientific realism. The situation is more complicated than this with machine learning. With the techniques we have discussed, the question is whether there is latent structure hidden in the source that can be uncovered, or whether the internal states are simply artifacts that are useful for making predictions. Zeiler and Fergus (2014) show that by deconvolving the

[19] For a discussion of the problem with respect to digital humanities see Alvarado and Humphreys (2017).

patterns in each layer, the representations in the hidden layers can be projected back on to the input space giving a 'natural' spatial representation of those patterns.

Humans made progress in the scientific realism case by developing scientific instruments that allowed us access to some of the theoretical entities. In the development of those instruments, methods had to be developed to provide interpretable representations of the unobservable entities. The situation is rather different in the machine learning situation. What is of interest is whether patterns of activation in the neural nets represent real but humanly unobservable properties in the source, whether that source is an image, a text, or some other type of object or whether they are just nonreferring bits of formalism construed as patterns. One answer is that the functional view we discussed earlier is correct B abstractly there is an unlimited number of patterns that could be extracted from the data – and that the patterns of activation within the hidden layers do refer but are for the most part opaque. A primary constraint on identifying patterns is that they be identifiable by humans as corresponding to a feature that humans can recognize as being nonarbitrary. This is why mapping the pattern on to a space that humans find familiar, as did Zeiler and Fergus, is so effective in distinguishing real from arbitrary features.

This question of arbitrariness is connected with the distinction between extensional and intensional representations. The familiar devices of vectors, lists, arrays, matrices, and the like are extensional representations in that they do not provide a compact term that captures what all the elements have in common, or what relations exist in the data. They simply provide N dimensional sets of data, where $N > 0$. The situation is complicated a little by the fact that everything is finite, so we cannot use a cardinality argument to show that there are extensional representations that do not have an intensional counterpart. Yet it should be clear that there are extensional representations that do not correspond to any existing visually or linguistically recognizable feature.

One of the central epistemological problems in this area thus concerns the difficulty for humans of understanding models oriented towards the needs of computational devices, or the lack of models altogether in dealing with very high dimensional data sets. Recalling our discussion of styles of representation, this is not entirely a new problem. Humans have previously addressed and solved a similar issue when mental representations were replaced with formal representations and those formal representations had interpretations that deviated significantly from everyday conceptual frameworks. Given previous successes in transitioning from one mode of representation to another, there is some hope that a transition could be made from explicit formal representations to other types of representations. The new challenge is in using representations that are tailored to computational efficiency, are often essentially statistical in nature, are often justified in terms of Shannon information transfer and mutual information measures, and that may not be translatable into more familiar, anthropocentric representations. At best these models are implicit and it may be that we should abandon the modeling approach entirely and replace it with information-based approaches. It may even be that the concept of a model is not the right representational framework for data centered science, at least not in the sense that we usually consider models. It was once claimed that given enough data, we can do without explicit models in science. That view has lost adherents but a variant on that view, or even a representation free approach, should be considered.

Reliabilism

Another concern may be that delegating the justification of our science to instruments requires relying on testimony rather than on direct evidence. The use of knowledge on the basis of authority has a long history, from using tables of logarithms and special functions, through the use of mathematical results in science that are taken on authority by scientists to have been established by mathematicians, to relying on the outputs of instruments we do not understand in full detail. These are all examples of relying on epistemic authorities. In terms of data, there are now open databases in astronomy, genomics, and other areas in which scientists can deposit their data for others to draw upon. The sources of a particular data set are important in deciding whether to use those data, so epistemic authority is playing a role in those cases. To address these situations we can appeal to reliabilist approaches to knowledge. A common form of reliabilism asserts that an individual S knows that p if and only if p is true, S believes that p, and a reliable process forms the belief that p, where p is a proposition and a reliable belief-producing process is one that produces a high proportion of true beliefs. This definition needs to be modified to apply to automated methods in the following way: An instrument I has the knowledge that F if and only if I contains a representation R of the entirety of F, R holds of the target, and a reliable process forms the representation R, where F is a fact and a reliable representation-producing process is one that produces a high proportion of accurate representations. Reliabilism in general, and this appeal to accurate representations in particular, makes good sense in cases in which the representations are accessible to humans. Such cases form the domain of most ordinary, anthropocentric epistemology as well as many cases of traditional representation-based AI. But reliabilism becomes problematical in situations in which the truth or the accuracy of a representation is unknown, because the representation is opaque. Here is where the worry from error is particularly pressing. One of the advantages of using artificial neural networks as an example rather than human brains is that we understand them better, even if that understanding is incomplete, and there is no conscious perceptual content in those networks.to confuse matters.

For supervised learning, there is some, albeit tentative and partial, evidence that the representations occurring within these neural nets are compositional and that the components used by successive layers correspond in a reasonably natural way to familiar concepts that are understandable by humans. For example, pixels are combined into edges, edges into shapes, shapes into objects. The representations are also atomistic in the sense that textons serve as the representational atoms from which more complex structures are built. Here, textons are "fundamental microstructures in natural images and are considered as the atoms of pre-attentive human visual perception" (Zhu et al. 2005, p. 121).

Despite this preliminary success, there remain difficult questions about how to interpret the correlations that underlie the correspondence claims, how generalizable the results are, whether this compositionality is a function of the number of layers in the network, and whether there are hidden anthropocentric assumptions in the methods. More generally, the question is how representative these methods are of future developments in automated science.

If we cannot find a way to justify reliabilist approaches in such cases, it may be necessary to fall back upon something like Tyler Burge's entitlement approach to computer assisted mathematics within which some sources of data, such as normal visual perception, are taken as primitive epistemic authorities (see Burge 1998).

Conclusion

Much, perhaps too much, has been made of the dangers of artificial intelligence. The concern about automation producing mass unemployment may or may not be genuine – we have survived McCormick's reaper, Ford's assembly line, steam shovels, and MOOCs – and there are more pressing concerns abroad today than the likelihood that malevolent robots will take over the world. But the automation of practical and theoretical knowledge together with its inscrutability is something genuinely new. If we humans cannot understand the representations used by machine learning, or precisely how they are arrived at, the prospect of unintended consequences or unpredictable consequences of the programs is raised considerably.

References

Alvarado, Rafael, and Paul Humphreys. 2017. Big Data, Thick Mediation, and Representational Opacity. *New Literary History* 48: 729–749.

Baldi, P., P. Sandowski, and D. Whiteson. 2014. Searching for Exotic Particles in High-Energy Physics with Deep Learning. *Nature Communications* 5: 4308.

Burge, Tyler. 1998. Computer Proof, Apriori Knowledge, and Other Minds. *Noûs* 32: 1–37.

Crombie, A.C. 1981. Philosophical Perspectives and Shifting Interpretations of Galileo. In *Theory Change, Ancient Axiomatics and Galileo's Methodology*, ed. J. Hintikka, D. Gruender, and E. Agazzi, 271–286. Dordrecht: D. Reidel Publishing Company.

Cybenko, George. 1989. Approximation by Superpositions of a Sigmoidal Function. *Mathematics of Control, Signals and Systems* 2 (4): 303–314.

Fleck, L. 1979. *Genesis and Development of a Scientific Fact*, ed. T. J. Trenn and R.K. Merton; trans. F. Bradley and T. J. Trenn. Chicago: University of Chicago Press.

Kononenko, Igor. 2001. Machine Learning for Medical Diagnosis: History, State of the Art, and Perspective. *Artificial Intelligence in Medicine* 23: 89–109.

Kourou, Konstantina, Themis P. Exarchos, Konstantinos P. Exarchos, Michalis V. Karamouzis, and Dimitrios I. Fotiadis. 2015. Machine Learning Applications in Cancer Prognosis and Prediction. *Computational and Structural Biotechnology Journal* 13: 8–17.

Gandy, Robin. 1980. Church's Thesis and Principles of Mechanisms. In *The Kleene Symposium*, ed. S.C. Kleene, J. Barwise, H.J. Keisler, and K. Kunen, 123–148. Amsterdam: North-Holland.

Hacking, Ian. 1992. 'Style' for Historians and Philosophers. *Studies in History and Philosophy of Science Part A* 23: 1–20.
Humphreys, Paul. 2004. *Extending Ourselves: Computational Science, Empiricism, and Scientific Method*. New York: Oxford University Press.
———. 2011. Computational Science and Its Effects. In *Science in the Context of Application*, ed. Martin Carrier and Alfred Nordmann, 131–142. Berlin: Springer.
———. 2013. What are Data About? In *Computer Simulations and the Changing Face of Experimentation*, ed. Eckhart Arnold and Juan Duran. Cambridge: Cambridge Scholars Publishing.
Leonelli, Sabina. 2016. *Data-Centric Biology: A Philosophical Study*. Chicago: University of Chicago Press.
Noë, Alva, and Evan Thompson. 2004. Are There Neural Correlates of Consciousness? *Journal of Consciousness Studies* 11: 3–28.
Rumelhart, David E., James L. McClelland, and PDP Research Group. 1987. *Parallel Distributed Processing*. Vol. 1. Cambridge, MA: MIT Press.
Shagrir, Oron. 2016. Advertisement for the Philosophy of the Computational Sciences. In *Oxford Handbook of Philosophy of Science*, ed. P. Humphreys, 15–42. New York: Oxford University Press.
Zarsky, Tal. 2014. Understanding Discrimination in the Scored Society. *Washington Law Review* 89: 1375–1412.
Zeiler, M.D., and R. Fergus. 2014. Visualizing and Understanding Convolutional Networks. In *European Conference on Computer Vision*, 818–833. Berlin: Springer.
Zhu, Song-Chun, Cheng-En Guo, Yizhou Wang, and Xu Zijian. 2005. What Are Textons? *International Journal of Computer Vision* 62: 121–143.

Instrumental Perspectivism: Is AI Machine Learning Technology Like NMR Spectroscopy?

Sandra D. Mitchell

Introduction

Philosophers of science explore and explain how scientists acquire knowledge of nature. Most have agreed that we must give up oversimplified accounts of direct experience of "the given" (which is the English translation of the Latin *datum* or *date*) and overambitious requirements that scientific knowledge be restricted to claims that are universally true and exceptionless. As a result, many factors that enter into scientific practice have been exposed as relevant to our understanding of how knowledge of nature is constructed, how it is judged, and how it is used. For example, which observations are judged to provide reliable data? What features of phenomena are represented in an explanatory model? In which contexts and for what purposes will an explanatory model be adequate? To be sure, science is a product of human activity, both *causally*, through experience and experiment and *inferentially*, though logic, calculation, and simulation. What is investigated and how it is investigated, is shaped by decisions which are themselves dependent on and constrained by human pragmatic goals, like curing diseases, or understanding the expanse of the universe.

The question, "Will science remain human?" is posed in response to a worry that AI machines will replace scientists, and that something crucial will be lost if that happens. Stark examples driving this worry are found in the proliferation of deep learning strategies of AI: AlphaGo beating the top ranked player of Go, DeepMind's application to problems in healthcare (Fauw et al. 2018), deep learning models for data reduction in high energy physics (Guest et al. 2018) and bias in autonomous

S. D. Mitchell (✉)
University of Pittsburgh, Department of History and Philosophy of Science, Pittsburgh, PA, USA
e-mail: smitchel@pitt.edu

M. Bertolaso, F. Sterpetti (eds.), *A Critical Reflection on Automated Science*, Human Perspectives in Health Sciences and Technology 1,
https://doi.org/10.1007/978-3-030-25001-0_3

systems (Danks and London 2017).[1] But are these new technologies really different from what we have come to see as legitimate extensions or instrumental replacements of human capacities by what we now accept as less threatening machines? In this paper my strategy is to explore in what ways machine learning is similar to other scientific instruments, taking the results of instrumental engagement as providing a useful non-human perspective on the phenomena. If AI is understood instrumentally, then it is clear we use it (or not) for our own, human, purposes. But when should we use it, and when not? When should we trust it, or why not? I will suggest that the same norms that govern judgments of other scientific instrumental reliability should be used to warrant the use of AI. My argument is in support of the norms to be applied, rather than an account of the success of any particular use of AI in practice.

Ever since the introduction of telescopes and microscopes humans have relied on technologies to extend beyond human sensory perception in acquiring scientific knowledge. Simple instruments relying on lenses present mediated images to the human observer. This constitutes an indirect causal interaction between the scientist and the phenomena studied. Contemporary scientific experiments, like x-ray crystallography, nuclear magnetic resonance spectroscopy (NMR), cryo-electron microcopy, and small-angle neutron scattering used for predicting the three-dimensional structure of proteins, involve more complicated causal interactions in order to detect and process information about the target phenomena. Scientists trust these detection instruments, from simple lenses to elaborate experimental equipment, to reveal features of nature. Indeed, we must trust them more than unaided human detection.

Recently developed artificial intelligence technologies appear to extend beyond human cognitive capacities. Are these forms of outsourcing cognitive aspects of scientific practice similar to instrument-mediated perception? If not, how do they differ and should we be worried about their increasing role in science? I will take up this challenge by investigating AI from an instrumental stance. Does AI provide just another instrumental perspective for humans to use in gaining scientific knowledge, like microscopes and NMR spectroscopy? Are the means by which we trust the results produced by these sensory technologies transferrable to the results produced by AI cognitive technologies? In this essay I will present reasons to think similar norms apply to both types of technology.

[1] AI, Machine Learning and Deep Learning are not identical. AI is a machine way of performing tasks that are characteristic of human cognition, but may or may not attempt to represent the way humans perform those tasks. Machine Learning is one set of practices to achieve AI, where the algorithm is not explicitly programmed to perform a task, but "learns" how to achieve a specified goal. Deep Learning is one form of Machine Learning that uses Artificial Neural Net structures, with many discrete layers (deep structure) of connected artificial neurons that implement a hierarchy of concepts.

Routes to Scientific Knowledge

Science aims to accurately characterize features of nature that permit explanation, prediction and intervention in order to further our human goals. Support and justification for scientific knowledge comes from experience (observation or experimentation) and reason (concepts, logic and inference). I argue that the theoretically and experimentally based models that result from well-executed scientific practices always encode a limited perspective. Much has been written about the perspectival character of scientific instruments and models of natural phenomena (e.g. Giere 2006; Van Fraassen 2008; Massimi 2012; Price 2007). Some appeal to the location of the "observer", i.e. the vantage point of a distance or scale from which structures can be detected. I argue that perspectivism follows from the partiality of representation itself. My argument rests on the claim that no scientific model, whether it is derived from more general theories or from the results of an experiment, can provide a complete account of a natural phenomenon.

What could be meant by model completeness? Completeness in formal systems, such as the propositional calculus, is tied to notions of proof and deductive inference. A set of axioms is complete if, given the rules of inference, all logical truths expressible in the theory are provable. But what could it mean for a scientific model of natural phenomena to be representationally complete? Weisberg (2013) suggests that model completeness is a representational ideal referencing the inclusiveness of the model ("each property of the target phenomenon must be included in the model," p. 106) and fidelity (models aim to represent "every aspect of the target system and its exogenous causes with an arbitrarily high degree of precision and accuracy," p. 106). As Weisberg acknowledges, this type of completeness is impossible to achieve.[2] Rather than reject completeness as a virtue of a model, Weisberg claims that we should treat it as a regulative ideal against which we judge the success of any given scientific model. Since no scientific model can satisfy the standard, we instead focus on how close or far from it a model comes. I disagree with this approach. As I have argued in the case of ideal, universal, exceptionless laws, (see Mitchell 2000, 2009), we should develop normative standards that track the character of what can be accomplished, that is, what scientists in fact do. More or less complete often will be unmeasurable when we are considering models that use different variables that do not stand in inclusive hierarchies. Since all scientific models are partial, and since many will represent differing features, how would we determine which one of them was "more" complete? Counting the number of variables clearly will not be adequate.[3]

[2] See also Madden 1967, p. 387: "The incompleteness of science arises from the impossibility of describing every detail of nature, whether the universe be conceived as infinite or finite in space and time, and from the fact that any explanatory deductive system depends upon assumptions which are themselves not explained."

[3] See also Craver 2006 who appeals to the continuum between a mechanism sketch and an "ideally complete" (p. 360) description of a mechanism. Craver and Kaplan 2018 endorse the norm of Salmon-completess which judges comparative completeness of explanations (in contrast to models) in terms of fewer or more relevant details.

Even if we came up with a way of measuring more or less complete models of natural phenomena, satisfying Weisberg's completeness ideal is neither necessary nor desirable for successful science. Not every describable feature of a system in every possible degree of precision is required for identifying features and relations that permit prediction, explanation, and intervention on that system.[4] Suppose we could meet a strong completeness standard whereby our model represents each property of the target phenomena (at all spatial and temporal scales) with the highest degrees of precision and accuracy. That representation would fail to constitute usable knowledge of the phenomenon; it would be a duplicate. It might be we want to rule this out as a representation at all, since representational models presuppose an agent (see van Fraassen 2008, Giere 2006) so it could be the case that no one would take a duplicate to be a representation. In addition, for the purposes of facilitating explanation, prediction, and intervention, a duplicate may be no better than engaging directly with the very system we are trying to understand. Of course if the model were not strictly complete, i.e. duplicated the original in scale, but not material, or in material but not scale, there could well be epistemic benefits to engaging with the model over the original. But then, I suggest, it would fail some ideal notion of completeness.

Model "goodness" should be judged by its *accuracy* with respect to existing empirical data, and its *adequacy* with respect to specific goals, not how close it comes to an unachievable and non-useful "ideal." The assumption that if we could represent everything then we can achieve any and all of our goals is undoubtedly the intuition supporting completeness as an ideal. However, given we cannot represent everything, including more details in a model can be detrimental to both its accuracy (e.g. by over parameterizing in ways that compound uncertainty) and its adequacy (e.g. by obscuring main factors whose manipulation might be sufficient to meet the goal).

A model represents relations that provide explanations and predictions. Since it cannot be complete in the sense of including all that could be described, it must be partial. Scientific models expose what is deemed causally relevant, what is salient, what is expressible in a particular framework, etc. What is represented and what is left out sometimes is guided explicitly by explanatory or pragmatic goals. Scientific representations also reflect limitations of the representational medium. Models are abstract and, even when accurate, are not complete. There are two consequences of this fact. First, the partiality of scientific models requires us to embrace model pluralism as essential to science achieving its goals. Second, the partiality of scientific models entails perspectivism.

- A useful representative model can specify only some aspects of its target phenomena.
- Therefore, aspects that might have been represented are omitted.

[4] Craver and Kaplan 2018 refer to this as the more details are better view, which the also reject.

- The omitted aspects could be (and typically are) included in other scientific models.
- Some sets of representational models are not incompatible with each other, nor intertranslatable/reducible, nor additive/mutually exclusive.
- *Therefore*: to explain, predict, and intervene on a given phenomenon, science requires a plurality of models to represent the features that are relevant in different contexts and for different purposes.

This account raises new questions about the relationships among the multiple models that are developed to represent the same phenomenon. While reduction, unification, and elimination are ways in which models of the same phenomenon may be related, I have argued that when there are substantial instances of compatible pluralism explanatory integration better accounts for model-model relationships (Mitchell 2009, 2019).

The flip-side of partiality of representation is perspectivism. One representational model, by leaving out some features or details explicitly and implicitly "selects" features to include. What is left in constitutes a perspective. Given there is no unique, complete representational model, there can be and frequently are multiple models that all satisfy the empirical demands science places on acceptance. The accuracy of a model is judged by its ability to account for accepted empirical data. Multiple models can provide equally accurate descriptions of a phenomenon, though they do so by referencing different features of it, perhaps at different scales, with different degrees of precision, etc. The adequacy of any model or set of models is judged by how well it or they serve some epistemic or non-epistemic purposes. Human purposes, as well as human capacities are reflected in both these judgments. How should we evaluate AI deep learning algorithms? Surely, AI provides new tools for developing models, which describe and predict features of nature. These models are partial and perspectival. As such, they cannot provide a complete model, though they may provide new perspectives that no human could produce.

The New Technologies

Machine learning includes a variety of computational algorithms that detect patterns, or make inferences from data, without applying explicit, human programmed rules specifically designed to solve the problem at hand. Rather they implement generic learning algorithms. From minimal information (a training set of input-output patterns bearing varying degrees of human-assigned labels) the machine builds its own models and new algorithms for making predictions from data in a specific domain. Artificial Neural Networks (ANNs), massively parallel systems with distributed computation were loosely modeled on simplified accounts of biological neural architecture in the brain. Research on ANNs has been a growth industry since the 1980s, due to, in part, the development of back-propagation learning algorithms for multilayer feed-forward networks that had widespread impact through Rumelhart and McClellan's 1986 book. While feed-forward networks are static, producing one set of output values for a given input, recurrent, or feedback network architectures are dynamic, where computed outputs modify response to subsequent inputs. ANN's learn from examples rather than explicit rules. This means that, among other things, connection weights between the neurons are adjusted by means of a learning rule. For example, a back-propagation algorithm can implement error-correction to train the network to minimize output error (Jain et al. 1996).

A human signature is ineliminable from all forms of machine learning. In so-called supervised learning, humans determine not just what the prediction problem is, but specify what counts as a correct answer, the target, that is used as part of the training process. The machine learns how to reach the target by calculating an error signal between the target and actual outputs and using that error to make changes in the weights in the algorithm. Unsupervised machine learning infers a function from unlabeled input to output that relies on hidden structure. In both cases, what data is presented, and what problem is to be solved is set by humans. Reinforcement machine learning employ goal-oriented algorithms, where the goal is specified by humans. The feature that makes artificial neural net machine learning (ANNs) a challenge is that the functions it infers to map input data into output patterns, and how the functions are acquired may not always be cognitively available or meaningful for humans. If AI machine learning algorithms do not learn the way humans learn, do not make inferences or patterns the way humans do, and we cannot see what it is doing, then how can we trust it is doing the right thing?

This seems to me to be parallel to the situation of causal detection instruments. They detect features of phenomena that we cannot detect, and they do it in ways that are different from how we do it. Yet we trust the results of such instruments. How is the perspective of AI machine learning different from causal experimental perspectives?

The Instrumental Stance

In commendation of ye Microscope

Of all the Inventions none there is Surpasses
The Noble Florentine's Dioptrick Glasses
For what a better, fitter guift Could bee
In the World's Aged Luciosity.
To help our Blindnesse so as to devize
A paire of new & Artificial eyes
By whose augmenting power wee now see more
Than all the world Has ever doun Before.

Henry Power, 1664 (Cowles 1934)

Power, one of the first scientists to be made a fellow of the Royal Society wrote the first book about microscopes (predating Hooke's *Micrographia* by two years).

Power's message that artificial eyes let us see more than anyone could have seen before continues to be true of the modern forms of spectroscopy. These, like X-ray crystallography and Nuclear Magnetic Resonance Spectroscopy, are less clearly extensions of human visual perception, than they are alternative detecting devices. Why do we trust the results of such instruments to provide data from which scientific inferences can be drawn? In what follows I will consider the use of these experimental instruments in the determination/prediction of protein structure. There are two components to trusting instrumental detection; one is reliability of the data output that result from the causal interaction with the target phenomenon; the other is the inferential or epistemic warrant that the measurements or models of data provide in supporting hypotheses about the phenomena.[5]

Briefly, proteins are the most common molecules found in living cells. They consist of one or more polypeptide chains which themselves are composed of amino acids. Proteins are coded for by DNA and produced through a process of transcription of RNA from DNA, then translation of the RNA on a ribosome to generate a chain of amino acids (the primary structure of a protein) that then folds into a functional conformation, of secondary, tertiary and sometime quaternary structure. Predicting the structure aims to identify the position of the atoms constituting the amino acids in their folded, functional form. Knowing the structure provides information about binding sites on a protein that are a clue to its function, since most proteins perform their biological function in response to or in conjunction with other molecules. This information can also aid in drug design to intervene on proteins for medical purposes. NMR and X-ray Crystallography are the two primary methods for experimentally ascertaining protein structure.

Obviously, humans cannot directly detect what is going on at the scale of atoms. NMR spectroscopy is an experimental alternative causal detection method that measures the way magnetic influences affect the behavior of the nuclei of atoms (e.g. hydrogen). Basically, a concentrated protein solution is placed in a strong magnetic field. Atomic nuclei, like hydrogen, have an intrinsic magnetization resonance or spin that is changed by the strong magnetic field. In the experiment, the initial alignment with the strong magnetic field is disrupted by a radio frequency (RF) electromagnetic pulse. As the hydrogen nuclei return to their aligned states, they emit RF radiation that is measured. The radiation emitted depends on the local environment so that excited hydrogen nuclei in other amino acids induce small shifts in the signals of close-by hydrogen nuclei (magnetization transfer). Given information about the protein's constituent amino acid sequence, the measurements provide information about where each atom of each amino acid is located in the 3-D structure of the protein.

As Hans Radder puts it "An experiment tries to realize an interaction between some part of nature and an apparatus in such a way that a stable correlation between a feature of that part of nature and a feature of the apparatus is produced" (Radder 2003). The first level causal output in NMR protein spectroscopy is the RF radiation associated with hydrogen nuclei in the various amino acids composing the protein

[5] See Bogen and Woodward 1988 for an important distinction between data and phenomena.

sample. From measuring the decay curves of the hydrogen atoms, the experimenter can recover information about the relative distance and rotational angles between atoms. From that, plus measurements of other types of atoms in the protein, an atomic structure of the protein can be inferred. What are our grounds for trusting the results of the instrument? We don't have any experience of "what is like to be a nuclear magnetic resonance spectrometer". Though we can decompose the causal process into more fine-grained steps, at some point we will not have any more direct perception. Thus, reliability of the causal process will be an inference we make – not a causal "experience" we have. That inference is made by appeal to theories, the "theory of the experiment" and the stability of the results of instrumental replication and multi-instrument convergence.

I will consider how this is achieved in the case of NMR experiments and question whether or not the same types of inference are available in the case of AI machine learning.

Theoretical Support

Instruments have perspectives, i.e. they selectively interact with some features of the target phenomena, not all, and in this sense, are causally biased (see Giere 2006). Measurements or some other meaningful representation of the causal effects of an instrument/phenomenon interaction encode theoretical assumptions and are not perfectly objective or given by experience. Numbers, graphs, and natural language represent the causal process in terms that express units, scales, and other content. As Tal (2017) clearly and correctly claims:

> To attain the status of an outcome, a set of values must be abstracted away from its concrete method of production and pertain to some quantity objectively, namely be attributable to the measured *object* rather than the idiosyncrasies of the measuring instrument, environment, and human operators. (Tal 2017, p. 35).

With respect to NMR experiments, we want to assign measurements (relative distances) to the atoms of the protein itself. How much, which and where theories are involved in detection matter to our judging the process and the outcomes as a reliable means to do that. In NMR determination of protein structure experiments, theories of how electromagnetic fields and pulses affect the behavior of atomic nuclei are required both for designing the experiments and interpreting their results. Additional theories are also required pertaining to the materials used in building the instrument, how the preparation of the sample might affect the target properties, the confounding influences of the environment in which the experiment is conducted and more. Theories are essential for both performing an experiment – the causal theory of the experiment – and for producing epistemically relevant information. We might require different causal operations to elicit measurements relevant to determining the structure of particular proteins, which are biologically functional only in conjunction with other molecules. We might need different sorts of information, and hence experimental access, to design drugs that will bind with a protein to negate a detrimental function it would otherwise perform.

The role of theoretical assumptions in experimental hypothesis testing long has been a subject of philosophical scrutiny. Concerns about theory-ladenness challenge the objectivity of experimental results for testing theories. But theories are required. Duhem points out that "The same theoretical fact may correspond to an infinity of distinct practical facts [...]. The same practical fact may correspond to an infinity of logically incompatible theoretical facts [...]" Duhem (1906/1962, p 152). Thus, for an experiment to be a test of a specific hypothesis or prediction, there needs to be some way to translate the causal outcome (practical fact for Duhem) into a claim relevant to the prediction of the hypothesis being tested (theoretical fact). There is no escaping *some* semantic infection from the theory being tested in describing the outcomes of an experiment.[6] There is no way to isolate an observation or measurement from the theoretical assumptions. But which theories are involved?

In its most extreme version, what is called Duhemian holism, it is claimed that every experiment implicates "a whole theoretical group" (Duhem 1906/1962, p. 183) and thus no isolated hypothesis can either be confirmed or falsified by an experimental result. There is always the possibility that a negative test result is due to one of the auxiliary assumptions, and not the theory under test. This has challenged the objectivity of experimental results to both accurately reflect the features of the phenomena and provide test for accepting or rejecting individual hypotheses.

Hasok Chang (Chang 2004) argues, in his examination of the history of the thermometer, that scientists can avoid the worst forms of holism by adopting a "principle of minimalist overdetermination". For Chang, overdetermination is the agreement by different methods on the measurement value ascribed to some phenomenon, e.g. a temperature determined by both calculation and measured by a mercury thermometer or measured by two different types of thermometer. The necessity of invoking auxiliary hypotheses in order to make predictions, build apparatus and interpret the results of an experiment is unavoidable, but by minimizing assumptions, Chang argues, the damage can be contained. "The heart of minimalism is the removal of all possible extraneous (or auxiliary) non-observational hypotheses." (Chang 2004, p. 94). Overdetermination by multiple experiments which make similar ontological assumptions about the nature of the phenomena, uses fewer assumptions than are required by predictions from high-level theories. His approach rests on the appeal for the most direct, or least theory-mediated correlation between what is in the world and the measurement. I understand this principle to require that the theories of the experiment should rely on *as few assumptions* about the function that associates the target feature with the measured feature as possible. Chang's principle acknowledges that there is no way to eliminate theoretical assumptions in generating experimental results, in contrast to others who have required something stronger, namely the independence of the theory of the

[6] There are different forms of theory ladenness. See Bogen's 2017 SEP article distinction of perception loading, semantic theory loading, and salience. On Bogen's classification, Duhem's claim is about semantic theory loading.

instrument and theory to be tested by the experimental results using the instrument. Complete independence is certainly too strong, since there needs to be some form of translation between experimental measurements and theoretical predictions in order for the former to be a test of the latter at all (Darling 2002). By minimizing, removing all possible auxiliary assumptions, Chang suggests we can provide the strongest grounds for taking experimental results to be confirmations or refutations of the hypothesis being tested.

The theory of the instrument does not need to be simple, and certainly is not in the case of NMR spectroscopy, but it should, if Chang is correct, rely on no more theory overlapping the hypotheses to be tested than is necessary. NMR spectroscopy relies on theories of nuclear spin and the effects on spin from magnetic influences (chemical shift). While the atoms in each functional protein are described by these theories, which particular arrangement of molecules occurs in a given protein is not. So, although NMR spectroscopy experiments may succeed or fail to correctly detect the atomic locations, the assumptions involved in addition to the theory of the instrument seem to be minimally overlapping with a hypothesis that protein P has conformational structure C. At least on Chang's account, it could be argued that NMR experiments are minimally overdetermined.

Chang's approach has the virtue of recognizing the ineliminabilty of influence of theoretical assumptions related to the hypothesis being tested in acquiring reliable measurements from any experimental set up. However, like Weisberg, his approach seems to take the impossible – complete independence – as a regulative ideal. Minimizing means using fewer assumptions, thus getting closer to independence. However, as in the case of Weisberg's ideal completeness ideal, reliance on the impossible-to-realize ideal norm of independence strikes me as an inappropriate goal. Instead, what matters is *what* is assumed in an experimental set up, not *how much*.[7]

In the case of NMR for protein structure prediction, assumptions invoke general theories about atomic nuclei and chemical steric constraints (what angles of rotation are possible, and that no two atoms occupy the same space at the same time) and a host of other theories. However, the specific location of atoms in the conformation of a functional protein in solution is not directly implicated in those theories. NMR can be trusted in testing rival structure predictions since NMR is not tuned to a specific protein structure but only to nucleic atomic behavior more generally. A more detailed explication of the theories involved in all the steps of generating measurements in NMR spectroscopy, akin to what Tal (2017) calls "white-box calibration" in contrast to "black-box calibration" would be required to convincingly make the case (see also Humphreys 2004). At some point, however, there will be no more sub-boxes between the phenomenon and detector that can be theoretically unpacked. There will just be input (some sample of the phenomenon) and an output. This is the rawest sense of data. With no theories left to support the veridicality of the correlation between phenomenal feature and instrument detection, we appeal to stability of the results to warrant claims of reliability.

[7] See Glymour 1980 for other ways to manage the theory-ladenness of experimental observations.

Replicability and Convergence

A theory or model of the experiment itself describes and predicts the causal relations between the target phenomenon and the output of the experiment. It involves theoretical constraints as well as sources of random and systematic error. Even if we have good grounds for accepting that the theory of the experiment is correct, there remains a question about if our experimental apparatus instantiates what the theory describes.

NMR spectroscopy based on theories of nuclear magnetization had been attempted in 1936 to testing Lithium compounds. However, the experiments were unable to produce any measurable signals. In 1941 it was reported that NMR signals of hydrogen in water had been detected, but there was insufficient replicability of the results. It wasn't until 1946 that the first successful results of NMR were reported by two labs, by Bloch at Stanford and Purcell at Harvard for which they were jointly awarded the Nobel Prize in 1952. There were many developments that refined the instrument; importantly Richard Ernst's 1964 introduction of the use of short, intense RF pulses to simultaneously excite all magnetic resonances and using Fourier transforms to computationally analyze the response (H. Pfeifer 1999). The RF technique increased the sensitivity of the instrument thus improving the signal to noise ratio. The first NMR spectrum of the protein ribonuclease was detected in 1957. In 1958 Kendrew and Perutz published the first high-resolution protein structure, using X-ray crystallography. In 1970 Wuthrich published a lecture showing that NMR spectroscopy and X-ray studies of proteins could yield similar spatial structures of protein under extremely different experimental conditions (NMR of proteins in solution and X-ray of proteins in a crystalline form). This is taken to be the pivotal moment for NMR in the history of protein experimentation. (Schwalbe 2003).

For an instrument to be reliable is for the output of its detection process to capture the targeted features of the phenomenon and thus permit accurate measurements that represent those features. But since we have no direct access to the phenomenon outside of some detection device, there is no way to directly compare the phenomenon with the instrumental response. Instead, instrumental reliability is supported by the stability of instrumental results in different places and times using the same experimental protocols (replicability) and stability of results among very different kinds of experiments (reproducibility or convergence). For NMR to reliably indicate the atomic structure of a protein is for the signals detected (chemical shift) and measured (decay curves) to reflect the distance and rotational angles between atoms present in the molecules constituting the protein.

The comparison of different instrumental techniques, like X-ray and NMR spectroscopy, permits cross-validation. If there is a designated standard technique with which to compare a new instrument, this is called calibration, if not, then multiple experiments each with different sources of systematic errors can be taken to be validating each other. Notice that just as there is no independent-of-theory test of a single measurement, neither is that any independent-of-theory calibration. Eran Tal (2017) has recently argued that successful calibration depends on two conditions.

First the measurement outcomes "are mutually consistent within their ascribed uncertainties" and second "the ascribed uncertainties are derived from adequate models of each measurement process" (2017:43). When the conditions are different, and the models are driven by different theories, or parts of a theory, Tal argues that the shared results are context invariant. This, Tal suggests, is what can be meant by the objectivity of the result. They are not theory independent but are independent of any particular theory. Replication under the same experimental protocol can support claims of internal stability – the phenomenal feature-instrument interaction is generating the signal measured, not a signal from some fluctuating or random extraneous source. Convergence can support a broader warrant. Tal suggests it permits prediction of experimental outcomes for other types of measurements with other sources of bias and uncertainty. As Tal claims, convergence accounts for "how it is possible to assess the reliability of measuring instruments despite the inaccessibility of 'true' quantity values, and despite the fact that measurement standards do not provide absolutely exact, infallible quantity values" (Tal 2017, p. 45). Stability across replication and convergence plays the same role as Chang's principle of minimalist overdetermination. However instead of minimizing the number of assumptions required for an experiment, on Tal's account, reliability is gained by varying the sources of uncertainty, no matter how many assumptions each source requires.

To sum up: trust in the reliability of the models inferred from the causal processes of experimental instruments derives from our theories of the instrument (how is nuclear spin affected by magnetic influences) and from the stability of the results across multiple trials (replication) and multiple instruments (convergence).

AI Instrumental Perspectives

I have presented a view that the inferences supporting the reliability of the causal capture of phenomenal features through experimentation is supported by theories of the experiment and by the stability of results in replication and convergence. Should AI machine-learning practices be subject to the same tests for reliability? How would they fare?

The instrumental stance focuses attention on the causal interaction between the target phenomena and the detecting device in the experiment that can output a measurement. Note that the measurement is a description that depends on interpretation and inference. In particular the measurement (or model of the data) involves distilling the signal from the noise in the initial data output. There are a variety of techniques to accomplish this in NMR protein experiments, from averaging out random error to 'correcting' for known systematic errors. In addition, the measured data must be represented in a way that can be related to the hypotheses at issue, i.e. decay curves in NMR experiments are translated into relative distance and rotation angles between atoms.

Instrumentally, machine-learning algorithms enable identification of associations and patterns in observed data, permitting scientists to build explanatory models and make predictions. The algorithms do not have explicit programmed rules that are applied directly to accomplish the pattern recognition, but rather "learn rules" based on past "experience". There are many different protocols for the type of learning implemented (reinforcement, supervised, unsupervised) and the character of the algorithms (linear regression, logistic regression, Bayes, nearest neighbor etc.). The point at issue is that how AI learns and then generates patterns and predictions seems to diverge from the way a human scientist learns and generates patterns and predictions. For some architecture of multiple layer (deep) neural networks, there may be no way in which a human can recover the rules that have been learned.[8] Thus, even though the architecture of ANNs is inspired by analogy to the way humans reason, it is challenging to determine if they actually reason in the ways humans reason. On the one hand there is some evidence (Dodge and Karam 2017) of significant differences between deep neural nets and humans in recognition performance of distorted images in success rates and in types of errors. On the other, Buckner 2018 argues that his account of machine learning in terms of generic transformational abstraction suggests there is alignment with human psychological processing.

In the case of AI, the "phenomena" are the data sets fed into the computer and the output is a pattern or a prediction. The output is more like a measurement or model of the data than a physical causal effect, in contrast to the case for the spectroscopy instruments. Perhaps what the learning algorithm does is eliminate noise and minimize error. In the NMR experiment the model of the data is based on both theories of the instrument and on the stability of repeated outputs.

> Most modern NMR experiments rely heavily on the reproducibility and stability of the spectrometer, because in order to select those signals that carry useful chemical information, it is necessary to cancel those that do not. [...]. In some experiments [...] the sensitivity of the technique is often determined not by the intrinsic signal-to-noise ratio of the instrument, but by its stability. The limiting factor is the ability to distinguish between the signals of interest and a background of unwanted signals derived from instrumental errors, rather than a background of random noise. (Morris 1992).

I suggest that the warrant of the output of machine learning is similar to the warrant of the measurement output of a causal detection device. Both rely on a theory of the instrument and the stability of the results. To assess reliability, one needs an analysis of the learning rules implemented in AI algorithm. We may ask, for example, what is the support for reinforcement learning vs. supervised learning relative to particular problem types? Just as in physical experiments, the theories behind the instrument in AI machine learning are subject to the norms of contrastive confirmation that apply across science. Which learning rules, representation protocols, etc. are theoretically warranted to solve specific kinds of problems can provide support for the reliability of AI.

In addition, it is clear that the stability of results through replication and convergence are also sources of warrant in the case of AI. Though different forms of stabil-

[8] See Buckner 2018 for an articulation of deep neural network AI processing as a form of transformation abstraction.

ity are signatures of reliable instrumental performance, they do not rule out the presence of stable multi-instrument bias. In a recent *Science* article (Hutson 2018), a "replication crisis" was announced for artificial intelligence. Scientists who wanted to test a new algorithm against a benchmark, a classic example of calibration, found they could not replicate the benchmark itself. Replication is a norm that AI scientists clearly adopt. Making explicit how a detector instrument is built, or a program is designed will accomplish two things. First, a replication can be attempted, checking the stability of the results across different iterations in space and time. Second, the assumptions implicit in the instrument or program can be made visible and the theories underlying them can be endorsed or challenged.

Other sources of failure to replicate include not having access to the training set upon which the algorithm learns the functions that execute predictions on new data input. Learning from different data sets can generate major differences in output predictions (Nishikawa et al. 1994). Other factors that are not central to the code may also influence the stability of the outcomes (see Islam et al. 2017). Factors in NMR experiments that are not directly in the source-signal-detector path also can influence outcomes, like input electricity source. Although there is no way to eliminate factors that disrupt the fidelity of the source features to the detected features, there are strategies in both NMR and in AI to identify and manage what may bias the results. Replication and convergence of results in each case can warrant the reliability of the instruments.

That the failure of reproducibility in AI is deemed a 'crisis' indicates that the sources of reliability and trustworthiness typical for extensions of our perceptual means of acquiring knowledge of nature, namely instrumented detection devises, also apply to the new extensions of our cognitive components of acquiring knowledge of nature, namely machine learning AI programs. In the case of NMR spectroscopy, the way it "sees" the world is not the same as the way we do and yet we can trust its results based on robust theories of the experiment and evidence of the stability of the output. The way AI machine learning detects patterns in data is also not necessarily the way we do it. Like the NMR case, artificial intelligence deploys capacities that are beyond our unaided human perceptual and cognitive abilities. Are there differences in how we can interrogate the causal instrument and the cognitive instrument to apply the norms required for judgments of reliability? Certainly yes. However, that the same general norms for reliability apply to both types of scientific practice supports taking an instrumental stance towards AI. It is a tool we can use, if reliable, in the same way we use NMR spectroscopy or other tested and trusted scientific instruments.

References

Bogen, James. 2017. Theory and Observation in Science. *The Stanford Encyclopedia of Philosophy* (Summer 2017 Edition), ed. Edward N. Zalta. https://plato.stanford.edu/archives/sum2017/entries/science-theory-observation/

Bogen, James, and James Woodward. 1988. Saving the Phenomena. *The Philosophical Review* 97: 303–352.

Buckner, C. 2018. Empiricism without magic: Transformational abstraction in deep convolution *al Neural Networks. Synthese* (12): 1–34. https://doi.org/10.1007/s11229-018-01949-1.
Chang, Hasok. 2004. *Inventing Temperature: Measurement and Scientific Progress*. Oxford: Oxford University Press.
Cowles, Thomas. 1934. Dr. Henry Power's Poem on the Microscope. *Isis* 21 (1): 71–80.
Craver, Carl F. 2006. When Mechanistic Models Explain *Synthese* 153: 355–376.
Craver, Carl F., and David M. Kaplan. 2018. Are more details better? On the norms of completeness for mechanistic explanations. *The British Journal for the Philosophy of Science*: axy015. https://doi.org/10.1093/bjps/axy015.
Craver, C.F., and D. Kaplan. 2018. Are More Details Better? On the Norms of Completeness for Mechanistic Explanations. *British Journal for the Philosophy of Science*.
Danks, David, and Alex London. 2017. Algorithmic Bias in Autonomous Systems. *Proceedings of the Twenty-Sixth International Joint Conference on Artificial Intelligence*: 4691–4697. https://doi.org/10.24963/ijcai.2017/654.
Darling, K.M. 2002. The Complete Duhemian Underdetermination Argument: Scientific Language and Practice. *Studies in History and Philosophy of Science* 33: 511–533.
De Fauw, J., J.R. Ledsam, B. Romera-Paredes, S. Nikolov, N. Tomasev, S. Blackwell, H. Askham, X. Glorot, B. O'Donoghue, D. Visentin, G. van den Driessche, B. Lakshminarayanan, C. Meyer, F. Mackinder, S. Bouton, K. Ayoub, R. Chopra, D. King, A. Karthikesalingam, C.O. Hughes, R. Raine, J. Hughes, D.A. Sim, C. Egan, A. Tufail, H. Montgomery, D. Hassabis, G. Rees, T. Back, P.T. Khaw, M. Suleyman, J. Cornebise, P.A. Keane, and O. Ronneberger. 2018. Clinically Applicable Deep Learning for Diagnosis and Referral in Retinal Disease. *Nature Medicine* 24: 1342–1350.
Dodge, Samuel, and Karam, Lina. 2017. A Study and Comparison of Human and Deep Learning Recognition Performance Under Visual Distortions. 2017 *26th International Conference on Computer Communication and Networks (ICCCN)*, pp. 1–7.
Duhem, Pierre. [1906] 1962. *The Aim and Structure of Physical Theory*. Trans. Philip P. Wiener. New York: Atheneum.
Giere, R. 2006. *Scientific Perspectivism*. Chicago: University of Chicago Press.
Glymour, Clark N. 1980. *Theory and evidence*. Princeton: Princeton University Press.
Guest, Dan, Kyle Cranmer, and Daniel Whiteson. 2018. Deep Learning and Its Application to LHC Physics. *Annual Review of Nuclear and Particle Science* 68: 1–22.
Humphreys, Paul. 2004. *Extending Ourselves: Computational Science, Empiricism, and Scientific Method*. New York: Oxford University Press.
Hutson, Matthew. 2018. Artificial Intelligence Faces Reproducibility Crisis. *Science* 359 (6377): 725–726.
Islam, R., Henderson, P., Gomrokchi, M., and Precup, D. (2017) Reproducibility of Benchmarked Deep Reinforcement Learning Tasks for Continuous Control. *ICML Reproducibility in Machine Learning Workshop*.
Jain, A.K., J. Mao, and K.M. Mohiuddin. 1996. Artificial Neural Networks: A Tutorial. *Computer* 29: 31–44.
Madden, Edward H. 1967. Book Review of Richard Schlegel. Completeness in science. *Philosophy of Science* 34: 386–388.
Massimi, M. 2012. Scientific Perspectivism and Its Foes. *Philosophica* 84 (2012): 25–52.
Mitchell, Sandra D. 2000. Dimensions of scientific law. *Philosophy of Science* 67 (2): 242–265.
———. 2009. *Unsimple Truths: Science, Complexity and Policy*. Chicago: University of Chicago Press.
———.2019. Perspectives, representation and integration. In *Understanding Perspectivism: Scientific challenges and methodological prospects*, ed. M. Massimi and C.D. McCoy, 178–193. Taylor & Francis.
Morris, G.A. 1992. Systematic Sources of Signal Irreproducibility and t_1 Noise in High-Field NMR Spectrometers. *Journal of Magnetic Resonance* 100: 316–328.

Nishikawa, R.M., M.L. Giger, K. Doi, C.E. Metz, F. Yin, C.J. Vyborny, and R.A. Schmidt. 1994. Effect of Case Selection on the Performance of Computer-Aided Detection Schemes. *Med. Phys.* 21: 265–269.

Pfeifer, H. 1999. A Short History of Nuclear Magnetic Resonance Spectroscopy and of Its Early Years in Germany. *Magnetic Resonance in Chemistry* 37: S154–S159.

Price, H. 2007. Causal Perspectivism. In *Causation, Physics, and the Constitution of Reality*, ed. R. Corry and H. Price. Oxford: OUP.

Radder, Hans. 2003. Toward a More Developed Philosophy of Experimentation. In *The Philosophy of Scientific Experimentation*, ed. Hans Radder, 1–18. Pittsburgh: University of Pittsburgh Press.

Rumelhart, David, James L. McClelland, and the PDP Research Group, eds. 1986. *Parallel Distributed Processing: Explorations in the Microstructure of Cognition*. Cambridge: MIT Press.

Schwalbe, H. 2003. Kurt Wüthrich, the ETH Zürich, and the Development of NMR Spectroscopy for the Investigation of Structure, Dynamics, and Folding of Proteins. *ChemBioChem* 4: 135–142.

Tal, E. 2017. Calibration: Modeling the Measurement Process. *Studies in History and Philosophy of Science* 65-66: 33–45.

van Fraassen, B. 2008. *Scientific Representation. Paradoxes of Perspective*. New York: OUP.

Weisberg, Michael. 2013. *Simulation and Similarity: Using Models to Understand the World*. Oxford: Oxford University Press.

How Scientists Are Brought Back into Science—The Error of Empiricism

Mieke Boon

Introduction

With the rise of A.I., expert-systems, machine-learning technology and big data analytics, we may start to wonder whether *humans as creative, critical, cognitive and intellectual beings* will become redundant for the generation and application of knowledge. And additionally, will the increasing success of machine-learning technology in finding patterns in data make *scientific knowledge* in the form of theories, models, laws, concepts, (descriptions of) mechanism and (descriptions of) phenomena superfluous?[1] Or can it be argued that human scientists and human-made scientific theories etc. remain relevant, even if machines were able to find data-models that adequately but *incomprehensibly* relate or structure data—for example, *data-models* that provide *empirically adequate* mapping relationships between data-input and data-output or determine *statistically sound* structures in data-sets.[2]

[1] In this chapter, 'theory' is taken in a broad sense, encompassing different kinds of scientific knowledge such as concepts, laws, models, etc. The more general term 'scientific knowledge' encompasses different kinds of specific *epistemic entities* such as theories, models, laws, concepts, (descriptions of) phenomena and mechanisms, etc., each of which can be used in performing different kinds of *epistemic tasks* (e.g., prediction, explanation, calculation, hypothesizing, etc.).

[2] On the terminology used in this chapter. In the semantic view of theories, *patterns in data* are also called *data-models* (see section "Empiricist epistemologies"), which are *mathematical representations of empirical data sets* (e.g., Suppe 1974; McAllister 2007). This chapter will adopt the term *data-model* in this very sense. In machine learning textbooks, data-models are also referred to as *mathematical functions*. Abu-Mostafa et al. (2012), for instance, speaks of the *unknown target function* f: X –> Y, where X is the input space (set of all possible inputs x), and Y is the output

M. Boon (✉)
Department of Philosophy, University of Twente, Enschede, The Netherlands
e-mail: m.boon@utwente.nl

M. Bertolaso, F. Sterpetti (eds.), *A Critical Reflection on Automated Science*, Human Perspectives in Health Sciences and Technology 1,
https://doi.org/10.1007/978-3-030-25001-0_4

Anderson (2008) indeed claims that the traditional scientific method as well as human-made scientific theories will become obsolete:

> This is a world where *massive amounts of data and applied mathematics replace every other tool* that might be brought to bear. *Out with every theory* of human behavior, from linguistics to sociology. Forget taxonomy, ontology, and psychology. Who knows why people do what they do? The point is they do it, and we can track and measure it with unprecedented fidelity. *With enough data, the numbers speak for themselves.* [...]. The big target here isn't *advertising*, though. It's *science*. The *scientific method* is built around testable hypotheses. These *models*, for the most part, are *systems visualized in the minds of scientists*. The models are then *tested*, and experiments *confirm* or *falsify* theoretical models of how the world works. This is *the way science has worked for hundreds of years*. [...]. *Scientists are trained* to recognize that *correlation is not causation*, that no conclusions should be drawn simply on the basis of correlation between X and Y (it could just be a coincidence). Instead, you *must understand the underlying mechanisms* that connect the two. Once you have a model, you can connect the data sets with confidence. *Data without a model is just noise*. [...]. But faced with massive data, this approach to science — *hypothesize, model, test — is becoming obsolete*. (Anderson 2008, my emphasizes).

Essentially, Anderson suggests that the meticulous work done by scientific researchers aiming at scientific concepts, laws, models, and theories on the basis of empirical data, will become superfluous because learning machines are able to generate data-models that represent relationships and structures in the data. Each set of data will be fitted by a unique data-model, which implies that we can give up on generalization and unification endeavors. Intermediate scientific concepts, laws, models, and theories, which are desired by humans for obvious metaphysical beliefs, and which are also practically needed to deal with the limitations of their intellect, can be bypassed if relating, structuring and simplifying data—which basically is what science does according to Anderson's quote—can be done by machines.

If let's say, scientists such as Boyle, Gay-Lussac, and Hooke, had fed their experimental data to a machine (e.g., data consisting of the measured pressure, volume

space (e.g., y_1 is 'yes' for x_1; y_2 is 'no' for x_2; etc.). The *machine learning algorithm* aims to find a *mathematical function* g that 'best' fits the data, and that supposedly approximates the unknown target function f. Abu-Mostafa et al. call the function g generated by machine learning 'the final hypothesis.' Alpaydin's (2010), on the other hand, uses the notion of model and function interchangeably. An example (Alpaydin 2010, 9) is predicting the price of a car based on historical data (e.g., past transaction). Let X denote the car *attributes* (i.e., properties considered relevant to the price of a car) and Y be the price of the car (i.e., the outcome of a transaction). Surveying the past transactions, we can collect a *training data set* and the *machine learning program fits a function* to this data *to learn Y as a function of X*. An example is when the fitted function is of the form $y = w_1.x + w_0$. In this example, the *data-model* is a *linear equation* and w_1 and w_0 are the *parameters* (weight factors) of which the values are determined by the machine learning algorithm to best fit the training data. Alpaydin (2010, 35) calls this equation 'a single input linear model.' Hence, in this example, the *machine learning algorithm* to fit the training data includes only one property to predict the price of a car. Notably, the machine learning program involves a learning algorithm, *chosen by human programmers*, that confines the space in which a data-model can be found – in this example, the *learning algorithm* assumes the linear equation, while the *data-model* consists of the linear equation together with the fitted values of the parameters (w_0 and w_1).

and temperature in a closed vessel, or the weights and extensions of different springs, respectively), the machine would have generated a data-model to connect these data, which could then be used to make predictions at new physical conditions. The Boyle/Charles/Gay-Lussac laws for gasses and Hooke's law for elasticity would not have existed. Taking this a step further, Anderson's claim implies that scientific concepts such as 'the ideal gas law', 'the gas-constant' (R), and 'the elasticity coefficient' (k) would be unnecessary. We would not even need related scientific concepts, such as 'gas-molecules,' 'the number of Avogadro,' 'collisions of molecules,' and 'reversible processes.'[3]

This short expose aims to raise the question whether a future is conceivable in which nobody needs to understand science any longer—a future in which the production and uses of scientific concepts, laws, models, mechanisms, theories etc. can be replaced by machine learning algorithms that produce *epistemically opaque* data-models[4] and networks stored in machines that will do all kinds of epistemic tasks for us—which would imply indeed that humans no longer need to learn theories etc. nor how to apply scientific knowledge in solving problems. Conversely, are there reasons to believe that scientific researchers still have a role to play?

The structure of this article is as follows. Section "Machine-learning" presents a brief overview of machine-learning technologies and applications. The different kinds of ways in which computers and machine-learning technologies may replace human experts and scientists are discussed. A list of epistemic tasks is drawn up, about which it can be reasonably assumed that machine learning will outperform humans.

In section "Empiricist epistemologies", I aim to make plausible that the abilities of computers and machine-learning technologies largely correspond with ideas in the empiricist tradition about the character of knowledge and ways of (deductive or inductive) reasoning on the basis of knowledge—and vice versa, about how general knowledge can be derived (inductively and statistically) from observations and data.

I will revisit accounts of empiricism at the beginnings of the philosophy of science, including (neo)positivism, because authors such as Mach and Duhem have articulated the basic assumptions of empiricism in a clear and straightforward manner. My aim is to first explain why epistemological and normative accounts of

[3]Current machine learning practices show that machine learning algorithms are dependent in varying degrees on our theoretical and practical background knowledge. Therefore, another option regarding Anderson's assumptions is that the current state of knowledge *suffices* for this purpose. Yet, in the context of this article, it will be assumed that he means to say that machine learning technology will eventually develop to the extent that such knowledge will become superfluous in the construction of machine learning algorithms.

[4]The notion of *epistemic opaqueness* of a process has been introduced by Humphreys (2009, 618): "a process is epistemically opaque relative to a cognitive agent X at time t just in case X does not know at t all of the epistemically relevant elements of the process. A process is essentially epistemically opaque to X if and only if it is impossible, given the nature of X, for X to know all of the epistemically relevant elements of the process."

science developed in the (neo-)positivist and empiricist tradition, make it very hard to articulate our intuitive discomfort about the suggestion that machines could take over and make human scientists virtually superfluous. I aim to make plausible that on an empiricist epistemology the elimination of any human contribution to scientific knowledge is in fact already built in as a normative ideal. Attempts to ensure the superiority of science seem to assume that the objectivity of epistemic results and of methods testing these results consists of some kind of algorithmic reasoning, be it deductive or (statistical) inductive. If this is so, it should not come as a surprise that we are forced to believe that data-models produced by machine learning algorithms are just better.

Three well-known ideas developed in the empiricist tradition will be discussed to show that a strict empiricist epistemologies indeed support the claim that objective, although opaque, data-models produced in machine learning processes can replace and may even be preferable to human-made scientific knowledge: (1) *Hempel's model of scientific explanation*, which supports the idea that the supposed laws and correlations operating in D-N and I-S explanation schema's can be interpreted as data-models constructed to represent input-output relationships in larger sets of observed or measured data; (2) The rejection of a *distinction between data and phenomena*, which supports the idea that (descriptions of) phenomena can be reduced to statistically sound data-models generated in machine learning processes; and (3) The *semantic view of theories*, which supports the idea that scientific knowledge in the form of theories or models does not add much to empirically adequate and/or statistically sound data-models to represent data.

Hence, several ideas central to empiricist epistemologies supports the belief that ultimately scientific knowledge is no longer needed, and show that the empiricist tradition offers hardly any possibilities for a more positive appreciation of the epistemic and cognitive roles of human scientists.

In the last section (section "Knowledge in the age of machine-learning technologies"), it will be argued that empiricist epistemologies are flawed, or at least too limited to understand the crucial role of scientific knowledge (theories, models, etc.) and human scientist in epistemic practices such as the engineering and biomedical sciences. It will be argued that a better understanding of knowledge in the age of machine-learning technologies requires widening our philosophical scope in order to include epistemological issues of *using* knowledge for all kinds of practical purposes. To that aim, philosophical accounts of science must start from the side of *epistemic tasks* and *uses* (e.g., Boon 2017c) and address questions such as, *how* science produces knowledge that can be *used*, and *how is it possible* that knowledge can be used anyway—for instance, in discoveries, technological innovations, 'real-world' problem-solving, and in creating and controlling functionally relevant phenomena by means of technology (e.g., Boon 2017a). Finally, on the basis of this analysis, many roles of scientists and of comprehensive scientific knowledge can be pointed out, which is how the human is brought back in science.

Machine-Learning

Machine-Learning Technologies

Machine-learning algorithms are increasingly used in dealing with complex phenomena or systems, aiming to detect, predict or intervene with the complex physical phenomena or systems in developing technological application such as in biomedical and healthcare contexts. Examples are machine-learning applications in the prediction and prognosis of chronic diseases (Kourou et al. 2015; Dai et al. 2015); drug discovery (Lima et al. 2016); brain imaging (Lemm et al. 2011); and genetics and genomics (Libbrecht and Noble 2015).

Other well-known machine-learning applications aim at automated pattern recognition in ways that replace human experts. For instance, recognizing *visual images*, which require the eye of an expert but not in-depth theoretical knowledge. Examples are automatic face recognition (Odone et al. 2009; Olszewska 2016); automated visual classification of cancer (Esteva et al. 2017); vision technologies for biological classification (Tcheng et al. 2016); and, forensics (Mena 2011). Another application of machine-learning concerns pattern-recognition in the sense of discovering *patterns*, *correlations* and *causal relationships* in (big) data sets, especially in the social sciences. Originally, these kinds of data-sets were analyzed by means statistical programs such as SPSS. Examples of machine-learning technologies drawing on finding patterns and structures in order to make proper predictions about specific cases situations, are: financial risk management (van Liebergen 2017); fraud detection (Phua et al. 2010); and manufacturing (Wuest et al. 2016). In these kinds of applications, machine-learning technologies develop towards more advanced strategies of finding patterns in data, e.g., by coupling data from different sources, and strategies such as network-based stratification to detect correlations or even causal structures (e.g., Hofree et al. 2013) that would be impossible through more traditional statistical programs.

Notably, machine learning is different from computer simulations, which utilize scientific knowledge to build mathematical models (e.g., sets of differential equations) that can be run on a computer—scientists use these simulation models, for instance to view dynamic processes and to investigate how changes in parameter values affects these processes. The machine-learning process does not draw on scientific models that are constructed by means of theories, laws, mechanisms and so forth. No theory or mechanism or law needs to be fed to the machine-learning process. Instead, the learning problem of the machine is to find a data-model that presents a correct mapping relationship between input and output data of a training-set (Alpaydin 2010; Abu-Mostafa et al. 2012). For example, in ML systems concerning face recognition, the relevant task is *classification* in which the inputs, which are the images of human faces, are classified into the individuals to be recognized, which are the outputs.

What Machines Can Do

Given the currently known examples, computers and machine-learning technologies can do different types of things for different uses, thereby taking over intellectual capabilities and types of reasoning that were previously carried out by experts and scientific researchers. Here, I propose a provisional categorization of *epistemic tasks* that can be performed by both humans and machine learning technologies, with the aim of making clearer how capacities of computers relate to those of humans:

(a) 'Match': Machine-learning technologies have the ability to learn to 'match' a visual images or data-strings (the input data) with a specific image or data-string somewhere sitting in a large data-set (e.g., automated face-recognition, finger-print recognition, matching of DNA profiles). Accordingly, ML technology is able to somehow mimic the human ability to recognize relevant similarities between visual images, or structural similarities in graphical pictures. It is often still possible to check (e.g., by and expert) whether the technology performs at least as good as the expert, but the ML-technology will outperform humans in speed. If images or data-strings get more complex, machines may perform more reliable or at a higher statistical precision (i.e., pointing out how reliable the outcome is).

(b) 'Interpret': Machine-learning technologies have the ability to learn to 'interpret' visual images as belonging to a specific type, in accordance with categories defined by humans. Accordingly, ML technology is able to take over the human ability to recognize or interpret the image as of a specific type of object, to belong to a specific category, or to subsume it under a specific concept (e.g., "that is an oak," "that is a car of brand Z," "that is *Picea mariana* rather than *Picea glauca*"). In these applications experts may have played a role in supervising the machine-learning process (e.g. Tcheng et al. 2016). Here as well, it is often still possible to check (by an expert) whether the technology performs at least as good as the expert, but the ML-technology will outperform humans in speed.

(c) 'Diagnose': Similarly, machine-learning technologies have the ability to learn to 'diagnose' data-strings as probably belonging to a specific class within pre-set categories, which may be generated by humans, but also by means of machines. Hence, ML technology is able to infer from limited information about a specific target that "it probably belongs to a specific category and therefore will probably also have several additional properties" (e.g., as in personalized advertisement of buyers; financial risk assessment of customers; and, in medical diagnosis of patients).

(d) 'Structure': Machine-learning technologies have the ability to learn to find patterns, correlations and causal relations in data, which is a task originally done by humans or by statistical programs. When data-sets get more complex (which can also be considered as 'richer'), the relationships will become more complex (which can also be considered as 'more refined'), which may then be accepted as empirically adequate but opaque structures in data. These structures, in turn, can be utilized in machines learning to 'match,' 'interpret,' or 'diagnose.'

(e) 'Discover': Additionally, structures found in data by ML technologies may point out, or point at (physical or social) phenomena, very similar to how human researchers infer from observed occurrences, causal relationships or measured regularities to (physical or social) phenomena. Yet, it will require human researchers to draw the relationship between computer outcomes and the real world, because the pattern does not speak for itself.

(f) 'Calculate': Machine-learning technologies are enabled by computers (the machine). Automated calculation was the first example of computers outperforming humans in accuracy and speed. Humans can check the calculations, and assume that the algorithm by which the computer performs the calculation somehow maps the algorithm as we understand it (e.g., adding up instead of multiplying).

(g) 'Simulate': Similarly, computer programs running complex simulations of dynamic processes outperform humans in accuracy and speed, as well as in handling complexity. Here, as has been briefly explained above, the adequacy of the computer program is firstly checked by how the scientific model (on the basis of which the computer-program was build) was constructed.

(h) 'Integrate': The performance of machine-learning technology will multiply if the mentioned abilities are combined. Natural language translation is an example, but also biomedical applications, for instance, as expressed in expectations regarding *personalized* and *precision* medicine.

This overview shows that, while computers already performed better than humans with regard to *deductive* reasoning in calculation and simulation—which basically consists of performing repetitive tasks guided by logical rules— they now also start to get better than humans in *recognizing* patterns and structure in data or pictures, for matching, interpreting and diagnosing purposes. Additionally, machine learning technology may contribute to the *discovery* of new theoretical concepts or categories, but in this case, the crucial role of humans is still to recognize the discovered structure (pattern, correlation or causal relationship) as a *representation* of something that is traceable or existent in reality, i.e., as a (physical or social) phenomenon.

One of the major *applications* of ML technology is their uses in making correct *predictions*. Computers were already widely used in their ability of *deductive inference*, thereby making deductively correct predictions—i.e., the prediction is logically correct, but may be empirically inadequate due to errors in the underlying models or the computational procedures. Machine-learning technology adds predictions that are based on *inductive inference*, which means that the algorithms (i.e., the correct mapping relationship in a learning set) is applied in new situations to predict statistically expected outcomes.

This vast range of machine-learning applications may suggest that scientific researchers and scientific knowledge become superfluous as learning from large data-sets, algorithms and data-models will be developed at a degree of complexity and adequacy far beyond the capacity of the human intellect. Yet, in section "Knowledge in the age of machine-learning technologies", it will be argued that scientists and scientific researchers still play a crucial role.

Empiricist Epistemologies

Basic Assumptions of Empiricism

The first claim of this paper is that, *if* we accept some of the fundamental presuppositions of empiricism, it becomes very difficult to argue against the idea that machines will ultimately perform better than human scientists. Presuppositions central to empiricist strands can be divided into two kinds, one normative and one epistemological. The normative ideal is driven by the desire to prevent superstition and abuse of power through knowledge, by requiring knowledge to be verifiable in principle, and is one of the reasons why objectivity plays a central role in science. Linked to this is also the explicit aim of avoiding metaphysical claims in science. This then requires an epistemology that explains how objectivity can be achieved while avoiding metaphysical content. Yet, empiricist epistemologies are not necessarily normatively motivated, but can also be determined by purely epistemological convictions. In order to substantiate my first claim, I will first outline the basic assumptions of empiricism by reference to Mach's Positivism and Logical positivism.

Central to Mach's positivism is the idea that the subject matter of scientific theories is phenomenal regularities.[5] Theories characterize these regularities in terms of *theoretical terms*, which need to be grounded in observation. Accordingly, theoretical terms in our theories and laws have to be explicitly defined in terms of phenomena, and are nothing other than abbreviations for such phenomenal descriptions. Additionally, Mach maintained that one must reject any *a priori* (or metaphysical) elements (such as causality) in the constitution of knowledge of things.

Logical positivism agreed that the subject matter of scientific theories is phenomenal regularities and that theories characterize these regularities in terms of theoretical terms being conventions used to refer to phenomena, and indeed added to positivism that a scientific theory is to be axiomatized in mathematical logic that specifies the relationships holding between theoretical terms.

The preliminary point I aim to make based on this brief overview, is that if these basic assumptions of a strict empiricism were correct, the theoretical terms (also called scientific concepts), mathematical relationships between them (also called scientific laws) and theories (also called axiomatic systems) generated in science by the meticulous efforts of scientists, are in fact quite arbitrary intellectual instruments to fit the data, which, in principle, can be replaced by the data-models

[5] Frederick Suppe (1974, Chapter One) presents a comprehensive outline on the historical background to the so-called Received View, which develops from positivism to logical positivism (e.g., Carnap) and logical empiricism (e.g., Hempel).

generated and executed in machine-learning technologies. Additionally, since machines can handle much bigger data-sets, and because machines are not confined by the kinds of idealizations and simplifications humans need to make in order to fit data into comprehensive mathematical formalisms, we may expect that machines will handle the inherent irregularity and complexity of data-sets more effectively than the human intellect ever could.

Scientific Explanation

Also Duhem's philosophy of science stands in the tradition of positivism and conventionalism of the late 19th and early 20th century. In accordance with the basic assumption of this tradition, Duhem denies that theories of physics present (causal) explanations. Instead, an explanation is a system of mathematical propositions, deduced from a small number of principles, which aim to represent as simply, as completely, and as accurately as possible a set of experimental laws. Experimental laws on this view, are simplified or idealized general descriptions of experimentally produced observable effects. Concerning the very nature of things, or the realities hidden under the phenomena described by experimental laws, a *theory tells us absolutely nothing*. On the contrary, from a purely logical point of view, there will always be a *multiplicity of different physical theories equally capable of representing a given set of experimental laws* (Duhem [1914] 1954; Craig 1998).[6]

Hempel's (1962, 1966) two models of explanation agree to the basic assumptions of empiricism as well. Although Hempel emphasizes that one of the primary objectives of the natural sciences is to *explain* the phenomena of the physical world, he defends that formal accounts of explanation should avoid the metaphysical notion of causality. Similar to Duhem, Hempel claims that: the explanation fits the phenomenon to be explained into a pattern of uniformities and shows that its occurrence was to be expected, given the specified laws and the pertinent particular circumstances. Explanations, therefore, may be conceived as deductive arguments whose conclusion is the explanandum sentence, E, and whose premise-set, the explanans, consists of general laws, L_1, L_2, ..., L_r, and of other statements C_1, C_2, ..., C_k, which make assertions about particular facts. Hempel calls explanatory accounts of this kind, explanations by deductive subsumption under general laws, or deductive-nomological (DN) explanations. The second model involves explanation of phenomena by reference to general laws that have a probabilistic-statistical form. In this case, the explanans does not logically imply the explanation, but involves inductive subsumption under statistical laws, called inductive-statistical (IS) explanation. In this case, the statistical laws make it only likely that the phenomenon was to be expected.

[6] A clarifying phrase "*to save the phenomena*" to capture the empiricist idea of how knowledge is obtained from data was originally introduced by Duhem (2015/1909) and later adopted by, among others, Van Fraassen (1977, 1980) and Bogen and Woodward (1988).

Similar to Duhem, Hempel's notion of explanation entails that an explanation only tells that, based on our empirical knowledge of the world so far, the phenomenon was to be expected—the phenomenon 'fits to', or 'can be subsumed under,' the regularities, patterns and correlations that have been found in observations and experimentally produced data.

Again, if, as empiricist epistemologies suggest, this is what science ultimately has to offer—that indeed, theories, models, laws and scientific concepts can be traced back to data, and are just helpful instruments that do not add anything to our knowledge about the world—then, it is to be expected that eventually machines will outperform human scientists. For, as especially Duhem's position suggests, there is no good reason to belief that the regularities, patterns and correlations in data found by humans would be better than the empirically adequate but opaque data-models found by a machines—and additionally, if empiricists are correct, there is no reason to doubt that machines will be capable to accurately fit a particular phenomenon into data-models stored in machines such as to predict that given a certain input a specific output is to be expected (with a specified probability).

There is a large literature on explanation that argues against Hempel's account of explanation, claiming that, although Hempel's theory may succeed in avoiding the (metaphysical) concept of causality, it is insufficient to account for the proper meaning of explanation. Well-known counter-examples, which meet Hempel's criteria of DN or IS explanations but are considered improper explanations, are: the barometer explaining the storm (which illustrates the problem of *common cause*); the length of the shadow of the flagpole explaining the length of the flagpole (which illustrates the problem of *symmetry*); and, taking the birth-control pill explaining why male do not get pregnant (which illustrates the problem of *explanatory relevance*). Conversely, counter examples that do not meet Hempel's criteria, but are considered proper explanations have been given, such as: the mayor's untreated syphilis explains why he got paresis (which illustrates the problem of *low probabilities*).

The briefly listed arguments against Hempel's logical empiricist account of explanation concern the meaning of explanation, assessed by what is commonly (and rather intuitively) taken as proper and improper (scientific) explanations. The listed arguments boil down to the idea that an explanation ought to be an answer to a *why* question, and therefore should refer to a relevant (physical) *cause*. But because reference to hidden causes is based on empirically untestable and thus metaphysical convictions, this is indeed what (strict) empiricism aims to avoid.

In the context of this article, the issue is *whether the opaque data-model generated by machine-learning technologies count as explanations* for the relationships found between input and output. As has been argued above, Duhem rejects (causal) explanations entirely, and may therefore agree that the possibility of empirically grounded algorithms produced by machines from which new conclusions can be derived, proves this even better. So, his point entails that we need no explanations anyway. Yet, contrary to Duhem, many of us will hold that we need explanations, and that an opaque data-model together with specific conditions producing an outcome—which basically is the logical or mathematical structure of an explanations on Hempel's account—is not a proper explanation for that outcome. But then the issue is, what 'being a (scientific) explanation' actually adds, and conversely, what is it that apparently is not provided by the opaque data-model. Does our resistance to the idea that an explanation in terms of an opaque data-model is not any better than an explanation in terms of theories and laws merely rely on deep 'scientific realist' intuitions, according to which—paraphrasing Van Fraassen (1980)—an explanation gives us a literally true story of what the world 'behind' the observable phenomena is like? Differently phrased, on a scientific realist view, opaque data-models do not provide explanations because genuine explanations *describe* or *represent* the unobservable (physical) causes (or mechanisms, processes, phenomena, or structures otherwise) that bring about the observed (physical) phenomena. In the last section, I will return to this issue, namely whether it is merely our metaphysical disposition, or whether genuine explanations are more than data-models that fit the data.

Data and Phenomena

The issue whether we eventually will need human-made explanatory laws and theories, rather than opaque data-models that merely fit the data, is at the heart of the question about *explanation* discussed in the previous section. Here, it will be laid out that the presuppositions of strict empiricism also challenge the distinction between data and phenomena as proposed by Bogen and Woodward (1988), because strict empiricism agrees to the idea that phenomena are nothing more than statistically justified mathematical structures in data.

Bogen and Woodward (1988) contest that there is a direct relationship between theories and data as assumed in strict empiricism. Instead, according to B&W, the notion of phenomena is crucial for understanding the relationship between data and theories. Therefore, different from the empiricist tradition, in particular Van Fraassen (1977) who builds on Duhem, a conceptual distinction is needed between data and phenomena. Loosely speaking, scientists infer to phenomena based on data, because data are idiosyncratic to particular experimental contexts and typically cannot occur outside them, whereas phenomena are objective, stable features of the world. Phenomena, therefore can occur outside of the experimental context, and are detectable by means of a variety of different procedures, which may yield quite different kinds of data, whereas data reflect the influence of many other causal factors, including factors that have nothing to do with the phenomenon of interest and instead are due to the measurement apparatus and experimental design (B&W 1988; Woodward 2011).

B&W's (1988) position, including some of the clarifications by Woodward (2011) and Bogen (2011), can be summarized as follows: (1) Phenomena are distinct from data, where data is what is directly observed or produced by measurement and experiment; (2) Often phenomena are unobservable, or at least, not observable in a straightforward manner; (3) B&W think of phenomena as being in the world, not just the way we talk about or conceptualize the natural order—i.e., phenomena exist independently of us, but beyond that B&W are ontologically noncommittal; (4) B&W don't want phenomena to be some kind of low level theories; (5) Phenomena are inferred from data; (6) Data produced by measurement and experiment serve as evidence for the existence or features of phenomena; and, (7) Theories aim at providing explanations of phenomena, whereas it is difficult to provide explanations of data from theory (even in conjunction with theories of instruments, non-trivial auxiliaries, etc.).

Bogen and Woodward's (1988) notion of phenomena has been criticized by several authors. McAllister (1997, 2011) assumes that B&W describe phenomena both as investigator-independent constituents of the world, and as corresponding to patterns in data-sets. He criticizes this view by arguing that there are always infinitely many patterns in any data-set, and so the choice of one as being a phenomenon is subjectively stipulated by the investigator, which make phenomena investigator-dependent. Also Glymour's (2002) criticizes on the point that B&W leave open the question of how scientists discern or discover phenomena in the first place. Are phenomena merely summaries of data? Or is there something more to phenomena than just patterns or statistical features. Glymour argues there is not. According to him scientists infer from data to patterns by means of statistical analysis, which does not add anything new to the data. This implies that phenomena coincide with patterns in data, and that no subjective grounds are involved. Accordingly, Glymour concludes that Bogen and Woodward are mistaken in thinking that a distinction between phenomena and data is necessary, while McAllister (1997) is mistaken in thinking that the choice about 'which patterns to recognize as phenomena' can only be made by the investigator on subjective grounds.[7]

Within a machine-learning context, we may start to wonder what B&W actually have in mind when distinguishing between data and phenomena. They take Nagel's example of the melting point of lead to explain this:

> Despite what Nagel's remarks seem to suggest, one does not determine the melting point of lead by observing the result of a single thermometer reading. To determine the melting point one must make a series of measurements. [...]. Note first that Nagel appears to think that the sentence 'lead melts at 327 degrees C' reports what is observed. But what we observe are the various particular thermometer readings—the scatter of individual data-points. [...]. So while the *true melting point is certainly inferred or estimated from observed data*, on the basis of a theory of *statistical inference and various other assumptions*, the sentence 'lead melts at 327.5 + 0.1 degrees C'—the form that a report of an experimental determination of the melting point of lead might take—*does not literally describe what is perceived or observed.* (Bogen and Woodward 1988, 308–309, my italics).

[7] McAllister (2007) presents an in-depth technical discussion of how to find patterns in data (i.e., data-models). He argues that "the assumption that an empirical data set provides evidence for just one phenomenon is mistaken. It frequently occurs that data sets provide evidence for multiple phenomena, in the form of multiple patterns that are exhibited in the data with differing noise levels" (Ibidem, 886). McAllister's (2007, 885) also critically investigates how researchers in various disciplines, including philosophy of science, have proposed quantitative techniques for determining which data-model is the best, where 'the best' is usually interpreted as 'the closest to the truth,' 'the most likely to be true,' or 'the best-supported by the data.' According to McAllister, these "[data-]model selection techniques play an influential role not only in research practice, but also in philosophical thinking about science. *They seem to promise a way of interpreting empirical data that does not rely on judgment or subjectivity*" (Ibidem, 885, my emphasis), which he disputes.

In this example 'the true value of the temperature at which lead melts' is considered to be the *phenomenon*, which, according to B&W is determined by statistical analysis of a set of data taken in measurements. Based on this example, one may be inclined to conclude that Glymour (2002) is correct in claiming that phenomena do not add anything to data.

In discussing this issue a bit further, I will use the notion 'physical phenomena' rather than just 'phenomena' to stress that phenomena in the sense of B&W are considered independently existing (physical) things (objects, properties or processes). Additionally, I will use the notion 'conceptions of phenomena,' to account for the fact, rightly pointed out by B&W, that phenomena are usually not observable in a straightforward manner, but need to be discovered and established. Hence, the notion of phenomena is connected to the notion of scientific concept, because a *scientific concept* can be considered *a conception of a physical phenomenon*, which, once the *meaning of the concept* is sufficiently established, becomes a *definition* in a dictionary or textbook. This definition gets the character of a (literal) *description* of the phenomenon (Boon 2012a).

The pressing question is whether *the formation of concepts of phenomena* (including establishing their definitions) will be still required once machine-learning technologies are able to find statistically justified patterns in data in the sense suggested by Glymour. More generally phrased, will *data-models* generated by statistical analysis of data make all other scientific knowledge superfluous, and will machine learning technology be able to generate these data-models?[8]

The Semantic View of Theories

Acceptance of scientific knowledge in empiricist epistemologies involve two important rules: knowledge must be *objective*, and it must be *testable*. Ideally, therefore, knowledge and the way in which it is *tested* must be *independent of specificities of human cognition*, and the measured data used for testing it must be independent of the knowledge to be tested. The so-called semantic view of theories, which in one or another version is held by authors such Suppes (1960), Van Fraassen (1980), Giere (1988, 2010), Suppe (1989), and Da Costa and French (2003), accounts for

[8] Affirmative answers to these questions can be taken as an interpretation of Anderson's position. Notably, even machine learning scientists and textbooks promote that knowledge of any sort related to the application (e.g., knowledge of concepts, of relevant and irrelevant aspects, and of more abstract rules such as symmetries and invariances) must be incorporated into the learning network structure whenever possible (Alpaydin 2010, 261). Abu-Mostafa (1995) calls this knowledge *hints*, which are the properties of the target function that are known to us independently of the training examples – i.e., hints are auxiliary information that can be used to guide the machine's learning process. The use of hints is tantamount to combining rules and data in the learning network structure – hints are needed, according to Abu Mostafa, to pre-structure data-sets because without them it is more difficult to train the machine. In *image recognition*, for instance, there are *invariance hints*: the identity of an object does not change when it is rotated, translated, or scaled.

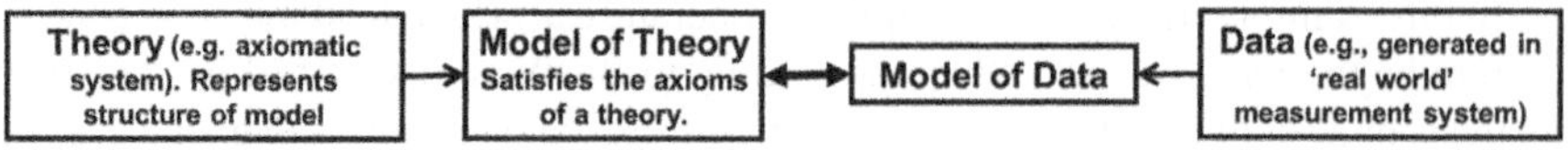

Schema 1 Semantical relationships between a theory and measured data according to the semantic view. *Theory acceptance* when (partial) isomorphism between model of theory and model of data

testing theories. It aims to account for the structure of theories and the *independent* relationship between theories and measurements, that is, between the outcomes predicted by the theory, and the outcomes of a measurement, by reducing the relationships between abstract theories, models and measured data to semantic relationships between abstract logical-, mathematical-structures and data-structures (see Schema 1).

Loosely speaking, the semantic view posits that a theory is a (usually deductively closed) set of sentences in a formal language, such as an abstract calculus, an axiomatic system, or a set of general laws (such as Newton's equations of motion), which enables to deduce logical consequences about particular types of physical systems (such as the model of a pendulum). The resulting model is a structure which is an *interpretation* (or *realization*) of the theory. Conversely, the theory *represents* the structure of the model. On this view, testing the *adequacy* of a theory only requires isomorphism (or similarity) between the model of the theory for a particular kind of system, and the measurement results called a model of the data. In brief, the semantic view explains how a theory is compared with measurements.[9] On Van Fraassen's (1980) version, *testing* whether a theory is *empirically adequate* means to assess (partial) isomorphism of a (mathematical) structures predicted by the theory (the models of the theory) and the structure in a set of measured data (the models of the data).

Obviously, the focus of the empiricist epistemology expressed in the semantic view is on the theory and how to test it. The question is not, for instance, whether the data-model is adequate. Conversely, in machine learning, the focus is on the data-model and how to test whether it is adequate. Therefore, from a machine learning perspective, someone may now ask 'why bother about the theory?' If machine learning technology can generate adequate data-models based on data, we do not need the theory any longer. Assume that a machine-learning technology has produced a data-model (although opaque and incomprehensive) that fits the data (see right part of Schema 1), and assume that the model of the theory is (partially) isomorphic with the data-model (see middle part of Schema 1), why would we need the left part of this schema anyway? Since, in empiricist epistemologies, the data and the data-model are taken as the solid ground of knowledge, the theory seems to be an unnecessary surplus. Hence, the *semantic view of theories* supports the idea that

[9]Notably, 'phenomena' in the sense of Bogen and Woodward (1988) do not occur in this view. Rather than *phenomena*, as B&W claim, the *model of data* mediates between the *measured data* and the *model of the theory*, which is a specific instantiation (interpretation, concretization) of the *theory* (see Schema 1).

scientific knowledge in the form of theories or models does not add much to empirically adequate and/or statistically sound data-models to represent data. Accordingly, it supports the belief that ultimately scientific knowledge is no longer needed. Again, the empiricist tradition offers hardly any possibilities for a more positive appreciation of scientific theories and the epistemic and cognitive roles of human scientists.

Knowledge in the Age of Machine-Learning Technologies

Empiricist Epistemologies: Theories Add Absolutely Nothing to Data-Models

In the previous section, it has been defended that a consequence of presuppositions and requirements of (anti-realist) empiricist epistemologies is that *explanations*, *phenomena*, and *theories* generated in science can (in principle, although maybe not yet in practice) be represented by, reduced to, or replaced with *data-models* generated by machine learning technologies. Empiricist epistemologies require that data-models adequately fit the data, but there are no specific epistemological reasons to require that data-models are intelligible for humans—that is, the fact that data-models generated by machines usually are opaque and incomprehensive for humans is not a problem in regard of the epistemic value of data-models. Additionally, referring to Duhem, and in his footsteps Van Fraassen, *theories tell us absolutely nothing about hidden realities*—rather, different theories may be equally capable of representing a given set of experimental laws. When taking *experimental laws*, in Duhem's wordings, to be *data-models*, this implies that no additional epistemic value is gained by theories over data-models, especially when data-models accurately represent large data-sets achieved by machine learning technologies. Hence, taking the semantic view of theories as a proper advancement of Duhem's ideas implies that the *epistemic* value of theories is to adequately represent *data-models*, where 'represent' means 'structural similarity,' i.e. being (partially) isomorphic. In turn, data-models *represent* the measured data. If we assume that representational relationships in science are transitive, this implies that from an epistemological point of view empirically adequate theories do not add anything to empirically adequate data-models[10]—as empirically adequate data-models already allow for adequate predictions of 'real-world' data, theories and models become unnecessary (see Schema 1). As a consequence, the claim that machine learning technologies will render human scientists and scientific knowledge superfluous accords with beliefs about the epistemic value of theories in anti-realist empiricist epistemologies.

[10] This claim only holds for anti-realist interpretations (as in Duhem and Van Fraassen) of the semantic view. Yet, the semantic view of theories also allows for realist interpretations of theories (e.g., Suppe 1989).

Empiricist epistemologies, therefore, support Anderson's (2008) claims cited in the introduction. Having this said, several arguments can be put forward against this conclusion.

Scientific Realism in Defense of Science

Anderson's (2008) claims can be more easily countered from scientific realist than from anti-realist viewpoints. Scientific realist positions are supported, at least in part, by the no-miracle argument: the successes of scientific theories would be a miracle unless we assume that theories truthfully describe or represent hidden realities behind the phenomena, which is why scientific realism is the best explanation for the successes of science. As a consequence, data-models, whether produced by human scientists or by machines, are epistemologically inferior to theories. Accordingly, contrary to the conclusion inferred from anti-realist empiricist epistemologies, scientific realists will believe that the successfulness of (approximately) true scientific theories cannot be superseded by data-models.

Additionally, scientific realists may argue that scientific theories have an *intrinsic* value, which has nothing to do with their epistemic or practical usefulness anyway. Many theories are not useful, at least not to begin with. One may even defend that the aim of science is not *useful* theories, but *true* theories. Science may be of epistemic and practical value to all kinds of applications such as in engineering and medicine, but this is a by-product of science, not its intended aim (also see Boon 2011, 2017c). Rather, science has an intrinsic *cultural* value in telling us what the world is like, which is a task that cannot be replaced by machine learning technologies whatsoever since incomprehensive, opaque data-models do not tell us anything meaningful about the world. Therefore, 'real science' and machine learning technologies operate in very different domains and must not be regarded as competing.

The Pragmatic Value of Scientific Knowledge in Epistemic Tasks

Empiricist epistemologies do not deny the pragmatic value of science and agree indeed that *pragmatic criteria* play a role in the *acceptance* of theories, but only deny that pragmatic criteria add to the epistemic value of theories (e.g., Van Fraassen 1980). In section "Empiricist epistemologies", it has been argued that machine-made data-models may become capable to perform better in regard of *epistemic criteria* (esp. empirical adequacy regarding the data) as compared to human-made scientific knowledge. In addition, it has been suggested that the generation and use of data-models for all kinds of *epistemic tasks* can be carried out by machine learning technology, which will in many cases perform better than human scientists who aim to generate and use scientific knowledge for similar tasks (see overview in section "Machine-learning"). It has also been argued that machine-made data-models usually are incomprehensible, opaque, and even inaccessibly 'sitting' in the machine, to the effect that they cannot be used by human epistemic agents. Therefore, machines do not produce scientific knowledge as in 'traditional' scientific practices— i.e., epistemic entities such as theories, laws, models and concepts that can be obtained from the machine and utilized in, say, self-chosen epistemic tasks by humans. Even if it were possible to obtain the data-models from the machine, they would be useless for epistemic uses by humans as these data-models do not meet relevant pragmatic criteria to enable such uses. The other way around, in order to be useful for humans in performing epistemic tasks, scientific knowledge must also meet pragmatic criteria.

The crux of pragmatic criteria such as *consistency*, *coherency*, *simplicity*, *explanatory power*, *scope*, *relevance* and *intelligibility* in generating and accepting scientific knowledge is to render scientific knowledge manageable for humans in performing epistemic tasks. I will leave unanswered whether machine learning technologies cannot offer this in principle. But if they cannot, a future without science would require machines to take over every epistemic task, which seems unlikely already regarding our daily interactions with the world.

Preparing the Data

Much needs to be in place before the machine-learning can even begin. Data-sets need to be generated, prepared and gathered, which requires epistemic activities by humans, such as designing experimental set-ups and measurement equipment (e.g., as in the experiments of Boyle, etc. in section "Introduction"). These epistemic tasks require scientific and background knowledge. As stated above, knowledge must meet specific pragmatic criteria to be manageable when performing epistemic tasks. For example, knowledge must be such that epistemic agents can see which real-world target-systems the knowledge is applicable to—for example, in order to recognize or explain specific phenomena in the data-generating experimental set-up. Conversely, it requires of scientists to have the cognitive ability to *think*, *theorize*, *conceptualize*, *explain*, *mathematize*, and *interpret* by means of scientific knowledge when performing epistemic tasks, not only when setting up the data-generating instrumentation and seeing to its proper functioning, but also in assessing and interpreting the data, drawing relationships between data from different sources, and for making the distinction between 'real' phenomena and artifacts. These crucial cognitive abilities go well beyond what empiricist epistemologies can explain, require, or allow in view of the requirements of objectivity.

The necessity to prepare data that are *about something in the real world* also implies that *phenomena* are crucial in scientific practices, even when only aiming at the generation of data for machine-learning processes. Harking back to the discussion above, the way in which Bogen and Woodward (1988) think about phenomena forces them to accept that *phenomena* coincide with *data-models*. However, this notion of physical phenomena is far too narrow regarding the uses of this notion in scientific practices, even if only practices aimed at measuring data. The description of a physical phenomenon such as 'lead melts at 327.5 + 0.1 degrees C' is not grasped by the number (i.e., the value) in this proposition. In contrast to the data-model that is statistically derived from the measurements in the way suggested by B&W, the described phenomenon can be analyzed in terms of a set of interrelated *heterogeneous* aspects, such as: the *observation* that substances (including lead) can melt; an *understanding* of the concept 'temperature' (also see Chang 2004); the *observed* regularity that a substance (including lead) always melts at approximately the same temperature; an *understanding* of the concept 'melting-point'; the *conception* that 'having a melting-point' is a specific characteristic of substances; the *regulative principle* that at the same (experimental) conditions the same effects will happen (Boon 2012b); the *assumption* that the temperature at which a substance melts (the melting-point) is an exact number; the *assumption* that the temperature can be measured at a pretty high accuracy; the *decision* or *assumption* that the observed fluctuations in the observed melting temperatures are due to (partially) unknown causes of the experimental set-up and measurement tools (e.g., Mayo 1996); and finally, an *understanding* of the workings of the measurement tools. In short, the full conception of the physical phenomenon consists of a collection of heterogeneous, mutually related but heterogeneous aspects, which are generated in a number of cognitive actions by human scientists, rather than being a statistically derived number only.

This elaboration of B&W's example shows that skills, knowledge, and understanding of scientists are required to establish both the physical phenomenon—which involves the experimental and measurement set-up, and also their proper operations to get a stable and reproducible measurement of the temperature at which lead melts—, as well as the *conception of the phenomenon*, even if the phenomenon under study is as simple as 'lead melts at 327.5 + 0.1 degrees C.' Additionally, this brief analysis shows that the formation of the concept of a phenomenon and physically establishing it in an experimental set-up, go hand-in-hand, and necessarily involve all kinds of basic assumptions that cannot be empirically tested (Chang 2004; Boon 2012a, b, 2015).

Expanding on this analysis, it can be argued that empiricist epistemologies are flawed in believing that the *theory-ladenness* of data is fundamentally problematic as it threatens the objectivity of science. More specifically, Bogen and Woodward are mistaken to hold that phenomena should not be some kind of low-level theories (claim 4). To the contrary, *theory-emptiness* of data fed to machine-learning processes would really be a problem. In actual scientific practices, the production of data representing supposed physical phenomena usually develops in a process of *triangulation* together with the development of the experimental set-up and measurement techniques and with the construction and application of scientific knowledge of all kinds. The data, phenomena and theory, as well as our interpretations of measurements and understanding of the working of instruments and experimental set-ups are intrinsically conceptually entangled (e.g., Chang 2004; Feest 2010; Van Fraassen 2008, 2012; Boon 2012a, 2017a; Van Fraassen 2008, 2012).

Epistemic Tasks in Engineering and Biomedical Sciences

In machine-learning-technologies, descriptions of (physical or social) phenomena are reduced to (and represented by) data-models, which is considered unproblematic within empiricist epistemologies. As sketched above, the data or data-model representing the phenomenon entail hardly any information relevant to epistemic tasks in dealing with phenomena, for instance, in aiming to *interact* with the targeted phenomenon in one or another way.

Yet, these kinds of epistemic tasks are at the center of the so-called applied sciences such as the engineering and biomedical sciences. These research practices aim at scientific knowledge about targeted (bio)physical phenomena, and about technological instruments that can possibly produce or control them—for the sake of the targeted phenomenon, not first theories, which are considered to be the focus of basic sciences.[11] These practices function in the way sketched above, which is to say that experimentally producing and investigating targeted phenomena (e.g., a phenomenon we aim to produce for a specific technological or medical function) is *entangled* with the generation of scientific knowledge and the development of technological instruments and measurement apparatus relevant to the phenomenon. Every tiny step in these intricate research processes involves epistemic tasks—e.g., to explain, interpret, invent, idealize, simplify, hypothesize, model, mathematize, design, and calculate—for which all kinds of practical and scientific knowledge is crucial and needs to be developed in the research process. Therefore, scientific knowledge needs to be *comprehensible* to the extent that it allows for these epistemic

[11] In other work, I have explained from a range of different philosophical issues, the crucial role of phenomena in the 'applied' research practices and what this means for our philosophical understanding both of scientific knowledge and of the aim of science (Boon 2011, 2012a, b, 2015, 2017a, c, forthcoming). The idea that these application-oriented scientific research practices aim at scientific knowledge in view of *epistemic tasks* aimed at learning how to do things with (often unobservable, and even not yet existing) physical phenomena has led to the notion of scientific knowledge as *epistemic tool* (Boon and Knuuttila 2009; Knuuttila and Boon 2011; Boon 2015; Boon 2017b,c; also see Nersessian 2009; Feest 2010; Andersen 2012). The original idea of scientific knowledge (or, originally more narrowly stated, 'scientific models') as epistemic tools, proposes to view scientific knowledge—such as descriptions, concepts, and models of physical phenomena—firstly as *representations of scientists' conceptions of aspects of reality*, rather than representations in the sense of a two-way relationship between knowledge and reality (as in anti-realist empiricist epistemologies as well as in scientific realism). The point of this (anti-realist) view is that someone can represent her conception (comprehension, understanding, interpretation) of aspects of reality by means of *representational means* such as text, analogies, pictures, graphs, diagrams, mathematical formula, and also 3D material entities. Notably therefore, scientists' conceptions of observable as well as unobservable phenomena arrived at by intricate reasoning processes (creative, inductive, deductive, hypothetical, mathematical, analogical, etc.), which employ all kinds of available epistemic resources, can be *represented*. By representing, scientists' conceptions become *epistemic constructs* that are public and transferable. Knuuttila and I have called these constructs *epistemic tools*, that is, conceptually meaningful tools that guide and enable humans in performing all kinds of different *epistemic tasks*.

tasks. Especially regarding these kinds of practically oriented scientific research practices in which human scientists aim at comprehensible scientific knowledge as well as epistemic and practical resources such as measurement instruments, technological procedures, and methods, it is inconceivable that machine-learning-technologies will make science and scientists superfluous.

The Error of Empiricism

Empiricist epistemologies insufficiently account for the types of epistemic tasks that are crucial for the development and use of epistemic and practical tools, which in turn are used in the development of, for instance, medical technologies. This shortcoming already applies to the generation of data that can be fed to machines that generate data-models for specific purposes. Empiricist epistemologies therefore miss out on crucial aspects of the uses and generation of scientific knowledge (theories, models, etc.) in intricate scientific processes taking place in application-oriented research practices like the engineering and biomedical sciences, and thus give room to beliefs such as defended by Anderson (2008). Rethinking the philosophical presuppositions of empiricist epistemologies that seem to force us to the view that science will become superfluous in the age of machine learning can help in gaining insights that bring the scientist back into science.

Acknowledgements This work is financed by an Aspasia grant (2012–2017 nr. 409.40216) of the Dutch National Science Foundation (NWO) for the project "Philosophy of Science for the Engineering Sciences." I want to acknowledge Koray Karaca for critical suggestions and fruitful discussions; I am indebted to him for introducing me to the machine learning literature, which I have gratefully used in writing the introduction to this theme. I also wish to acknowledge Marta Bertolaso and her team for organizing the expert meeting (March 5–6, 2018 in Rome) on the theme "Will Science remain Human? Frontiers of the Incorporation of Technological Innovations in the Biomedical Sciences," and participants of this meeting for their fruitful comments, especially by Melissa Moschella.

References

Abu-Mostafa, Y. 1995. Hints. *Neural Computation* 7: 639–671. https://doi.org/10.1162/neco.1995.7.4.639.

Abu-Mostafa, Y. S., Magdon-Ismail, M., and Lin, H.- T. (2012). *Learning from data*. AMLbook.com. ISBN 10:1-60049-006-9, ISBN 13:978-1-60049-006-4

Alpaydin, E. 2010. *Introduction to Machine Learning*. Cambridge: The MIT Press.

Andersen, H. 2012. Concepts and Conceptual Change. In *Kuhn's The Structure of Scientific Revolutions Revisited*, ed. V. Kindi and T. Arabatzis, 179–204. Routledge.

Anderson, C. (2008). The End of Theory: The Data Deluge Makes the Scientific Method Obsolete. *Wired Magazine,* June 23. Retrieved from: https://www.wired.com/2008/06/pb-theory/

Bogen, J. 2011. 'Saving the Phenomena' and Saving the Phenomena. *Synthese 182* (1): 7–22. https://doi.org/10.1007/s11229-009-9619-4.

Bogen, J., and J. Woodward. 1988. Saving the Phenomena. *The Philosophical Review 97* (3): 303–352. https://doi.org/10.2307/2185445.

Boon, M. 2011. In Defense of Engineering Sciences: On the Epistemological Relations Between Science and Technology. *Techné: Research in Philosophy and Technology 15* (1): 49–71. Retrieved from http://doc.utwente.nl/79760/.

———. 2012a. Scientific Concepts in the Engineering Sciences: Epistemic Tools for Creating and Intervening with Phenomena. In *Scientific Concepts and Investigative Practice*, ed. U. Feest and F. Steinle, 219–243. Berlin: De Gruyter.

———. 2012b. Understanding Scientific Practices: The Role of Robustness Notions. In *Characterizing the Robustness of Science After the Practical Turn of the Philosophy of Science*, ed. L. Soler, E. Trizio, Th. Nickles, and W. Wimsatt, 289–315. Dordrecht: Springer: Boston Studies in the Philosophy of Science.

———. 2015. Contingency and Inevitability in Science – Instruments, Interfaces and the Independent World. In *Science as It Could Have Been: Discussing the Contingent/Inevitable Aspects of Scientific Practices*, ed. L. Soler, E. Trizio, and A. Pickering, 151–174. Pittsburgh: University of Pittsburgh Press.

———. 2017a. Measurements in the Engineering Sciences: An Epistemology of Producing Knowledge of Physical Phenomena. In *Reasoning in Measurement*, ed. N. Mößner and A. Nordmann, 203–219. London/New York: Routledge.

———. 2017b. Philosophy of Science in Practice: A Proposal for Epistemological Constructivism. In *Logic, Methodology and Philosophy of Science – Proceedings of the 15th International Congress (CLMPS 2015)*, ed. H. Leitgeb, I. Niiniluoto, P. Seppälä, and E. Sober, 289–310. College Publications.

———. 2017c. An Engineering Paradigm in the Biomedical Sciences: Knowledge as Epistemic Tool. *Progress in Biophysics and Molecular Biology* 129: 25–39. https://doi.org/10.1016/j.pbiomolbio.2017.04.001.

———. forthcoming. Scientific methodology in the engineering sciences. Chapter 4. In *The Routledge Handbook of Philosophy of Engineering*, ed. D. Michelfelder and N. Doorn. Routledge. Publication scheduled for 2019.

Boon, M., and T. Knuuttila. 2009. Models as Epistemic Tools in Engineering Sciences: A Pragmatic Approach. In *Philosophy of Technology and Engineering Sciences. Handbook of the Philosophy of Science*, ed. A. Meijers, vol. 9, 687–720. Elsevier/North-Holland.

Chang, H. 2004. *Inventing Temperature: Measurement and Scientific Progress*. Oxford: Oxford University Press.

Craig, E. 1998. Duhem, Piere Maurice Marie. In *Routledge Encyclopedia of Philosophy: Descartes to gender and science.*, vol. 3, 142–145. London/New York: Routledge.

Da Costa, N.C.A., and S. French. 2003. *Science and Partial Truth. A Unitary Approach to Models and Scientific Reasoning*. Oxford: Oxford University Press.

Dai, W., et al. 2015. Prediction of Hospitalization Due to Heart Diseases by Supervised Learning Methods. *International Journal of Medical Informatics* 84: 189–197.

Duhem, P. 1954/[1914]. *The Aim and Structure of Physical Theory*. Princeton: Princeton University Press.

———. 2015/[1908]. *To Save the Phenomena: An Essay on the Idea of Physical Theory from Plato to Galileo*. Trans. E. Dolland and C. Maschler. Chicago: University of Chicago Press

Esteva, A., B. Kuprel, R.A. Novoa, J. Ko, S.M. Swetter, H.M. Blau, and S. Thrun. 2017. Dermatologist-level Classification of Skin Cancer with Deep Neural Networks. *Nature 542*: 115. https://doi.org/10.1038/nature21056.

Feest, U. 2010. Concepts as Tools in the Experimental Generation of Knowledge in Cognitive Neuropsychology. *Spontaneous Generations 4* (1): 173–190.

Giere, R.N. 1988. *Explaining Science*. Chicago/London: The University of Chicago Press.

———. 2010. An Agent-Based Conception of Models and Scientific Representation. *Synthese* 172 (2): 269–281. https://doi.org/10.1007/s11229-009-9506-z.

Glymour, B. 2002. Data and Phenomena: A Distinctions Reconsidered. *Erkenntnis 52*: 29–37.

Hempel, C.G. 1962. Explanation in Science and Philosophy. In *Frontiers of Science and Philosophy*, ed. R.G. Colodny, 9–19. Pittsburgh: University of Pittsurgh Press.
———. 1966. *Philosophy of Natural Science*. Englewood Cliffs: Prentice-Hall.
Hofree, M., J.P. Shen, H. Carter, A. Gross, and T. Ideker. 2013. Network-based Stratification of Tumor Mutations. *Nature Methods 10*: 1108. https://doi.org/10.1038/nmeth.2651. https://www.nature.com/articles/nmeth.2651#supplementary-information.
Humphreys, P. 2009. The Philosophical Novelty of Computer Simulation Methods. *Synthese* 169 (3): 615–626. https://doi.org/10.1007/s11229-008-9435-2.
Knuuttila, T., and M. Boon. 2011. How Do Models Give Us Knowledge? The Case of Carnot's Ideal Heat Engine. *European Journal for Philosophy of Science 1* (3): 309–334. https://doi.org/10.1007/s13194-011-0029-3.
Kourou, K., et al. 2015. Machine Learning Applications in Cancer Prognosis and Prediction. *Computational and Structural Biotechnichnology Journal* 13: 8–17.
Lemm, S., et al. 2011. Introduction to Machine Learning for Brain Imaging. *NeuroImage* 56: 387–399.
Libbrecht, M.W., and W.S. Noble. 2015. Machine Learning Applications in Genetics and Genomics. *Nature Reviews Genetics* 16: 321–332.
Lima, A.N., et al. 2016. Use of Machine Learning Approaches for Novel Drug Discovery. *Expert Opinion on Drug Discovery* 11: 225–239.
Mayo, D.G. 1996. *Error and the Growth of Experimental Knowledge*. Chicago: University of Chicago Press.
McAllister, J.W. 1997. Phenomena and Patterns in Data Sets. *Erkenntnis 47* (2): 217–228. https://doi.org/10.1023/A:1005387021520.
———. 2007. Model Selection and the Multiplicity of Patterns in Empirical Data. *Philosophy of Science 74* (5): 884–894. https://doi.org/10.1086/525630.
———. 2011. What do Patterns in Empirical Data Tell Us About the Structure of the World? *Synthese 182* (1): 73–87. https://doi.org/10.1007/s11229-009-9613-x.
Mena, J. 2011. *Machine Learning Forensics for Law Enforcement, Security, and Intelligence*. Boca Raton: CRC Press.
Nersessian, N.J. 2009. *Creating Scientific Concepts*. Cambridge, MA: MIT Press.
Odone, F., M. Pontil, and A. Verri. 2009. Machine Learning Techniques for Biometrics. In *Handbook of Remote Biometrics. Advances in Pattern Recognition, ch. 10*, ed. M. Tistarelli, S.Z. Li, and R. Chellappa. London: Springer.
Olszewska, J.I. 2016. Automated face Recognition: Challenges and Solutions. In *Pattern Recognition – Analysis and Applications*, ed. S. Ramakrishnan, 59–79. InTechOpen. https://doi.org/10.5772/62619.
Phua, C., et al. (2010). A comprehensive survey of data mining-based fraud detection research. https://arxiv.org/abs/1009.6119
Suppe, F. (1974). *The Structure of Scientific Theories* (1979 second printing ed.). Urbana: University of Illinois Press.
———. 1989. *The Semantic Conception of Theories and Scientific Realism*. Urbana/Chicago: University of Illinois Press.
Suppes, P. 1960. A Comparison of the Meaning and Uses of Models in Mathematics and the Empirical Sciences. *Synthese* 12: 287–301.
Tcheng, D.K., A.K. Nayak, C.C. Fowlkes, and S.W. Punyasena. 2016. Visual Recognition Software for Binary Classification and Its Application to Spruce Pollen Identification. *PLoS ONE 11* (2): e0148879. https://doi.org/10.1371/journal.pone.0148879.
Van Fraassen, B.C. 1977. The Pragmatics of Explanation. *American Philosophical Quarterly 14*: 143–150.
———. 1980. *The Scientific Image*. Oxford: Clarendon Press.
———. 2008. *Scientific Representation*. Oxford: Oxford University Press.
———. 2012. Modeling and Measurement: The Criterion of Empirical Grounding. *Philosophy of Science* 79 (5): 773–784.

Van Liebergen, B. 2017. Machine Learning: Revolution in Risk Management and Compliance? *The Capco Institute Journal of Financial Transformation* 45: 60–67.
Woodward, J.F. 2011. Data and Phenomena: A Restatement and Defense. *Synthese 182* (1): 165–179. https://doi.org/10.1007/s11229-009-9618-5.
Wuest, T., et al. 2016. Machine Learning in Manufacturing: Advantages, Challenges, and Applications. *Production & Manufacturing Research* 4 (1): 23–45.

Information at the Threshold of Interpretation: Science as Human Construction of Sense

Giuseppe Longo

Introduction: The Origin of Sense

We may dream of the origin of the human construction of sense in the early gestures by which our ancestors, while inventing human communicating communities, tried to interpret the totally meaningless spots of light in the night's sky. They probably pointed out to each other configurations of stars by interpolating them with lines, a founding mathematical gesture. Language allowed to name them: here is a lion, there a bison, or whatever. This surely had also some use, in orientation, in recognizing seasons… but the shared meaningful myth was probably at the origin of this early invention of configurations of sense in a meaningless dotted sky.

Structuring, organizing reality, which in turn resists and canalizes these actions of ours, these are the primary gestures of human knowledge. We single out and qualify fragments of reality by setting contours, by giving names and associating properties with them, and correlating unrelated fragments.

These cognitive gestures are not to be reduced to, but have their condition of possibility in our biological activity. Recall that Darwin's first principle of heredity in evolution is "reproduction with modification". But also allopatric speciation, by individual or populations' movement, and, in general, hybridizing migrations, are at the core of his earliest analysis: the finches of Galapagos he observed are the result of moving populations. Thus "motility" must be integrated as the other founding principle for the understanding of biological evolution (Soto et al. 2016). In other words, *agency*, in the broadest sense, is essential for life, from motility and action in space to cognitive phenomena, since "motility is the original intentionality"

G. Longo (✉)
Centre Cavaillès, République des Savoirs, CNRS et École Normale Supérieure, Paris, France

School of Medicine, Tufts University, Boston, MA, USA
e-mail: Giuseppe.Longo@ens.fr

M. Bertolaso, F. Sterpetti (eds.), *A Critical Reflection on Automated Science*, Human Perspectives in Health Sciences and Technology 1,
https://doi.org/10.1007/978-3-030-25001-0_5

(Merleau-Ponty 1945). Constraints interactively impose limits and canalize motility and organismal dynamics, from cells and organisms (Montévil and Mossio 2015) to human activities.

Agency is at the origin of significance, since a signal, a disturbance, a collision that impacts a moving amoeba is "meaningful" for the amoeba according to the way it affects this original intentionality – it may favor or oppose the ongoing protensive action, the amoeba's anticipations (Saigusa et al. 2008) – a thesis hinted in (Bailly and Longo 2006).

It is always an organism, as a unity, that acts and moves, that attributes meaning to an incoming signal or perturbation. The "correlated changes" induced by a hit, a paraphrase of Darwin's "correlated variations" within organisms and within an environment, are made possible only by the organismal unity and its history in an ecosystem. The formation of sense is thus historical, beginning with the phylogenetic and then the ontogenetic history, including embryogenesis for multicellular organisms.

As for comparing organisms and machines, including computers, note that embryogenesis is obtained by *reproduction with differentiation* from one organism, the zygote. At each step of embryogenesis, cells' reproduction preserves the organism's unity. Differentiation has nothing to do with the assemblage of parts, elementary and simple ones, by which we construct any artifact. A child is not obtained by attaching a leg, gluing a nose, introducing an eye in a hole, but by reproductive differentiation of individual cells. Thus, at each step, the unity of the organism is preserved and is a subject of sense: sensations and reactions to sensation are meaningful at the proper level of the forming individual, they are interpreted. This is radically different from the piecewise assembly of any machine whatsoever, which precludes the constitutive formation of meaning as interpretation of the "deformation" of a self-constructed, differentiating and self-maintained biological unity and of its correlation to an environment. An "interpretation" is always the result of a historical and contextual construction of sense: different histories, such as embryogenesis and assemblage, produce different meanings.

Of course, there is a huge distance, actually several "critical transitions", between this biological formation of sense, as action and reaction of an organism in an ecosystem, and the meaningful constructions of human knowledge. Yet, the evolutionary and historical dynamics of knowledge formation have their roots, and their *conditions of possibility*, in these primary aspects of life: the moving eukaryote cell, the embryogenesis of a multicellular organism, whose active unity and motility allow to interpret, actually force to attribute meaning to its deforming "frictions" within the ecosystems. By a long evolutionary and historical path, this original formation of sense, as *deformation of an active unity*, becomes knowledge construction in the human communicating community, with its proper levels of interaction, thus its proper unity.

The challenge we have today concerns a newly invented observable, information, and its relation to sense and knowledge construction. Of course, "information" exists since long, or as soon as an animal brain forms the invariants of action: the retention of what matters in an ongoing action and its selective recalling for a protensive activity jointly single out what may be considered the "relevant information" to be retained and later used for further action. This relative independence from the context of the retained action underlies the construction of the invariant, a trajectory, a sound … and may be analyzed as a primary form of information. That is, the constitution of cognitive invariants w.r.to a context may be viewed as the founding practice of the subsequent notion of information that only language and, even more so, writing allows to properly single-out and define. Yet, only the modern invention of machines for transmission and elaboration of information, independently of meaning, as strings of dots and lines in Morse alphabet, 0 s and 1 s in today's computers and in Shannon's theory, fully detached information from other observables and, in particular, from meaning. In conclusion, once the invariants resulting from animal and human cognitive activities were made independent from action and stabilized as (electronic) signs,[1] transmission and elaboration of information became a mathematical science, actually at least two very relevant sciences (Turing's and Shannon's). Then, information could be formally analyzed, elaborated and transmitted, independently of any interpretation.

The Modern Origin of Elaboration of Information as Formal Deduction: Productivity and Limits of 'Nonsense' in the Foundational Debate in Mathematics

It may be fair to say the powerful promotion of meaningless formalisms as information carriers has its modern origin in the formal-axiomatic approach to the Foundations of Mathematics, at the end of the nineteenth century. Historians may better specify the predecessors of this "linguistic turn" in Logic, such as Leibniz's "lingua characteristica", as well as whether this turn was just a consequence or a symptom or an actual promoter of the new vision of human knowledge that will mark the twentieth century, as formal elaboration of signs. In any case, the foundational issue in Mathematics brought to the limelight the formal-linguistic perspective in the clearest way. This issue is at the origin, in the '30s, of the various mathematical notions of computation, such as Church's lambda-calculus (1932) and Turing's "Logical Computing Machine", as he called it (1936). In short, whatever was the general cultural role of the debate internal to Logic, computers, as mechanical devices for elaborating information, and their networks that are changing our lives were invented within that debate on the foundations of Mathematics.[2] In order

[1] We use "sign", in reference to a priori meaningless strokes, letters, 0 s and 1 s …, thus we use the expression "sign pushing" instead of "symbol pushing" widely used in Computer Science. "Symbol" retains the Greek etymology of "sym-ballein" or "to bring together" or to unify different acts of experience or synthesize meaning.

[2] Note that computability and its machines were invented in order to prove the *limits* of the formal-linguistic approach. As hinted above, in order to prove undecidability and incomputability, Gödel,

to criticize the abuses of the computational approaches to the notion of information, which we can find in many, vague, un-scientific, commonsensical references in biology and even in other sciences, let's briefly survey its robust origins within Mathematics.

The issue at stake at the end of the nineteenth century was to save Mathematics from the major foundational crisis induced by the invention of non-euclidean geometries: the reliability and certainty of the mathematical edifice, guaranteed by the perfect correspondence between physical space and Euclid's geometry in Cartesian spaces, was suddenly lost. Hilbert (1898), with extreme rigor, proposed formal axiomatic approaches for the various existing geometries, where certainty was no longer based on "meaning", as reference of the geometric properties to actual physical space, but only on the exact axiomatic presentation and on the internal coherence of the theories so defined. The information had to be entirely formalized in the axioms and deduction developed independently of meaning, in particular as reference to a "meaningful" geometric structure of space.[3] The absence of logico-formal contradiction, that is of a purely formal deduction of a proposition *A* and of its negation, *not A*, was the only quest and guarantee for certainty. By an analytical encoding of all the mathematical theories he had axiomatized, into Arithmetic, that is in the Formal Theory of Integer Numbers, Hilbert transferred the burden of coherence into Arithmetic: a formal, "potentially mechanisable" proof of its coherence could reassure the "absolute certainty" of existing Mathematics, non-euclidean geometries first – this was one of the famous 23 open problems he posed in 1900. Gödel, Church, and Turing's negative answers required first … the positive definitions of "formal deduction or computation", for example by the construction of the Logical Computing (Arithmetic) Machine by Turing, that is of a rule based formal system. By their negative results they did not set a limit to thinking, as the analytic view of mind as a "sign pushing" and information-carrier device makes us believe, since the actual limit is elsewhere and cannot be *thought*, surely not from outside. Yet, they set a limit only to the formal-linguistic manipulation of signs, and this from *inside* – as often forgotten: by smart diagonal techniques on strings of signs, they formally showed that certain strings of signs, as formal definitions of formulae or of functions, cannot be proved nor negated. As it was later shown, even in mathematics, actually in Arithmetic, we do think (and prove) beyond that limit set to formal axioms and deductions, so that we can now look at them from outside – see below and (Longo 2011a, b).

Church and Turing (1936) had first to give precise definitions of "computable function" and "formal derivation", thus of formal computing/arithmetic machines, the mathematical foundation of modern digital computers – see (Longo 2010) for the relevance of these and other *negative results*.

[3] H. Weyl, Hilbert's "best student", recalls his supervisor's philosophy by quoting: "a typically Hilbertian manner: "It must be possible to replace in all geometric statements the words point, line, plane by table, chair, mug." In [Hilbert's] deductive system of geometry the evidence, even the truth of axioms, is irrelevant …" (Weyl 1953).

In spite of the limitations thus proved (the existence of undecidable sentences, the formal unprovability of coherence, the construction of incomputable functions), the mechanical notion of deduction is still now setting a paradigm for knowledge, well beyond its role as a remarkable (yet incomplete) mechanical *tool* for knowledge. And coded "sign pushing" has been even extended to an understanding of biology (DNA as "an encoded program") and, by some, of physics. Moreover, even when considered just as a tool for knowledge, it should be clear that no tool is neutral: it organizes, it shapes the object of investigation.

Reconquering Meaning

As for human cognition, the formal, digital elaboration of information was soon to be identified with "knowledge construction" and more generally "intelligence". This excludes what Alessandro Sarti calls "imagination of configurations of sense": those primary gestures of knowledge that give meaning even to nonsense, by tracing, interpolating, structuring the world, from connecting the bright dots in the sky to Euclid's notion of the "line with no thickness". The invention of this founding mathematical structure, spelled out in Euclid's books definition beta, is at the origin of western mathematics (Longo 2015a, b). This line is a pure length, a constructed border of a figure, a trajectory – all organizing features of action in space.

As a further answer to the formalists myths of mathematics as sign pushing, note that any relevant proof requires the invention of new mathematical structures, from Euclid's lines to … Grothendieck's toposes, or even the insight into provably non-formalizable properties, as we will hint. A mathematician goes nowhere without the imagination of new concepts and structures, often "deformations" of existing ideas or invention of new correlations. These may be, sometimes, but not always, only a posteriori fully formalizable. But, what does *a posteriori* mean? Proving a theorem, in general, in Mathematics, is not proving an already given formula, the formalist parody – this is very exceptional, such as the quest for "Fermat's last theorem", i.e. the proof of a simple formula that required four centuries to be given.[4] Normally, proving a theorem is answering a question, which may later lead to "writing a formula". For example, the reader should try to tell which is the sum of the first *n*

[4] Even Wiles' 1993 proof of this easy to state arithmetic property, Fermat's last theorem, required the invention of new deep mathematical ideas, a complex blend of advanced algebraic geometry, the "vision" of a path that linked homology to the analysis of elliptic curves, leading to the totally unrelated new ideas and techniques of the "modularity lifting theorem" (Cornell et al. 2013). Checking formally the proof, a posteriori – once it has been given, may be very useful, but it is a different matter, (Hesselink 2008).

integers, for *n* generic integer. If you know the formula, a computer may easily prove it by induction, but … where is the formula? Its invention requires an insightful game of symmetries, i.e. the imagination of a geometric configuration of numbers, (Longo 2011a, b).[5]

The "vision" here concerns an order in space and its inversion, that is, young Gauss' proof in the footnote, a typical mathematical construction, is the "imagination of a configuration of sense", a sense given by the human *aim* of the proof and the *geometrical structuring* of numbers, as an order. Interpolating, ordering stars and giving them names of animals or gods is a similar invention. None of these constructions is an "elaboration of (pre-given?) information", surely not in the formalist or mechanical sense. There is no miracle in this, but *active* insight and conceptual bridges based on the *meaning* of mathematical structures – geometric/algebraic meaning as ways to organize scattered points in space or concepts in mathematics. The Greek word "theorem" has the same root of "theater". In "acting out" a theorem, though, a mathematician moves possibly original pieces in the game, (s)he makes organizing "*gestures*". The notion of gesture is a hard to define; it is at the core of Chatelet's close analysis of the foundations of the nineteenth century's physics and mathematics, when mathematical gestures invented a new physics and motivated new mathematics (Chatelet 1993). More recently, Grothendieck's ideas and geometric "insights", probably one of the most original and broad mathematical work of the last 70 years, reinvented algebraic geometry; a philosophical account of his organizing mathematical gestures may be found in (Zalamea 2012), see (Longo 2015b) for a review.

In summary, the formalist myth of pure calculi of signs (formal handling of information) was born or enhanced by the analysis of the foundations of mathematics. This is why a reflection on these foundations may greatly help in the discussion, in particular in setting the limits and specifying the alternatives to formal rule-based thinking, viewed as the only access to rationality and even to … biological dynamics. This view was promoted by the same formalist trend that enabled the invention of digital computers, from Turing Machines to von Neuman or Crutchfield automata, that we will mention, and today's computers i.e. the possibility of moving signs according to formal rules prescribing how to write and re-write (manipulate) them without any reference to meaning. These are all Discrete State Machines, implementing a dynamic of information as strings of signs, handled by rules, both signs

[5] In short, following a proof allegedly given by Gauss at the age of 7, place

1	2	…	n	on a row
n	(n-1)	…	1	on the next row (an audacious mirror symmetry of the usual order of number writing), then obtain:

(n + 1) … (n + 1) by adding the columns. Thus, the sum of the first *n* integers is $(n + 1)n/2$, which is easy now to *check* by induction; yet, we had first to produce the formula, by this or other similar constructions. The rule-based formalist approach confuses *proving*, in mathematics, which includes inventions like here, with a posteriori *proof checking*, a remarkable technique in Computer Science, that we will mention below.

and rules encoded by 0 s and 1 s. How can this rule based computation replace knowledge construction, if it cannot invent the formula that computes the sum of the first *n* integers or, even worse, fully encode the proofs of some recent interesting *Arithmetic* statements, well beyond Gödel's fantastic diagonal trick – the Liar's inspired "this formula is not provable"? As a matter of fact, some "concrete" (meaningful) propositions of Arithmetic are provably *unprovable* within Arithmetic – this is much stronger than Gödel's diagonalization. As a matter of fact, different sorts of "geometric judgments" unavoidably step into their proofs, see (Longo 2011a, b) for a technical and philosophical introduction. In these cases, even a posteriori proof-checking is badly defined. The "everything is information" fans should then clarify whether their notion of information does or does not include sense construction, gestures, and frictions on reality, often of a pre-linguistic nature and at the origin of interpretation.

A close mathematical analysis, that we only hinted here, can help to better specify the vague references to "insight" and "practices" in human reasoning used by many authors (see the survey in Nickles 2018). In mathematics, one may "see" the gestures that correlate, organize possibly new structures by interpolating points, constructing borders, by symmetries, by ordering .... As we will stress, a human body in a history, beginning with embryogenesis, is needed for this, embedded in an evolutionary, linguistic and historical community. In particular, the largely pre-linguistic shared action in space enables the human universality of mathematics: we all share the monkey's practice of trajectories and surfaces, symmetries and ordering, that, once made explicit by human language, founded mathematics, before any formalization (see (Longo 2011b) for an analysis of the symmetries that underlie Euclid's axioms). Symmetries are one of our active organizational interpretations of the world; they precede and give meaning to geometric constructions, from Euclid's axioms to the founding diagrams of Category Theory, e.g. the "Natural Transformations" that motivated Eilenberg and MacLane's work, (Mac Lane 1970), (Asperti and Longo 1991) and Grohendieck Toposes (Zalamea 2012).

In the development of "intelligent" machines', though, a major turn happened recently, by the invention of multi layered neural nets, better known as Deep Learning. This is based on ideas that go well beyond the rule-based logical formalisms of classical Artificial Intelligence and it will be discussed below. The debacle of classical AI's view of the brain as a formal-deductive 0–1 machine (see below) parallels a paradigm shift in biology and, perhaps, also in (analytic) philosophy. In the latter case, "information" seems to stand, now and at least for a few, for the "construction of a perspective", of a theoretical frame or *interpretation* of natural phenomena. Two more "ways out" from the debacle of the formalist philosophy and the linguistic turns, from Logic to Biology, will be worth mentioning next. A discussion about them may further help in clarifying what one means by the fashionable reference to information: is this dehumanized, formal sign pushing, like in digital machines, or interpreted information, that is the explicit proposal of an interpreting perspective for knowledge construction?

The Role of 'Interpretation' in Programming, as Elaboration of Information

The religion of life and intelligence, and even of physical dynamics, as elaboration and transmission of information, as moving strings of bits and bytes, has its own limits within Programming Language Theory, first. Do programs, our invention, "stand alone"? Do they work correctly when piled up in complicated interactions or do they run into bugs and inconsistencies? A simple consequence of Gödel's and Turing's negative results shows that there is no general algorithmic way to prove the correctness of programs (Rice Theorem, (Rogers 1967)). Decades of Software Engineering, Programing Language Theory etc. have been dedicated to produce rigorous programming frames, in order to avoid bugs and inconsistencies and to invent methods for proving partial correctness – see the broad literature on Model Checking (Sifakis 2011), on the Abstract Interpretation (Cousot 2016), on Type Theory (Girard et al. 1989) and a lot more. Two of the many amazing successes of the Science of Programming are … one, the Internet, whose functioning is also made possible by geometric tools in Concurrency (Aceto et al. 2003) and in networks (Baccelli et al. 2016), and, the other, Embedded Computing and their Nets, such as Aircraft and Flight Control, following, among others, the approach in (Sifakis 2011).

In short, also in order to design sufficiently robust programming languages and analyze the correctness of formal computations, we human scientists have to *interpret* programs for elaborating information in "meaningful contexts" of mathematics, often of a topological nature, (Scott 1982) – for a survey see (Longo 2004)). Otherwise, sufficiently complicated or long programs produce bugs, inconsistencies, loopholes etc.… Formal computations do not stand alone: geometric meaning frames their design, helps checking their correctness. Recall, say, the dramatic improvement of abstract algebra since Argand and Gauss (end of eighteenth century) interpreted the imaginary number/letter i, a purely algebraic-formal sign, on the Cartesian plane: its meaning in space renewed analysis and geometry and provided a paradigm for the enriching interplay of formal syntax and semantics.

Programming is our modern and even more abstract alphabetic invention, as formal writing of signs that are modified according to rules written as signs, that is by replacement rules: current elaboration of information is a formal "term writing and re-writing system" (Bezem et al. 2013). We, the people of the alphabet, invented it, together with cryptography – a way to encode letters by numbers and/or other letters. Letters and numbers are not in *nature*, though. Ideograms or hieroglyphics are completely different inventions for writing, thus stabilizing and making visible the invisible sounds of language in their own way: none of these human techniques is "already there", in the world.

Moreover, in order to associate a number or a letter to a physical process, we need to make the difficult operation of measurement, i.e. chose an observable, a metrics, construct an instrument etc. and perform a measurement. By programming, we invented machines that move meaningless letters or 0–1 pixels (which is the same) according to written instructions: the alphabet or the 0 and 1's, by replacement of one by the other, move on our screens. This is such a fascinating invention that mystics believe that it is already in nature: nature formally elaborates information, like our computers, it moves signs, 0 and 1 in the squares of a (huge) Cellular Automaton (Wolfram 2013). Instead, if computation refers to the robust science of term-rewriting and, thus, of computable functions over integer numbers, as it should since the beautiful use of these notions summarized in (Rogers 1967) and (Barendregt 1984), nature *does not compute*. We, humans, associate numbers to natural processes by measurement and embed them in more or less successful theories, where mathematics allows also to approximately compute – but first to understand, by organizing phenomena in very rich structures (see section "From geodetics to formal rules and back again"). Besides, computations are on integers or on computable real numbers: are the fundamental constants of physics, c, h, G, α … integer or computable reals? A question I keep asking to computationalists who project Turing Machines onto the world, against Turing, see (Longo 2018c).

In summary, uninterpreted information, as strings of moving signs, does not stand alone. Sufficiently secure programming as well as an understanding of determination (and randomness, see the footnote) need an interpretation within a relevant, meaningful mathematical, physical or biological theory. Dehumanizing science by subtracting its notions to human interpretation and theorizing does not work. It all depends then on whether the notion of information an author refers to crosses or not the threshold of interpretation.

Which Information Is Handled by a Magic Demon?

Shortly after the birth of Computability Theory as the mathematical foundations of "*elaboration* of information", (Shannon 1948) set the basis of the theory of "*communication*" (of information). Once more, as Shannon stresses in several places, the transmission of strings of meaningless signs is at the core of it. The 'informational content' is only given by the inverse of the probabilities of the appearance of a sign in a pre-given list of possible signs (receiving a rare string or sign is more informative than a frequent one, see Lesne 2014 for a recent survey). The formula used to express this property happens to be the same, by the use of logarithms of probabilities, as Boltzmann expression of entropy, with an opposite sign and modulo a coefficient (Boltzmann's *k*, a dimensional constant, see Castiglione et al. 2008). Now, in physics, the same mathematical formula may have very different interpretations, a fortiori when modifying coefficients and constants. For example, Schrödinger's diffusion equation in Quantum Mechanics is a "wave" equation, with a complex coefficient *i* and a constant, Planck's *h*. These coefficient and constant are crucial and yield a quantum-state function, representing the dynamics of a amplitude of probability or the probability of obtaining a value at measurement. That is, if, in Schrödinger's equation, *h* is replaced by a different dimensional constant or the imaginary number *i* by a dimensional number or by a real number, the mathematics refers to a different physical phenomenon and one obtains different diffusion equations, such as the equation for the diffusion of heath or … of water waves. Conflating the meaning of these different writings of an equation is a physico-mathematical abuse.

In the case of entropy, one may give a conversion equation towards "information", by setting bit = $k$log2, where k is Boltzmann's constant (Landauer 1991). This quantity, $k$log2, is indeed, a fundamental physical invariant: $k$log2 = *the least amount of entropy produced by any irreversible process.*

Calling "bit" this quantity, whose dimension is energy divided by temperature (that is, entropy), may be legitimate, if one knows what it refers to, as most physicists do (see the very insightful experiments in (Lutz and Ciliberto 2015)), otherwise it is a word-play. Indeed, on the grounds of a formal analogy and of this "conversion" (or dimensional "forcing"), an abuse immediately started: information is *the same as* entropy, with a negative sign. Since entropy is produced everywhere energy is transformed, information is everywhere. This understanding has even been attributed to (Schrödinger 1944), chap. 6. Yet, Schrödinger refers to negative entropy as the extraction of *order* from the environment by the organism, while increasing or maintaining its own order, a different concept. Moreover, he opposes the approach in chap. 6 to the previous one by "forgetting at the moment all that is known about chromosomes" as *code-script* (Schrödinger never mentions "information" in his book). In (Bailly and Longo 2009) we developed that approach by Schrödinger in biology by the notion of "anti-entropy" as phenotypic complexity in organisms, which differs from information, at least for its dimensionality (and more: it is a geometric notion, since metric properties, dimensions and coding do matter – in contrast to digital information). We then applied it to a mathematical analysis of an idea in (Gould 1996) as for the increase of organisms' complexity along evolution, as a random, not oriented, but asymmetric diffusion. The situation then is far from obvious and conceptual short-cuts may misguide knowledge: both information and organization may oppose to entropy production, in the broad sense of "disorder", yet they yield different notions, also dimensionally, and, more deeply, as tools for understanding, life in particular (see (Longo 2018a)).

Brillouin is more consistently credited for conflating information and negative entropy in physics, by an argument worth recalling (Brillouin 1956). Maxwell's demon was an inventive game, playfully opposing Boltzmann's dramatic vision of the end of the Universe by the entropic final state of equilibrium: a sufficiently smart and fast demon could decrease entropy, by separating mixed gas particles according to a measurable property of some of them, see (Leff and Rex 1990). Brillouin's relevant remark is that such a demon needs to transform energy in order to detect and let pass only certain particles through the separating door. As any energy transformation process, this would increase entropy. "We have, nevertheless, discovered a very important physical law in Eq. (13.10): every physical measurement requires a corresponding entropy increase, and there is a lower limit, below which the measurement becomes impossible" (Brillouin 1956). "We cannot get anything for nothing, not even an observation", Brillouin continues.

In Quantum Mechanics, measurement produces a new object, irreversibly, by the interaction of the classical measurement instrument and quantum "reality" (this corresponds to the irreversible projection of the state vector in Schrödinger's equation). In Classical Physics, measurement was considered to be "for free". Instead, Brillouin shows that measurement has a cost, it transforms energy, hence producing entropy – a remarkable observation. Thus, measurement, the only way we have to access the world, is irreversible, also classically. In spite of the interpretative abuses (the dimensions of constants and the amusing game-play), "We have, *nevertheless*, discovered …" says Brillouin. Many focused more on the abuses than on the discovery of an "important physical law".

The key point is that entropy production is associated to *all* irreversible processes and *k*log2 quantifies the least amount of entropy production and this can be experimentally checked (just beautiful). Now, measuring, communicating and elaborating information as well require energy and thus produce entropy, at least in that amount, e.g. even when erasing a bit of information – Landauer's Principle (Landauer 1991), see above. By measurement we do produce information, a number, to be transmitted to colleagues, but that number (actually, an interval) is not "already there", as it is the result of the choice of an observable, the construction of a measurement tool etc. Thus, information *is* physical in the sense that it requires quantifiable physical transformations, but physics is not information, nor information is intrinsic to inert matter. We invented this fantastic *new* observable, information, independent from its material realization, a beautiful invariance property, whose irreversible treatment does require some energy – and this is measurable.

In our view, the key issue is the irreversibility of physical processes, even in classical (and relativistic) frames where time is usually considered to be reversible: measurement, at least, forces, *always,* an arrow of time, or equivalently, it produces entropy. And, I insist, we have no other way to access to reality but by measurement. This equivalence, irreversibility of time and entropy production, seems to be a core, inter-theoretic physical property.[6]

Unfortunately, the anthropomorphic reference to a demon encourages the use of the word information and its "projection onto the world": the demon looks and measures and decides – he actually "transforms information into negative entropy" (Brillouin 1956, p.168). Yet, as Brillouin observes, "we may replace the demon by an automatic device … an ingenious gadget", with no need to refer to information nor to an observing eye, but just to the use of some physical energy to reduce a form of entropy, while producing more entropy (due to the physical work done). That is, a physical or an engineering gadget would do, by a purely mechanical activity. The reference to an impossible, playful (and inessential) demon, though, allows to revisit

[6] In (Longo and Montévil 2014), we added to irreversibility of time and to entropy production the coexistence of a "symmetry breaking" and of a random event: these four phenomena seem to be correlated in all existing physical theories. In our work, this correlation extends to proper biological dynamics, such as embryogenesis and evolution, where increasing organization as well (anti-entropy production) produces entropy.

Maxwell's game on Boltzmann's approach and may suggest an informational agent acting on the world.

Along this line of thought, Brillouin's wording may be interpreted as a philosophy of nature (Bühlmann et al. 2015). The *human* description or "encoding" of natural phenomena requires the production of information, by measurement first. It has a cost, which can be even measured, as negentropy, a remark consistent with Brillouin's insight and the quantification above. I understand similarly the philosophical reflection by many, for example in (Ladyman and Ross 2008): the construction of scientific knowledge requires a subject performing first the non-trivial act of measurement. Theory building is thus production of information. Consequently, it has the dimension of information and may be measured as a physical observable, whose analysis may unify science.

This understanding of information may depart from the meaningless "sign pushing", that is, from the sign transformation or communication proper to formal elaboration or transmission of information by input-output piecewise assembled machines: it may ground (and it is grounded on) human "interpretation" of natural phenomena, which begins with measurement. The latter requires a protensive gesture, starting with the choice of an observable, then a metrics, the construction of a tool for measurement etc. by the knowing subject. It is an active friction on reality, along the lines of the moving amoeba above, which is also "measuring", in its own way, its Umwelt. Of course, as stressed in (von Uexküll 1934): "Behaviors are not mere movements or tropisms, but they consist of perception (Merken) and operation (Wirken), they are not mechanically regulated, but meaningfully organized". Multicellular embryogenesis and a long evolutionary history add on top of the movements and tropisms of a unicellular eukaryote, but all these biological processes organize life very differently from mechanical assemblage and regulation and set the protensive grounds or condition of possibility for meaning.

As for Brillouin, he seems to have contradicting views, sometimes crossing, sometimes not, the threshold of information as pure sign pushing or as resulting also from interpretation. On one side, one may find, in his writings, an interpreting role of human observation and experience that may be described as production of meaningful information; on the other, in many examples, such as the analysis of a game of cards in the first chapter of his book, information is presented as the formal analysis of a collection of signs and its mechanical transmission.[7] As a matter of fact, Brillouin aims to lay the foundation of a scientific discipline of information, by "eliminating the human element" (Introduction, p. viii). Can we actually construct an invariant notion of information that encompasses theory building, and is, on one side, based on the activity of a knowing human subject, while, on the other, is independent of the *historical formation of sense*?

[7] The information content is the logarithm of the probability, as a frequency, of the chosen card. This example opens the way to conflating Boltzmann's logarithmic formula for entropy to Brillouin's quantification of information, modulo a negative sign and a differing constant, as pointed out above, also by its dimensionality – a further major difference.

Today we need a rigorous clarification of this issue, in view of the role that mechanically elaborated and transmitted information is having on our lives. We may improve by this both human knowledge and machines, while working towards the invention of the next machine: in spite of the impact on our society and on science, by its speed and memory size, this digital, exact, iterating machine is rather boring, per se. Typically, the mathematics of the new AI of multi-layers neural nets in continua finds a tool but also a bottleneck when implemented in discrete state machines.

The Biology of Molecules, Well Before the Threshold of Biological Meaning

"A theorem by Einstein or a random assembly of letters contain the same amount of information as long as the number of letters is the same" observes A. Lwoff in "Biological order" (1969). His co-winners of the 1965 Nobel Prize in Biology, F. Jacob and J. Monod, in several writings, set the theoretical basis of molecular biology along the same lines. "Since about twenty years, geneticists had the surprise that heredity is determined by a message written in the chromosomes, not by ideograms, but by a chemical alphabet" (Jacob 1974). "The program represents a model borrowed from electronic computers. It assimilates the genetic material of an egg to the magnetic tape of a computer" (Jacob 1970). One should ask first where the operating system, the compiler or the interpreter is… In Computer Science, compilers translate one language into another (a lower level one, usually) and allow the operating system to handle the mechanical computation at the level of machine language, as a rule-based transformation of 0 s into 1 s and vice versa. Interpreters do the same job, but are written in the same language of the source language: lambda-calculus is the mathematical paradigm for this (Barendregt 1984). In this context, thus, "interpreters" have nothing to do with the notion of interpretation that we are using here as "human proposal for the construction of meaning". As a matter of fact, within Computer Science, a clear distinction is made. On one side, "denotational semantics" is the mathematical interpretation of formal computation over meaningful, possibly "geometric", structures (Scott 1982), (Goubault 2000), (Longo 2004) – we recalled above an early example of this sort of interpretation: Argand-Gauss invention of the geometric meaning of the formal imaginary *i* over the Cartesian plane (Islami and Longo 2017). On the other side, "operational semantics" remains a form of sign pushing, that is the rule-based job done by compilers, interpreters and operating systems within computers: meaning is reduced to the formal operations carried on by these different programs in order to have the machine work. This second notion of "semantics" is a legitimate abuse, as long as one knows what it refers to: "sign pushing" as the internal, purely formal-mechanical, handling of programs by programs (compilers, interpreters, operating systems).

In biology, lactose operon (Jacob and Monod 1961) provided an early experimentally insightful example of operational control of genetic information by the genes themselves. This suggested a "microscopic cybernetic" in (Monod 1970). However, Wiener's cybernetic, as non-linear control-theory in continua (Rugh 1981), (Sontag 1990), is well beyond the programming-alphabetic approach proposed by the founding fathers of molecular biology, an exact Cartesian Mechanism, says Monod. The feed-back or circular diagrams in (Monod 1970) are perfectly handled in programming, in particular by recursion or by impredicativity, the strongest features of lambda-calculus (Barendregt 1984), (Girard et al. 1989), (Asperti and Longo 1991). Indeed, lambda-calculus is a most advanced reference for DNA as a computer program (see below for more).

Since then, the lactose operon mechanism was proposed as a fully general paradigm and the Central Dogma of Molecular Biology (Crick 1958) was reinforced in its strongest sense: DNA contains the complete information for proteins' formation and, thus, for ontogenesis. Waddington's or McClintock's work on the epigenetic control of gene expression was forgotten and no longer quoted for many years (Fox Keller 2000). Thus, once the human genome was decoded, it was going to be possible, according to Gilbert (1992), to encode it on a CD-Rom and say: "Here is a human being, this is me". The decoding was completed by 2001.[8]

Following the early theorizing in Jacob's "Linguistic model in Biology" (1974), quoted above, (Danchin 2003, 2009) stress the relevance and clarify the terminology of this linguistic turn in biology: the cell is a computer that may generate another computer – a feature easily implemented in lambda-calculus, as soundly observed also in (Danchin 2003, 2009). Then, evolutionary novelty may be due to the genetic implementation of Gödel's formal method, which diagonalizes over strings of letters and numbers (see the 2018 version of (Longo 2010) for a critique). The organism is an avatar (Gouyon et al. 2002) of the selfish gene computing evolution (Dawkins 1976), (Chaitin 2012). The discovery of many fundamental mechanisms in Molecular Biology, that must be acknowledged, was mostly embedded in this cascade of computational follies or vague metaphors.[9]

The consequences have been immense. As discussed in (Longo 2018a), the informational approach on discrete data types (the alphabet) diverts the causal analysis in biology, typically in the etiology of cancer, a major issue today. Still now, most cancer research in biology is guided by the Central Dogma, interpreted as "any phenotype has its antecedent in the genotype": the cancer phenotype must then be studied at the genetic level, in spite of evidence to the contrary (Sonnenschein and Soto 2011, 2013), (Versteg 2014), (Adjiri 2017). An acknowledgement of the diffi-

[8] See "We Have Learned Nothing from the Genome" (Venter 2010), written by the leader of the team that first decoded a human DNA, and (Longo 2018a) for some references to the amazing promises made in early 2000s as for cancer's genetic diagnosis, prognosis and therapy, see also (Weinberg 2014), (Gatenby 2017).

[9] In a rare attempt to clarify the different role of Turing-Kolmogorov vs Shannon-Brillouin approaches in biology, (Maynard-Smith 1999) confuses, in the explanatory examples, the dual correlation that entropy and complexity have in the two theories, see (Longo et al. 2012), (Perret and Longo 2016) for details.

culties of this approach to an increasingly spreading and deadly disease have been made also by some of the founding fathers of cancer biology, (Weinberg 2014), (Gatenby 2017). Yet, the focus on the molecular level, where the alphabetic program of life may be found, still dominates. The hypothesis of the completeness of genotypic information justifies the focus on (*promises* of) genetic therapies for cancer, neglecting the analysis of carcinogenes, thus prevention – a surprising and overwhelming impression to the mathematician stepping into the cancer literature (Longo 2018a). The search for the "magic bullet" that would reprogram the oncogene or the "proto-onco-gene" or compensate for the lack of "onco-suppressor-gene", gave and gives an incredibly minor role to the analysis of ecosystemic *causes* of cancer. Instead, as Weinberg (2014) acknowledges, "most carcinogenes are not mutagenic". What is then the causal chain of cancer formation and progression? A paradigm shift (Sonnenschein and Soto 1999; Baker 2014, 2015) has found major obstacles, in spite of the evidence against the focus on genetic drivers of cancer (Kato et al. 2016), (Brock and Huang 2017). The entrenched molecular-alphabetic-informational vision of the organism forbids other approaches, including the analyses of the tissue-organism-ecosystem interactions and their role in the control of cell proliferation. In our joint work in organismal biology, following (Sonnenschein and Soto 1999) and … Darwin, we consider cell proliferation with variation and motility as the "default state" of all cells, including within an organism, see (Soto et al. 2016). In particular, thus, if this Darwinian approach is correct, carcinogenes would interfere, at various levels of organization, not just molecular, with the control of cell proliferation – mutations follow, an evolutionary-developmental reaction of cells that are less constrained and/or under stress. This reaction may be present both in pathological developments and in "healthy" aging tissues (see the references above and in (Longo 2018a)).

More generally, the need to explain how information is elaborated and transmitted in an organism forced the genocentric approach and the idea that, in a cell, macro-molecular interactions are exact, like in a "boolean algebra" … thus "evolution is due to noise" (Monod 1970); "biological specificity … is entirely … in complementary combining regions on the interacting molecules" (Pauling 1987). Their enthalpic random oscillations in a quasi-turbulent and complex energetic environment were disregarded (Onuchic et al. 1997). Randomness, a key component of diversity production, thus of adaptivity of life, was and still is identified with "noise", (Bravi and Longo 2015), a term related to information. The production of biological diversity is then considered a pathology opposing the rule-based norm (Ramellini 2002). Along these views on the exact combinatorics of macro-molecules, that would follow the formal rule of a program or Chomsky's grammatical rules (Searls 1992), the "key-lock" paradigm of the perfect fit for too long dominated the analysis of "transmission of information", for example through cellular receptors. Thus, little attention has been paid to low probability affinities, that do matter in time, a major issue for understanding endocrine disruptors and their relevance in carcinogenesis (Soto and Sonnenschein 2010, 2017). In general, the constructive role of stochasticity in gene expression and macromolecular interactions, observed since (Kupiec 1983), only recently found its way into the literature (Elowitz et al.

2002; Paldi 2003; Fromion et al. 2013; Marinov et al. 2014). Of course, it is hard to conceive elaboration and transmission of information in a largely stochastic macromolecular soup, such as the proteome of a eukaryote cell.[10] Against the informational monomania, the view of biological *constraints acting on* and *canalizing* these physical dynamics and, more generally, organizing biological functions in an organism is slowly maturing (Deacon et al. 2014), (Montévil and Mossio 2015), (Soto et al. 2016) and many others.

Finally, observe that Genetically Modified Organisms (GMOs) are the direct children of the Central Dogma, in its strict interpretation: by modifying the DNA encoded information, a "blue-print" of the organism, one may completely pilot the plant in the ecosystem. This is false. For example, the microbial communities in the roots and soil are heavily affected by the use of GMOs, while they contribute to plants phenotypes in an essential way (Kowalchuk et al. 2003). However, the medium-long term disaster may partly be balanced out, in the short term, by enough chemical fertilizers – see the recent Bayer-Monsanto fusion.

In conclusion, under the slogan "biology is information", specified under the form of alpha-numeric information, thus genetic information, we witnessed a major distortion of knowledge construction. Yet, we should not be distracted by this. DNA is an amazingly important physico-chemical trace of evolution, used by the cell according to the context (tissue, organism, ecosystem). It is an *internal* constraint to development (Montévil and Mossio 2015). Its biological *meaning* lies in its use as a constraining template for the production of molecules, in ways largely incompatible with Turing's and Shannon's theories of information. In particular, it strictly depends on dimensionality and its specific physico-chemical matter, on torsions and pressures on chromatin (Cortini et al. 2016) etc., far away from the uni-dimensional encoding and hardware independent theories of elaboration and transmission of information.

Besides the increasingly acknowledged role of stochastic gene expression mentioned above, the very notion of "gene" is being deeply revised, as it has been several times in the twentieth century (Fox Keller 2000): alternative splicing, overlapping genes,[11] the "Physics of Epigenetics" (Cortini et al. 2016) … totally

[10] Even more radically "proteins never do fold into a particular shape, but rather remain unstructured or 'disordered' […]. In mammals, about 75% of signaling proteins and half of all proteins are thought to contain long, disordered regions, while about 25% of all proteins are predicted to be 'fully disordered' […]. Many of these intrinsically unstructured proteins are involved in regulatory processes, and are often at the center of large protein interaction networks" (Gsponer and Madan Babu 2009). See also the increasingly acknowledged important role of long non-coding RNAs (Hadjiargyrou and Delihas 2013). Dogmas on the exact mechanisms inspired by the need to transmit and elaborate alpha-numeric information are collapsing one after the other (Mouilleron et al. 2016).

[11] Overlapping genes are parts of a given DNA or RNA region that can be translated in at least two different reading frames, yielding very different proteins. In fact, a shift in the reading frame of a nucleotide sequence, by one-two bases, totally changes the resulting protein. Discovered in the late '70s in a small viral genome (Barrell et al. 1976), there are coming to the limelight only recently

modify our understanding of DNA and bio-molecular dynamics in their physical and geometric context. The reader who absolutely cares of the informational terminology at least in mathematical modeling, should at least look at the novel approaches in Geometry of Information, where the group-theoretic notion of "reduction of ambiguities" provides a beautiful interpretation of some dynamics in terms of continuous symmetries and their breaking, (Barbaresco and Mohammad-Djafari 2015). So far, this geometric approach to information is ignored by the tenants of the rule-based world, where agents must follow the formal, alpha-numeric norm, possibly independently of meaning, from biology to economy, as we shall argue.

From Geodetics to Formal Rules and Back Again

Galileo's inertia is a principle of intelligibility. It is a limit principle, since no physical body moves like a point-mass on a straight line. Yet, it allows to understand all classical movements and to focus on what modifies them: gravitation and frictions, both closely analyzed by Galileo. Einstein unified inertia and gravitation, as the latter may be understood as inertial movement in a curved Riemannian space. Conservation principles, such as inertia (conservation of momentum) and energy conservation, are the main unifying assumptions encompassing even incompatible physical theories (Quantum Mechanics, Relativity Theory, Hydrodynamics, etc.). These principles may be described as *continuous* symmetries in equations (Kosman-Schwarback 2010). Thus, one may say: I do not know the result of this coin flipping nor where that river will go exactly, but they will all move along geodetics (optimal paths), *for symmetry reasons.*[12] Falling stones, coins, rivers … will never "go wrong": their unpredictability, if any, is a matter of non-linearity of our mathematical approach to their dynamics jointly to physical measurement that is always an interval – and a minor fluctuation below the best possible measurement may give a very different geodetics. The point is that, in general, we do not obtain a number from physical measurement, but an interval; this allows to *prove* and even evaluate,

since many studies have shown they are not restricted to viruses (Chirico et al. 2010), but they are present as well in cellular organisms, man included (reviewed in Pavesi et al. 2018). M. Granero and A. Porati, among the pioneers in this field, nicely described the phenomenon in Italian: by a one letter shift CARABINE MICIDIALI becomes ARABI NEMICI DI ALI, and GAS INODORO becomes ASINO D'ORO (A. Vianelli, personal communication). This sort of shifts in 'reading' is not recommended in linguistic analysis nor in programming, where it is avoided by the use of parenthesis, typically – like in lambda-calculus, a consistently evoked reference for the genetic program, see (Danchin 2003, 2009). Actually, even in genomes (at least bacterial ones) there might be more STOP off-frame codons than expected to avoid unneeded reading frame shifts (Seligmann and Pollock 2004; Abraham and Hurst 2018). What is a gene, then?

[12]As observed (and promoted) by H. Weyl (1949), physics moved "from causal lawfulness to the structural organization of time and space (structural lawfulness), nay, from causal lawfulness to intelligibility by mathematical (geometric) structures", see also (Bailly and Longo 2006).

in some cases (by Lyapounov exponents, (Devaney 1989)), the unpredictability of non-linear dynamics and to *understand* quantum non commutativity (up to Planck's h). The discrete replaces measurement and the enumeration of acts of measurement, proper to theories over continuous manifolds, with solely enumeration: in the discrete, one can only count – a beautiful remark in (Riemann 1854). And continuous deformations of Riemannian manifolds and their relation to the metrics (thus measurement) found Relativity Theory. Riemann's distinction, measuring and counting vs just counting, allows to pose explicitly the challenges in computer modeling.

Note first that randomness is unpredictability relative to the intended theory (cf. classical vs quantum randomness), and its analysis increases intelligibility, (Calude and Longo 2016). That is, the understanding of determination in terms of symmetry principles and, by a fine analysis, of the nature of randomness (from Poincaré's non-linear dynamics to indeterminism and non-commutativity in Quantum Mechanics) may not imply predictability. On the contrary, predicting, given a theory, is a further issue: it is computing a number, actually an interval, that is a future possible evolution of a "trajectory" in the broadest sense. We understand very well the dynamics of dice and rivers in complicated landscapes, with little or no predicting power – and a rather limited one even for Planets (Laskar 1994). However, we are theoretically sure that, much like a falling stone, they never go wrong.[13]

A computational approach to theorizing in physics conflates two notions: understanding and predicting. This is particularly absurd in biology: Darwin's evolution is a bright light for knowledge, but it has little to do with predicting. If theorizing boils down to creating Turing or Shannon information, over discrete databases,[14] a deterministic theory would always compute an exact value, thus predict, even in the most chaotic non-linear case (see below). In non-linear dynamics, instead, given a continuous solution of a system of equations, any discrete time and space implementation "soon" diverges from the continuous description proposed by the theory – usually written on the grounds of a conservation principle (write the Hamiltonian, says the physicist, as continuous symmetries). At most, under some reasonable mathematical assumptions, the discrete time and space implementation of equations given in continua is approximated by a continuous trajectory (Pilyugin 1999), *not the opposite*. That is, one can find a continuous trajectory that approximates the discretized one, *but not a discrete one approximating the continuous trajectory*, except, sometimes, for a "short" time-space trajectory. The point is that

[13] This is in blatant contrast with biological organisms, which "go wrong" very often or most of the time – but, in absence of a pre-given phase space for biological dynamics, it is hard to pre-define "wrong/right": very rarely, but crucially, hopeful monsters may be "right" in evolution, and contribute to speciation, as we may observe *a posteriori* (Longo 2017). The point is that ecosystemic compatibility and viability is not optimality, in particular because the ecosystem is not pre-defined, but co-constructed with/by the organism.

[14] "… due to the inherent limitations of scientific instruments, all an observer can know of a process in nature is a discrete time, discrete-space series of measurements. Fortunately, this is precisely the kind of thing — strings of discrete symbols, a "formal" language—that computation theory analyzes for structure." (Crutchfield 1994).

measurements do not produce discrete series of integer (or rational) numbers (see the footnote), but series of interval and a fluctuation below those intervals of approximation yields (classical) unpredictability, in non-linear or similar systems that would amplify that fluctuation – this makes no sense in discrete dynamics. Missing this point is passing by "the fundamental aporia of mathematics" (Thom), discrete vs continuum, already singled out by Riemann.

Thus, one understands, frames and discusses falling stones, planets' trajectories, river's paths, hurricanes, Schrödinger's diffusion etc. by writing equations that are believed pertinent for good reasons (general conservation principles). If the system is sufficiently "complex" (non-linearity describes "interactions", usually, the least required to express "complexity"), then predictability is possible at most for short time scales. Computers enhanced immensely our predictive power by increasing approximations and computing speed. Yet, the theoretical challenge should be clear: digital computers do not follow continuous, mathematical, nor actual dynamics. Thus, the transfer of intelligibility principles, given as continuous symmetries, into discrete sign pushing presents major epistemological and modeling problems (Lesne 2007), (Longo 2018a). And here, we discuss exactly the move from principles of intelligibility to computational rules, as formal norms to follow. Of course, computers' implementation is essential to modern science, but exactly because of this, a close analysis is needed, in order to improve the use of these fantastic computational tools, beyond myths.

Computations as Norms

By the previous remarks, if theorizing is identified with active production of information, a major shift is implicitly produced. On the one side, a clear and important role is (at last) given to the knowing subject: he/she has to fix observables and parameters, measure etc. in order to produce information. In Weyl's terms, scientific objectivity is the result of the passage from "subjective-absolute" judgments on space and time to the "objective-relative" construction of invariants.[15] That is, relativizing knowledge is the production of invariants w.r.to changing reference frames and knowing subjects, with his/her measurement tools. This relativization yields objectivity. In this sense, we do actively produce "information" in the broadest sense and radically depart from "science as the search for absolute or intrinsic truth" proper to naive philosophies of knowledge, which place the subject in an absolute position.

On the other side, if theorizing is production of information and this refers to Turing's or Shannon's elaboration or transmission of information, we move from intelligibility to *formal normativity*. Typically, we understand all trajectories, including the trajectory of an amplitude of probabilities in Schrödinger's equation, as

[15] "subjective-absolute and objective-relative seems to me to contain one of the most fundamental epistemological insights that can be extracted from natural sciences" (Weyl 1949).

geodetics in suitable phase spaces. These locally maximize a gradient and the integral of these gradients yields the optimal path. However, this is just a principle of intelligibility, as mentioned above: in no way can we, the observers, know exactly and even less force where the object will actually go. The actual path depends also on local fluctuations possibly below measurement. This is very different from a computational rule, which describes *and* prescribes: go from point (5,3,7) to (8,1,3) in space, say. This is exactly what a program does on your screen, even when simulating the wildest non-linear dynamics. Nothing moves on a digital computer's screen. The formal rule-based implementation, a term re-writing system, prescribes exactly how signs/pixels must be written and then re-written, that is switched on-off, in order to implement movement.[16]

By contrast, we understand how falling objects go by the gradient principle, and this according to the context (local frictions, fluctuations …), often with little predictability. Instead, the computer implemented image of that very object will follow exactly the pixel by pixel rule: frictions, fluctuations … are either excluded or explicitly formalized in the rules. The writing of the equations does not norm, but describes. The writing of a program describes *and* norms. Moreover, equations, as sequences of signs, do not move; instead, programming is a writing and re-writing sequences of signs, including the programs themselves, the extraordinary idea underlying Turing's invention – and fully expressed in lambda-calculus by programs acting on themselves and on their types, see (Barendregt 1984), (Girard et al. 1989), (Asperti and Longo 1991), (Kreisel 1982). Compilers, operating systems, interpreters … are all written in sequences of 0 s and 1 s, like all programs, in one dimension, a line of discrete entities: they are implementations of Turing's Universal Machine, the beautiful invention of a major mathematical invariant – the class of computable functions. Mystics transformed it into an absolute.

Again, the focus on the production of information while doing science, first by measuring, is one thing – it may help to clarify and unify in different sciences the role of the knowing subject (Rovelli 2004), (Zeilinger 2004). Claiming instead that theorizing is sign pushing on discrete data types (Crutchfield 1994) is another: the structures of intelligibility are then transformed into normative frames. Of course, one reaches the bottom line when claiming that a body falls because it is programmed to fall: "We can certainly imagine a universe that operates like some behaviour of a Turing machine" (Wolfram 2013).[17]

[16] Note that also declarative or functional languages are "imperative": they apply reduction rules to get to "normal forms" or implement the equality (and this on the grounds of fundamental theorems, such as Normalization (Girard et al. 1989) and Church-Rosser (Barendregt 1984)).

[17] Unfortunately, this attitude is shaping minds. A student in mathematics, in my top institution in higher education (ENS, Paris), asked me not long ago: "how can the Universe compute at each instant where it will go in the next instant?". I answered: "there are plenty of very small Turing Machines, hidden everywhere, that do the computations!". And I wrote a letter to Alan Turing (Longo 2018c) – he explicitly and radically departed from these views (the brain … the Universe, are surely not discrete state machines, i.e. Turing Machines, says he – see references in my letter).

One more issue. It should be clear that in theoretical universes of discrete signs there is no classical randomness. Discrete data types, where one can only count, are given exactly, and access to data is exact. Classical randomness requires the interplay of *non-linearity*, or some form of mathematical chaos, and *approximated measurement* as an interval (Devaney 1989). Of course, Quantum Physics adds the intrinsic indetermination of measurement, up to Planck's h, and of the discrete spin-up or down of a quanton, say. But, Turing Machines and von Neuman's Cellular Automata (CA) or similar devices are treated classically.[18] Now, in classical discrete state machines there is no unpredictability: they just "follow the written rule". Typically, if one iterates (restarts) a programmable discrete state machine, including CAs, on the same initial data, one obtains *exactly the same trajectory*. Instead, a necessary property of random processes, in physics, is that, when iterated in the "same" initial/border conditions (the same intervals of measurement), in general, they do not follow the same trajectory. In digital machines there is no randomness, at most noise, and this is successfully eliminated. Even in networks of computers, given in space-time continua where noise is massively present, it is largely reduced, it is "do not care" (Longo et al. 2010) – a remarkable feat of network and concurrent computing (Aceto et al. 2003). Finally, "Indeterministic" Turing Machines are just "one input – many outputs" machines: once the latter is encoded in one number, they become faster, but perfectly deterministic machines (Longo et al. 2010).

As a consequence, there is no "emergence" in deterministic discrete state machines, including CAs. Emergence should, at least, imply unpredictability, thus randomness (Calude and Longo 2016): instead, restart your CA, on the same, exact, discrete, initial data, and you get exactly the same complicated "emergent" shape – noise and statistic must be introduced on purpose, from outside. Moreover, the definition of randomness as irreducibility in (Crutchfield 1994) and (Wolfram 2002) is ill-defined in two dimensional CAs, as a mathematical invariant: it depends on coarse graining (Israeli and Goldenfeld 2004) and it is subject to "speed up" – in more than one dimension there is no lower bound in program size, by using concurrent processes, see (Collet et al. 2013). Finally, even when incompressibility is well defined, as it is on one dimensional, finite strings of numbers *à la* Kolmogorov, it does not correspond to physical randomness.[19]

[18] Quantum Computing formally introduces a major feature of Quantum Mechanics, entanglement (Zorzi 2016). By this, it forces a connection between hardware and software, a major computing and engineering challenge, beyond Turing's fruitful split, hardware vs software, perhaps along the way of a major revolution in computing.

[19] Many projected into nature Kolomogorof's effective notion of "incompressibility" for finite strings, as the dual component of predictive determination, i.e. as randomness. A finite sequence or string of letters/numbers w, say, is *incompressible* if no program to generate it is shorter than w (Calude 2002). Today, this is a crucial notion since our machines need to compress strings. However, it does not describe physical randomness, as thought by many, yet another abuse of a (remarkable) computational invention. Only asymptotically, infinite random sequences, as defined by Martin-Löf, relate to classical and quantum randomness (Calude and Longo 2016). This is to be expected: limit constructions already unified, by Boltzmann work, random particle trajectories and

In short, von Neuman universes of exact formal rules handling information on discrete data types are the grounds for von Neuman's nightmare games in life, economy and wars, a formally normed Informational Universe that still dominates the collective imaginary: just follow the meaningless rule on discrete data and iterate exactly – any classical dynamics on discrete space and time is perfectly predictable, it is normed once and for all, it is computable, it is thus doomed to iterate identically on the same data, for ever. Fortunately, these nightmares are slowly fading away even inside one of the areas where the de-humanization of knowledge construction was first brought to the limelight, AI.

Back to Geodetics in Artificial Intelligence and to Sense Construction

Hilbert's attitude to mathematics often wavered between a strict formalist credo, as recalled by H. Weyl (see the footnote above) and a practice of inventive and meaningful constructions, in actual work. In a sense, he wanted to get rid of the foundational issue for good ("a definitive solution"): once consistency and completeness of major formal theories were proved, one could work freely in "Cantor's paradise", rich in infinities, intuition and sense (Hilbert 1926). In formalist mathematical circles, formal deduction is rarely extended to an understanding of human general cognition and the threshold of interpretation and meaning is often crossed. Yet, the myth of intelligence as sign pushing has been at the core of Classical Artificial Intelligence, of which (Turing 1950) is considered the founding paper. Note that Turing explicitly says that "The nervous system is certainly not a discrete-state machine" (p. 451). He thus invents an "imitation game" whose aim is to simulate human intelligence independently of how the brain could actually be organized, but in such a way that an interrogator would not be able to make a difference between his machine and a woman – in 30% of the cases, says Turing, for a game no longer than 5 min and … by year 2000 (an enormous literature is available on this paper, I dare to refer to my personal letter to Turing also for references, (Longo 2018c)).

Soon after 1950, Hilbert's deductive formalisms and Turing's cautious imitation of cognitive capacities were transformed in a vision of intelligence as formal operations on signs, since "a physical symbol system has the necessary and sufficient means of general intelligent action." (Newell and Simon 1976). The biological

thermodynamic entropy (Chibbaro et al. 2015). In the finite, all sufficiently long strings are compressible, by Van der Waarden theorem (Calude and Longo 2017). Thus, in terms of incompressibility, there would be no long random sequences …. And a string to be generated, in the *future* (next year lottery drawings), is blatantly not random if one can compute, *now*, just one element of it, independently of its 'a posteriori' compressibility or not. Once more, there is no way to analyze such a meaningful notion for physics, randomness for finitely many events, in abstract computational terms: one has first to specify the intended theory, as an interpretation and organization of (a fragment of) nature, and then define randomness relatively to it (Calude and Longo 2016).

structure of the brain had no interest in this perspective that dominated AI for decades. In contrast to this approach, a few tried to "model", not just imitate, brain activities as abstract "nets of neurons", (McCulloch and Pitts 1943), (Rosenblatt 1958). Mathematical neural nets function by continuous deformations and reinforcement of connections, following the ideas in (Hebb 1949). This *connectionist* modeling was rather marginalized for decades, or its two dimensional nets were considered only for their computational power: as for input-output functions on integer numbers, they compute no more than a Turing Machine.[20]

In the late '80s, it was observed that an animal brain is actually in three dimensions. Multi-layer neural networks were then invented for recognizing patterns by filtering and back propagation algorithms over the layers (Boser et al. 1991). Very refined gradients methods entered this new, radically different AI, including difficult "wavelets" techniques from mathematical physics (Mallat and Hwang 1992), (Mallat 2016). By these methods, mostly in continuous structures, invariants of vision may be constructed: roughly, after inspecting thousands, millions of images of the same object under different perspectives, a sort of "optimal" result or fundamental invariant structure of that object is singled out – a geodetic that results from the depth of the many layers (thus the well selling name, Deep Learning). The mathematical challenge is a major one as well as the implementation on discrete state machines – the usual problems of approximation and shadowing pop out, as these methods are mostly non-linear (see above). Moreover, they are generic, in the sense that, so far, they uniformly work for the recognition of images, sounds, language, etc.

It is interesting to note that the more these techniques advance and work (there exist already fantastic applications), the more the geometric and topological structures and methods they use radically depart from the brain structure. The visual cortex is the best known, and its "neuro-geometry" is better understood as a complex manifold of multidimensional hyper-columns, far from multi-layered networks (Petitot 2017). Moreover, different brain functions are realized by other, much less known, yet very different structuring of neurons (Cant and Benson 2003). Finally, emotions, (Violi 2016), or pregnances, (Sarti et al. 2018), that is meaning, essentially contribute to the shaping and the sensitivity of the brain: that is, its dynamic structuring depends also on emotions and meaning, as its activity takes place only in the brain's preferred ecosystem, the skull of a material living body, in interaction with an environment and on the grounds of an evolutionary and individual history that gives meaning to its frictions with the world.

[20] A red herring, the so-called *extended* Church Thesis, often blurred computational novelties: any physical finite structure computes at most Turing computable functions. Now, mathematical neural nets are not meant to compute number-theoretic functions. Of course, in this and other cases, if one forces a physical dynamic to take a digital input and then produce at most one digital output and formalizes the dynamics "*à la* Hilbert", one can prove that that formal system computes no more than a Turing Machine. This says nothing about the proper expressivity and possible functions of a continuous dynamics of networks (see below), whose job is not to compute functions on integers (see below for more). A similar trivializing game has been played also with Quantum Computing and Concurrent Networks (Aceto et al. 2003).

Input-Output Machines and Brain Activity

In a time when fantastic discrete state machines for elaborating and transmitting information are changing science and life, we need to focus with rigor on the notion of information. The limits of the arithmetising linguistic turn, which spread from the foundations of mathematics to biology and cognition, has been shown from inside several times. In Logic, by purely formal methods, Gödel, Church and Turing's (very smart) diagonal tricks proved its incompleteness from within arithmetic itself, (Longo 2010, 2018c) – we can do better today, (Longo 2011a). Biology is, slowly, undergoing a major paradigm shift away from the myth of the completeness of the alphabetic coding of the organism in DNA – the decoding of DNA greatly helped in this, as hinted above. Classical sign pushing AI is being silently replaced by non-trivial mathematical methods in continua etc. where "information" is elaborated in a totally different way, before being passed over to digital computers.

As for the latter issue, note that an input-output machinery underlies both classical AI and Deep Learning. Also multi-layers neural nets are *a priori* static. They receive inputs and then output invariants of vision, audition, language etc. Moreover, so far, their continuous dynamics must be encoded in discrete state machines.

The animal brain works in a totally different way. First, it is always active, indeed, super-active. The friction with the world, by the body, canalizes, constrains, selects its permanent re-structuring and activities. Neuron's continuous and continual deformations of all sorts include electrostatic critical transitions, that are too poorly schematized as 0 or 1, as well as moving synaptic connections. In case of sensory deprivation (perfect darkness and silence, lack of skin sensations … a form of torture), one gets crazy by the increasing chaoticity of neural activities. Even the formation of the visual cortex, the best known so far, goes by, first, an explosion of connections, later selected by their activity (Edelman 1987). Action on the world, beginning with the restless uncoordinated activities of the newborn, is at the origin of the construction of meaning, as hinted in section "Introduction: the origin of sense": meaning results from an active friction with the world. Moreover, as analyzed in (Violi 2016) as for humans, the mediation of the mother contributes to the earliest form of interpretation of the environment: the baby feels or stares at the mother in order to make sense, in the interaction, of a new event. Meaning, in humans, requires and accompanies the historical constitution of the individual. Emotions and the material bodily structure contribute to it: this skin, smell, cells' membranes, chemical structure of DNA ….

And this is a crucial issue. All theories of information mentioned above, are based on an essential and radical separation of signs from their material realization: the great idea by Turing to formalize the split hardware vs software, practiced since Babbage and Morse or earlier, is at the core of computing as well as of Shannon's transmission of information. The same signs may be transmitted and elaborated by drums, smoke signs, electric impulses, wave frequency modulations … valves, diodes, chips, etc. In general, we use gradients of energy or matter for this purpose. The claim that in nature, in cellular chemical exchanges for examples, there is "information" any time the variation of a gradient matters more than an energy or matter flow, is an amazing anthropomorphic projection. In the perspective of the immateriality and uni-dimensionality of discrete information, the radical materiality of life, its intrinsic space dimensionality, is lost in vague abstract notions of information, rarely scientifically specified. Some of the consequences are hinted in (Longo 2018a).

It may be appropriate now to ask whether the construction of invariants of vision, for example, in Deep Learning, soundly models the brain's behavior, once observed that it does not model its architecture nor biological activity. Memory or, more generally (pre-conscious) retension, selects the relevant invariants by forgetting the details that do not matter, both at the moment of the memorized action and when memory is used to act in a new context. Retension exists for the purpose of protension (Berthoz 1997), (Longo and Montévil 2014), e.g. for capturing the prey by preceding its trajectory. Forgetting is crucial to animal memory: the child must remember what matters of the trajectories of a ball in order to grip it, not the color of the ball, say. We recognize a friend 20 years later because we remember what mattered to us, his/her smile, the expression, a movement of her/his eyes. These are fundamental cognitive constructions of invariants of action and intentions. Intentionality and pre-conscious protension are at the core of them, while they are based on and constitute meaning: because of the protensive nature of action, the friction on the world interferes and give sense to its deformations. With no emotion nor protensive affection, without the joy of gripping a ball thrown by the father or the need to capture a prey, the selection of what matters for an aim is hard to conceive: what is interesting for action and what may be relevant to recall for a new activity is largely related to the meaning we attach to an object of desire. It is hard to see anything of this nature, surely not a model, in the gradients' or wavelets' methods used by the excellent mathematics of multi-layered neural nets. Yet, they may provide increasingly effective imitations of key cognitive activities. As for the biological brain, note finally that its continuous material deformations are not implemented by a digital computer "in the background", in contrast to the non-linear mathematics of Deep Learning.

A Societal Conclusion

Besides a hint to the Geometry of Information and to neural nets, I only quoted the main mathematical frames for elaboration and transmission of information on *discrete* data types. By a historical simplification, I named them from Turing and Shannon. Wiener's approach (1948) belongs more to, and greatly enhanced, the theory of continuous control, based on the calculus of variations and other areas of optimization theory, (Rugh 1981), (Sontag 1990). The harmony of a noiseless world is obtained by smooth feed-back control of a dynamics, with a prefixed goal. Besides the construction of analogical machines, the social aim is explicit: a harmonious society governed independently of ideologies and passions, adjustable, by a fine tuning in continua, to the smooth governance of any political system, once its goals are given. A world where each society would move along a unique, mathematically defined geodetics, beyond the painful frictions caused by the democratic debate and opposing goals.[21]

There is a similar commitment, but even more "dry" and radical, in the myth of digital information as ruling nature. Smooth control is replaced by lists of "formal instructions", of pitiless "follow the rule" orders, exactly like in von Neuman's Automata. As often in science, from classical Greece to the Italian Renaissance, a vision of nature is also (or first) a perspective on the human condition. This is for example clear when rule-based classical AI is (was) considered an imitation or even a model of the human brain: it was first an image of what human cognition is, a rule-based sign pushing (see the quotations and references above).

The formalization of information as a new observable by the new fantastic sciences of elaboration and transmission of data has had a role in this. In order to be implemented in machines, information was separated from its use as meaningful symbolic human exchange, from our biological body and its historicity. It now seems to provide tools that float in between the two possible abusive transfers to the governance of human societies, I discuss below. Their common effect or, perhaps, aim is to subtract the economic, thus social and political decisions, to the democratic debate, anchored on possibly different interpretations. It is now beginning to modify human justice, increasingly affect by automated tools (Garapon and Lassègue 2018), while the human and historical "interpretation of the law" is crucial.

One of these abuses is the transfer of the mathematics of physical equilibrium processes to economic equilibria, since (Walras 1874), now enriched by the informational terminology.[22] Control theory is a variant of it and some more general versions in non-linear mathematics pretend to be applicable to all sorts of social dynamics, an approach also inaugurated by (Wiener 1954). The other abuse is the new role of information sciences in finance. Non-trivial mathematics, see (Merton 1990), (Bouleau 2017), (Biasoni 2020), joins sophisticated algorithms in elaborating abstract market information. Largely automatized information processing even more radically excludes human interpretation and social meaning from the financial dynamics that drive economy.[23]

[21] See (Supiot 2017) for the difference between democratic government and governance, and (Longo 2018b) for the entanglement of science and democracy.

[22] "Optimal pricing of goods in an economy appears to arise from agents obeying the local rules of commerce" (Fama 1991), quoted in (Crutchfield 1994), where it is stressed: "global information processing plays a key role".

[23] Statistical analysis of "abstract" data govern finance and, thus, economy, (Bouleau 2017). These analyses are independent from the underlying assets, that is from any reference to "tables, chairs, mugs", as Hilbert would put it (footnote 3).

In either case, a supposed (mathematical, numerical) objectivity leaves no alternative: when freely moving along a geodetics, the unique optimal path, with no human/economic meaning nor political friction, economy and finance cannot go wrong. Its implementation by formal rules, in digital machines, even further dehumanizes the government of the common house (the Greek oiko-nomos) as well as human conducts. In my view, the new hegemony of this autonomous universe of senseless financial data, in the governance of the world, plays a key role in encouraging the unbounded diffusion of the informational and computational terminology, superposed to all phenomena, independently of meaning and interpretation. The cultural hegemony of the "all is computation and digital information" fashion is further reinforced by today's general role of our extraordinary information networks and their computing machines. For example, what do Ladyman and Ross (2008) mean when claiming that "what cannot be computed cannot be thought" (p. 209)? Are they strictly referring to computing as what can be implemented in networks of digital computers? Or do they refer to some broader notion of computing that would vaguely include the *thinkable meaning* produced by a dancer, a painter or by the "geometric judgments" needed to prove the formally unprovable statements of number theory (Longo 2011a)? They seem to lean more towards the first interpretation and, thus, say: the construction of sense that cannot be produced/computed by the machines currently owned/developed by Google, Apple, Microsoft etc. *cannot be thought*. Perhaps, this is more a "normative" statement than a scientific analysis; that is, in view of the role of these corporations, this claim is meant to become a norm: you are not allowed to think what the GAFAM cannot compute – even more so: the digital information they handle "is" the world.

We need to oppose this trend and analyze and work at meaningful knowledge constructions, as the result of an historical formation of sense. An analysis of the historicity of biological evolution may be one of the possible links of science to the humanities, with no subordination. A modest attempt is carried on in the interplay we hinted to in (Koppl et al. 2015), by stressing the role of changing spaces of possibilities (phase space) and rare events, further analyzed in biology in (Longo 2017); this role is shared by theorizing both in evolution and in the historical humanities, economy in particular. Some mathematics is being invented on these and on semeiotic grounds in (Sarti et al. 2018) – see (Montévil et al. 2016) as for variability and diversity production in biological morphogenesis. These approaches to "heterogenesis", as called in (Sarti et al. 2018), are a tentative way to depart from an analysis and a governance of biological processes and of our communicating humanity by optimal paths in pre-given spaces of possibilities, towards pre-given goals, possibly by mechanical rules.

We may then conclude by quoting Simondon (1989, p.272): "La machine peut se dérégler et présenter alors les caractéristiques de fonctionnement analogues à la conduite folle chez un être vivant. Mais elle ne peut se *révolter*. La révolte implique en effet une profonde transformation des *conduites finalisées*, et *non un dérèglement de la conduite*".[24]

[24] "The machine may be out of order and present the operating characteristics similar to a mad behavior of a living being. But it cannot revolt. The revolt indeed implies a profound transformation of the finalized behavior, and not a malfunction of the conduct."

Acknowledgments Inigo Wilkins made a critical and competent reading of a preliminary version of this paper.

References[25]

Abraham, L., and L.D. Hurst. 2018. Refining the Ambush Hypothesis: Evidence that GC- and AT-Rich Bacteria Employ Different Frameshift Defense Categories. *Genome Biology and Evolution* 14: 1153–1173.

Aceto, L., G. Longo, and B. Victor, eds. 2003. The Difference Between Concurrent and Sequential Computations, Special issue. *Mathematical Structures in Computer Science* 13: 4–5. Cambridge University Press.

Adjiri, A. 2017. DNA Mutations May Not Be the Cause of Cancer. *Oncol Therapy* 5 (1): 85–101.

Asperti, A., and G. Longo. 1991. *Categories, Types and Structures*. Cambridge, MA: MIT Press.

Baccelli, F., H.M. Mir-Omid, and A. Khezeli. 2016. Dynamics on Unimodular Random Graphs. In arXiv:1608.05940v1 [math.PR].

Bailly, F., and G. Longo. 2006. (Trans. 2011) *Mathematics and the Natural Sciences: The Physical Singularity of Life*. London: Imperial College Press, (original French version, Hermann, 2006).

———. 2009. Biological Organization and Anti-Entropy. *Journal of Biological Systems* 17 (01): 63–96. https://doi.org/10.1142/S0218339009002715.

Baker, S. 2014. Recognizing Paradigm Instability in Theories of Carcinogenesis. *British Journal of Medicine and Medical Research* 4 (5): 1149–1163.

———. 2015. A Cancer Theory Kerfuffle Can Lead to New Lines of Research. *Journal of the National Cancer Institute* 107: dju405.

Barbaresco, F., and A. Mohammad-Djafari, eds. 2015. *Information, Entropy and Their Geometric Structures*. Basel/Beijing: MDPI.

Barendregt, H. 1984. *The Lambda-Calculus: Its Syntax, Its Semantics*. Amsterdam: North-Holland.

Barrell, B.G., G.M. Air, and C.A. Hutchison 3rd. 1976. Overlapping Genes in Bacteriophage φX174. *Nature* 264: 34–41.

Berthoz, A. 1997. *Le Sens du Mouvement*. Paris: Odile Jacob (English version, 2000).

Bezem, M., J.W. Klop, and R. Roelde Vrijer. 2013. *Term Rewriting Systems*. Cambridge: Cambridge University Press.

Biasoni, S. 2020. *De la précarité épistémique des systèmes complexes en transition critique étendue. Vers une épistémologie ouverte des processus de marchés*. Thèse de Doctorat, sous la direction de J. Lassègue et G. Longo, Ens, Paris (publication attendue: 2020).

Boser, B., E. Sackinger, J. Bromley, Y. LeCun, and L. Jackel. 1991. An Analog Neural Network Processor with Programmable Topology. *IEEE Journal of Solid-State Circuits* 26 (12): 2017–2025.

Bouleau, N. 2017. *Wall street ne connait pas la tribu borélienne*. Paris: Spartacus IDH.

Bravi, B., and G. Longo. 2015. The Unconventionality of Nature: Biology, from Noise to Functional Randomness. In *Unconventional Computation and Natural Computation*, LNCS 9252, ed. Calude and Dinneen, 3–34. Springer.

Brillouin, L. 1956. *Science and Information Theory*. New York: Academic.

Brock, A., and S. Huang. 2017. Precision Oncology: Between Vaguely Right and Precisely Wrong. *Cancer Research*. https://doi.org/10.1158/0008-5472.CAN-17-0448. Published December.

Bühlmann, V., L. Hovestadt, and V. Moosavi. 2015. *Coding as Literacy*. Basel: Birkhäuser.

Calude, C. 2002. *Information and Randomness*. 2nd ed. Berlin: Springer.

[25] Papers (co-)authored by Giuseppe Longo are downloadable here: http://www.di.ens.fr/users/longo/download.html

Calude, C., and G. Longo. 2016. Classical, Quantum and Biological Randomness as Relative Unpredictability. Special issue of *Natural Computing* 15 (2): 263–278, Springer, June.

———. 2017. The Deluge of Spurious Correlations in Big Data. *Foundations of Science* 22 (3): 595–612.

Cant, N.B., and C.G. Benson. 2003. Parallel Auditory Pathways: Projection Patterns of the Different Neuronal Populations in the Dorsal and Ventral Cochlear Nuclei. *Brain Research Bulletin* 60 (5–6): 457–474.

Castiglione, P., M. Falcioni, A. Lesne, and A. Vulpiani. 2008. *Chaos and Coarse-Graining in Statistical Mechanics*. New York: Cambridge University Press.

Chaitin, G. 2012. *Proving Darwin: Making Biology Mathematical*. Pantheon: Random House.

Chatelet, G. 1993. *Les enjeux du mobile*. Paris: Seuil.

Chibbaro, S., L. Rondoni, and A. Vulpiani. 2015. *Reductionism, Emergence and Levels of Reality: The Importance of Being Borderline*. Berlin: Springer.

Chirico, N., A. Vianelli, and R. Belshaw. 2010. Why Genes Overlap in Viruses. *Proceedings of the Biological Sciences* 277 (1701): 3809–3817. https://doi.org/10.1098/rspb.2010.1052. Epub 2010 Jul 7. PubMed PMID: 20610432.

Collet, P., F. Krüger, and O. Maitre. 2013. Automatic Parallelization of EC on GPGPUs and Clusters of GPGPU Machines with EASEA and EASEA-CLOUD. In *Massively Parallel Evolutionary Computation on GPGPUs*, Natural Computing Series Book Series, 35–59. Berlin: Springer.

Cornell, Gary., Joseph H. Silverman, and Glenn. Stevens. 2013. *Modular Forms and Fermat's Last Theorem* (illustrated ed.). Springer.

Cortini, R., M. Barbi, B. Caré, C. Lavelle, A. Lesne, J. Mozziconacci, and J.M. Victor. 2016. The Physics of Epigenetics. *Reviews of Modern Physics* 88: 025002.

Cousot, P. 2016. Abstract Interpretation. In *SAVE* 2016, Changsha, China, 10–11 December.

Crick, F.H.C. 1958. On Protein Synthesis. In *Symposia of the Society for Experimental Biology*, Number XII: The Biological Replication of Macromolecules, ed. F.K. Sanders, 138–163. Cambridge: Cambridge University Press.

Crutchfield, J. 1994. The Calculi of Emergence: Computation Dynamics, and Induction. Special issue of *Physica D*, Proceedings of the Oji International Seminar "From Complex Dynamics to Artificial Reality".

Danchin, A. 2003. *The Delphic Boat. What Genomes Tell Us*. Cambridge, MA: Harvard University Press.

———. 2009. Bacteria as computers making computers. *FEMS Microbiology Reviews* 33 (1). https://onlinelibrary.wiley.com/doi/full/10.1111/j.1574-6976.2008.00137.x.

Dawkins, R. 1976. *The Selfish Gene*. Oxford: Oxford University Press.

Deacon, T.W., A. Srivastava, and J.A. Bacigalupi. 2014. The Transition from Constraint to Regulation at the Origin of Life. *Frontiers in Bioscience – Landmark* 19: 945–957.

Devaney, R.L. 1989. *An Introduction to Chaotic Dynamical Systems*. Reading: Addison-Wesley.

Edelman, J. 1987. *Neural Darwinism: The Theory of Neuronal Group Selection*. New York: Basic Books.

Elowitz, M.B., A.J. Levine, E. Siggia, and P.S. Swain. 2002. Stochastic Gene Expression in a Single Cell. *Science* 297: 1183–1186.

Fama, E. 1991. Efficient Capital Markets II. *The Journal of Finance* 46: 1575–1617.

Fox Keller, E. 2000. *The Century of the Gene*. Cambridge, MA: Harvard University Press.

Fromion, V., E. Leoncini, and P. Robert. 2013. Stochastic Gene Expression in Cells: A Point Process Approach. *SIAM Journal on Applied Mathematics* 73 (1): 195–211.

Garapon, A., and J. Lassègue. 2018. *Justice digitale*. Paris: PUF.

Gatenby, R.A. 2017. Is the Genetic Paradigm of Cancer Complete? *Radiology* 284: 1–3.

Gilbert, W. 1992. A vision of the Grail. In *The Code of Codes: Scientific and Social Issues in the Human Genome Project*, ed. Daniel J. Kevles and E. Leroy. Cambridge, MA: Harvard University Press.

Girard, J.Y., Y. Lafont, and R. Taylor. 1989. *Proofs and Types*. Cambridge: Cambridge University Press.

Goubault, E., ed. 2000. *Geometry in Concurrency*, Special issue, *Mathematical Structures in Computer Science*, Cambridge University Press., vol.10, n. 4.

Gould, S.J. 1996. *Full House*. New York: Three Rivers Press.

Gouyon, P.H., J.P. Henry, and J. Arnoud. 2002. *Gene Avatars, The Neo-Darwinian Theory of Evolution*. New York: Kluwer.

Gsponer, J., and M. Madan Babu. 2009. The Rules of Disorder or Why Disorder Rules. *Progress in Biophysics and Molecular Biology* 99: 94–103.

Hadjiargyrou, M., and N. Delihas. 2013. The Intertwining of Transposable Elements and Non-coding RNAs. *International Journal of Molecular Sciences* 14 (7): 13307–13328. https://doi.org/10.3390/ijms140713307.

Hebb, D. 1949. *The Organization of Behavior*. New York: Wiley.

Hesselink, Wim H. 2008. Computer Verification of Wiles' Proof of Fermat's Last Theorem. www.cs.rug.nl. Retrieved 2017-06-29.

Hilbert, D. 1898. *The Foundations of Geometry*. Trans. 1901. Chicago: Open Court.

———. 1926. Über das Unendliche. *Mathematische Annalen* 95 (1): 161–190.

Islami, A., and G. Longo. 2017. Marriages of Mathematics and Physics: A Challenge for Biology. In *The Necessary Western Conjunction to the Eastern Philosophy of Exploring the Nature of Mind and Life*, ed. K. Matsuno et al. Special Issue, *Progress in Biophysics and Molecular Biology* 131: 179–192.

Israeli, N., and N. Goldenfeld. 2004. Computational Irreducibility and the Predictability of Complex Physical Systems. *Physical Review Letters* 92 (7): 074105.

Jacob, F. 1970. *La logique du vivant*. Paris: Gallimard.

———. 1974. Le modèle linguistique en biologie, *Critique*, XXX, 322.

Jacob, F., and J. Monod. 1961. Genetic Regulatory Mechanisms in the Synthesis of Proteins. *Journal of Molecular Biology* 3: 318–356.

Kato, S., S.M. Lippman, K.T. Flaberty, and R. Kurrock. 2016. The Conundrum of Genetic "Drivers" in Benign Conditions. *JNCI Journal of the National Cancer Institute* 108 (8).

Koppl, R., S. Kauffman, G. Longo, and T. Felin. 2015. Economy for a Creative World. *Journal of Institutional Economics* 11 (01): 1–31.

Kosman-Schwarback, Y. 2010. *The Noether Theorems: Invariance and Conservation Laws in the Twentieth Century*. Berlin: Springer.

Kowalchuk, G.A., M. Bruinsma, and J. van Veen. 2003. Assessing Responses of Soil Microorganisms to GM Plants. *Trends in Ecology and Evolution* 18: 403–410.

Kreisel, G. 1982. *Four Letters to G. L.* http://www.di.ens.fr/users/longo/files/FourLettersKreisel.pdf.

Kupiec, J.J. 1983. A Probabilistic Theory for Cell Differentiation, Embryonic Mortality and DNA C-Value Paradox. *Speculations in Science and Technology* 6: 471–478.

Ladyman, J., and D. Ross. 2008. *Every Thing Must Go, Metaphysics Naturalized*. Oxford: Oxford University Press.

Landauer, R. 1991. Information is Physical. *Physics Today* 44: 23.

Laskar, J. 1994. Large Scale Chaos in the Solar System. *Astronomy and Astrophysics* 287: L9–L12.

Leff, Harvey S., and Andrew F. Rex, eds. 1990. *Maxwell's Demon: Entropy, Information, Computing*. Bristol: Adam-Hilger.

Lesne, A. 2007. The Discrete vs Continuous Controversy in Physics. *Mathematical Structures in Computer Science* 17 (2): 185–223.

———. 2014. Shannon Entropy: A Rigorous Notion at the Crossroads Between Probability, Information Theory, Dynamical Systems and Statistical Physics. *Mathematical Structures in Computer Science* 24 (3). Cambridge University Press.

Longo, G. 2004. Some Topologies for Computations. In *Géométrie au XX siècle, 1930–2000*. Paris: Hermann.

———. 2010. Incompletezza. In *La Matematica*, vol. 4, 219–262, Einaudi. Revised and translated: 2018. Interfaces of Incompleteness. In *Systemics of Incompleteness and Quasi-Systems*, ed. G. Minati, M. Abram, and E. Pessa. New York: Springer.

———. 2011a. Reflections on Concrete Incompleteness. *Philosophia Mathematica* 19 (3): 255–280.

———. 2011b. Theorems as Constructive Visions. In *ICMI 19 conference on Proof and Proving*, Taipei, Taiwan, May 10–15, 2009, ed. Hanna, de Villiers. Springer.

———. 2015a. The Consequences of Philosophy. *Glass-Bead*, Web Journal. http://www.glass-bead.org/article/the-consequences-of-philosophy/?lang=enview.

———. 2015b. Conceptual Analyses from a Grothendieckian Perspective: Reflections on Synthetic Philosophy of Contemporary Mathematics by Fernando Zalamea. In *Speculations* Web Journal. https://www.urbanomic.com/speculations_lo/.

———. 2017. How Future Depends on Past Histories and Rare Events in Systems of Life. *Foundations of Science*: 1–32.

———. 2018a. Information and causality: Mathematical reflections on Cancer biology. *Organisms. Journal of Biological Sciences* 2 (1): 83–103.

———. 2018b. Complexity, Information and Diversity, in Science and in Democracy. *To appear* (see https://www.di.ens.fr/users/longo/download.html for a prelimnary version).

———. 2018c. Letter to Alan Turing. In *Theory, Culture and Society*, Special Issue on Transversal Posthumanities, ed. Füller, Braidotti.

Longo, G., and M. Montévil. 2014. *Perspectives on Organisms: Biological Time, Symmetries and Singularities*. Dordrecht: Springer.

Longo, G., Catuscia Palamidessi, and Thierry Paul. 2010. Some Bridging Results and Challenges in Classical, Quantum and Computational Randomness. In *Randomness Through Computation*, ed. H. Zenil, 73–92. World Scientific.

Longo, G., P.A. Miquel, C. Sonnenschein, and A. Soto. 2012. Is Information a Proper Observable for Biological Organization? *Progress in Biophysics and Molecular Biology* 109 (3): 108–114.

Lutz, E., and S. Ciliberto. 2015. Information: From Maxwell's Demon to Landauer's Eraser. *Physics Today* 69 (30).

Lwoff, A. 1969. *L'ordre biologique*. Paris.

Mac Lane, S. 1970. *Categories for the Working Mathematician*. New York: Springer.

Mallat, S. 2016. Understanding Deep Convolutional Networks. *Philosophical Transactions of the Royal Society A* 37: 20150203.

Mallat, S., and W.L. Hwang. 1992. Singularity detection and processing with wavelets. In *IEEE Transactions on Information Theory*, vol. 32, no. 2, March.

Marinov, G.K., B.A. Williams, K. McCue, G.P. Schroth, J. Gertz, R.M. Myers, and B.J. Wold. 2014. From Single-Cell to Cell-Pool Transcriptomes: Stochasticity in Gene Expression and RNA Splicing. *Genome Research* 24: 496–510.

Maynard-Smith, J. 1999. The Idea of Information in Biology. *The Quarter Review of Biology* 74: 495–400.

McCulloch, W., and W. Pitts. 1943. A Logical Calculus of Ideas Immanent in Nervous Activity. *Bulletin of Mathematical Biophysics* 5 (4): 115–133.

Merleau-Ponty, M. 1945. *Phénomenologie de la perception*. Paris: Gallimard.

Merton, R.C. 1990. *Continuous-Time Finance*. New York: Oxford University Press.

Monod, J. 1970. *Le Hasard et la Nécessité*. Paris: PUF.

Montévil, M., and M. Mossio. 2015. Closure of Constraints in Biological Organisation. *Journal of Theoretical Biology* 372: 179–191.

Montévil, M., M. Mossio, A. Pocheville, and G. Longo. 2016. Theoretical Principles for Biology: Variation. *Progress in Biophysics and Molecular Biology* 122 (1): 36–50. Soto, Longo & Noble eds.

Mouilleron, H., V. Delcourt, and X. Roucou. 2016. Death of a Dogma: Eukaryotic mRNAs Can Code for More Than One Protein. *Nucleic Acids Research* 44 (1): 14–23.
Newell, A., and H.A. Simon. 1976. Computer Science as Empirical Inquiry: Symbols and Search. *Communications of the ACM* 19 (3): 113–126.
Nickles, T. 2018. Alien Reasoning: Is a Major Change in Scientific Research Underway? *Topoi*. https://link.springer.com/article/10.1007/s11245-018-9557-1.
Onuchic, J., Z. Luthey-Schulten, and P. Wolynes. 1997. Theory of Protein Folding: The Energy Landscape Perspective. *Annual Review of Physical Chemistry* 48: 545–600.
Paldi, A. 2003. Stochastic Gene Expression During Cell Differentiation: Order from Disorder? *Cellular and Molecular Life Sciences* 60: 1775–1779.
Pauling, L. 1987. Schrödinger Contribution to Chemistry and Biology. In *Schrödinger: Centenary Celebration of a Polymath*, ed. Kilmister. Cambridge: Cambridge University Press.
Pavesi, A., A. Vianelli, N. Chirico, Y. Bao, R. Belshaw, O. Blinkova, A. Firth, and D. Karlin. 2018. Overlapping Genes and the Protein they Encode Differ Significantly in Their Sequence Composition from Non-overlapping Genes. *PLOS ONE* 13 (10): e0202513.
Perret, N., and G. Longo. 2016. Reductionist Perspectives and the Notion of Information. *Progress in Biophysics and Molecular Biology* 122 (1): 11–15. Soto, Longo and Noble eds.
Petitot, J. 2017. *Elements of Neurogeometry. Functional Architectures of Vision*, Lecture Notes in Morphogenesis. Cham: Springer.
Pilyugin, S. 1999. *Shadowing in Dynamical Systems*. Berlin: Springer.
Ramellini, P. 2002. L'informazione in biologia, 39-48. In *La Nuova Scienza*, La società dell'informazione, ed. U. Colombo and G. Lanzavecchia, vol. 3. Milano: Libri Scheiwiller.
Riemann, B. 1854. *On the Hypothesis Which Lie at the Basis of Geometry* (Engl. by W. Clifford, Nature, 1973).
Rogers, H. 1967. *Theory of Recursive Functions and Effective Computability*. New York: McGraw Hill.
Rosenblatt, F. 1958. The Perceptron: A Probabilistic Model for Information Storage and Organization in the Brain. *Psychological Review*. 65 (6): 386–408.
Rovelli, C. 2004. *Quantum Gravity*. Cambridge: Cambridge University Press.
Rugh, W.J. 1981. *Nonlinear System Theory: The Volterra/Wiener Approach*. Baltimore: The Johns Hopkins University Press.
Saigusa, T., A. Tero, T. Nakagaki, and Y. Kuramoto. 2008. Amoebae Anticipate Periodic Events. *Phyiscal Review Letters*. 100: 018101.
Sarti, A., G. Citti, and D. Piotrowski. 2018. Differential Heterogenesis and the Emergence of Semiotic Function. *Semiotica*, in press.
Schrödinger, E. 1944. *What is Life? The Physical Aspect of the Living Cell*. Cambridge: Cambridge University Press.
Scott, D. 1982. Domains for Denotational Semantics. In *Proceedings ICALP 82*, LNCS 140. Springer.
Searls, D.B. 1992. The Linguistics of DNA. *American Scientist* 80: 579–591.
Seligmann, H., and D.D. Pollock. 2004. The Ambush Hypothesis: Hidden Stop Codons Prevent Off-Frame Gene Reading. *DNA and Cell Biology* 23: 701–705.
Shannon, C. 1948. A Mathematical Theory of Communication. *Bell Systems Technical Journal* 27: 279–423, 623–656.
Sifakis, J. 2011. Modeling Real-Time Systems-Challenges and Work Directions. In *EMSOFT01*, Tahoe City, October 2001. Lecture Notes in Computer Science.
Simondon, G. 1989. *L'individuation psychique et collective*. Paris: Aubier, 2007.
Sonnenschein, C., and A.M. Soto. 1999. *The Society of Cells: Cancer and Control of Cell Proliferation*. New York: Springer.
———. 2011. The Death of the Cancer Cell. *Cancer Research* 71: 4334–4337.
———. 2013. The Aging of the 2000 and 2011 Hallmarks of Cancer Reviews: A Critique. *Journal of Biosciences* 38: 651–663.
Sontag, D. 1990. *Mathematical Control Theory*. New York: Springer.

Soto, A.M., and C. Sonnenschein. 2010. Environmental Causes of Cancer: Endocrine Disruptors as Carcinogens. *Nature Reviews Endocrinology* 6 (7): 363–370.

———. 2017. Why Is It That Despite Signed Capitulations, the War on Cancer is Still on? *Organisms. Journal of Biological Sciences* 1: 1.

Soto, A.M., G. Longo, and D. Noble, eds. 2016. From the Century of the Genome to the Century of the Organism: New Theoretical Approaches. Special issue. *Progress in Biophysics and Molecular Biology* 122 (1): 1–82.

Supiot, A. 2017. *Governance by Numbers*. New York: Hart Publishing.

Turing, A. 1936. On Computable Numbers, with an Application to the Entscheidungsproblem. *Proceedings of the London Mathematical Society. (Series 2)* 42: 230–265.

———. 1950. Computing Machinery and Intelligence. *Mind* 50: 433–460; also in Boden 1990, *Collected Works* (Volume 1).

Venter, C. 2010. We Have Learned Nothing from the Genome. *Der Spiegel*, July 29.

Versteg, R. 2014. Cancer: Tumours Outside the Mutation Box. *Nature* 506: 438–439.

Violi, P. 2016. How our Bodies Become Us: Embodiment, Semiosis and Intersubjectivity. *Journal of Cognitive Semiotics* IV (1): 57–75.

von Uexküll, J. 1934. A Theory of Meaning, reprinted in *Semiotica* (1982) 42–1, 25–82.

Walras, L. 1874. *Théorie mathématique de la richesse sociale*. Paris: Corbaz.

Weinberg, R. 2014. Coming Full Circle – Form Endless Complexity to Simplicity and Back Again. *Cell* 157: 267–271.

Weyl, H. 1949. *Philosophy of Mathematics and of Natural Sciences,* 1927, English Trans. Princeton: Princeton University Press.

———. 1953. Axiomatic Versus Constructive Procedures in Mathematics (a cura di T. Tonietti) *The Mathematical Intelligence,* vol. 7, no. 4. New York: Springer, (reprinted) 1985.

Wiener, N. 1948. *Cybernetics or Control and Communication in the Animal and the Machine*. Cambridge, MA: MIT Press.

———. 1954. *The Human Use of Human Beings: Cybernetics and Society*. London: Da Capo Press.

Wolfram, S. 2002. *A New Kind of Science*. Champaign: Wolfram Media Inc.

———. 2013. The Importance of Universal Computation. In *A. Turing, His Work and Impact*, ed. Cooper. Waltham: Elsevier.

Zalamea, F. 2012. *Synthetic Philosophy of Contemporary Mathematics*. Cambridge: Urbanomic/Sequence Press.

Zeilinger, A. 2004. Why the Quantum? 'It' from 'Bit'? A Participatory Universe? Three Far-Reaching Challenges from John Archibald Wheeler and Their Relation to Experiment. In *Science and Ultimate Reality: Quantum Theory, Cosmology and Computation*, ed. J. Barrow, P. Davies, and C. Harper, 201–220. Cambridge: Cambridge University Press.

Zorzi, M. 2016. Quantum Lambda Calculi: A Foundational Perspective. *Mathematical Structures in Computer Science* 26 (7): 1107–1119.

Mathematical Proofs and Scientific Discovery

Fabio Sterpetti

The Method of Mathematics and the Automation of Science

This chapter aims to disentangle some issues surrounding the claim that scientific discovery can be automated thanks to the development of Machine Learning, or some other programming strategy, and the increasing availability of Big Data (see e.g. Allen 2001; Colton 2002; Anderson 2008; King et al. 2009; Sparkes et al. 2010; Mazzocchi 2015). For example, according to Anderson, the "new availability of huge amounts of data, along with the statistical tools to crunch these numbers, offers a whole new way of understanding the world" (Anderson 2008). In this perspective, we "can throw the numbers into the biggest computing clusters the world has ever seen and let statistical algorithms find pattern where science cannot" (Ibidem). Sparkes and co-authors state that the advent of computers and computer science "in the mid-20th century made practical the idea of automating aspects of scientific discovery, and now computing is playing an increasingly prominent role in the scientific discovery process" (Sparkes et al. 2010). This increasing trend to automation in science has led to the development of so-called robot scientist, which allegedly, "automatically originates hypotheses to explain observations, devises experiments to test these hypotheses, physically runs the experiments by using laboratory robotics, interprets the results, and then repeats the cycle" (King et al. 2009). Some authors have gone, if possible, even further by claiming that the process of theory formation in pure mathematics can be automated. For example, Colton writes that theory formation in mathematics "involves, amongst other things, inventing concepts, performing calculations, making conjectures, proving theorems and finding counterexamples to false conjectures," and that computer programs "have been written which automate all of these activities" (Colton 2002, p. 1).

F. Sterpetti (✉)
Department of Philosophy, Sapienza University of Rome, Rome, Italy
e-mail: fabio.sterpetti@uniroma1.it

M. Bertolaso, F. Sterpetti (eds.), *A Critical Reflection on Automated Science*, Human Perspectives in Health Sciences and Technology 1,
https://doi.org/10.1007/978-3-030-25001-0_6

Unlike other researches that deal with automated discovery, I will not focus here on assessing some specific and recent achievements in computer science to verify whether computer machines were really able to autonomously make the scientific discoveries that are credited to them. Rather, I wish to underline how the very idea that scientific discovery can be automated derives, at least in part, from a confusion on the nature of the method of mathematics and science, and how this confusion affects the judgement over whether machines provide genuine discoveries. The origin of this confusion may be due to the fact that the mathematical tools developed to deal with computability originated in the first half of the twentieth century, in what can be called a formalist and axiomatic cultural 'environment'. From a formalist point of view, the method of mathematics is the axiomatic method, according to which, to demonstrate a statement one starts from some given premises, which are supposed to be true, and then deduces the statement from them. Hilbert, whose ideas were very influential at the time, viewed the axiomatic method as the crucial tool for mathematics and scientific inquiry more in general (Rathjen and Sieg 2018). For example, Hilbert writes that he believes that everything that can "be object of scientific thinking in general [...] runs into the axiomatic method and thereby indirectly to mathematics. Forging ahead towards the ever deeper layers of axioms [...] we attain ever deepening insights into the essence of scientific thinking itself" (Hilbert 1970, p. 12). The mathematical agenda of the first decades of the twentieth century was set by Hilbert by fixing, in his famous address to the International Congress of Mathematicians held in Paris in 1900, the most relevant mathematical problems, and by putting forward his formalist program (Zach 2016). Hilbert's formalist program pursued two main goals: (1) the formalization of all of mathematics in axiomatic form, together with (2) a proof that this axiomatization of mathematics is consistent. In this view, the consistency proof itself was to be carried out using only finitary methods, in order to avoid any reference to infinitary methods, which were regarded as problematic because of their being related to a problematic and disputed concept such as that of infinity.

The theoretical success achieved by logicians and mathematicians in dealing with computability led someone to think that the view on the nature of mathematical knowledge implied by the axiomatic method was in some way vindicated by such extraordinary achievements. But the assessment of some results in computability theory and the inquiry on what is the most adequate way to describe what we do when we do mathematics are distinct issues, and they should be kept distinct (Longo 2011). And whether we keep those issues distinct affects how we assess the claim that science can be automated.

The way in which the confusion between mathematical successful developments in computability theory and the justification of axiomatic ideas originated, can be summarized as follows: the axiomatic view led to the rigorous study of computability. The study of computability led to the development of computer machines. Computer machines proved extraordinary useful and effective. This confirmed that computability theory was correct. And this led many to conclude that the axiomatic view, from which computability theory seemed to stem, should had been the correct view on what we do when we do mathematics, namely deducing consequences from given axioms. Now, mathematics has always been regarded as one of the highest achievements of human thought. So, the argument goes, if doing mathematics means to deduce consequences from given axioms, and machines are able to deduce consequences from given axioms thanks to the mathematical results developed by assuming that doing mathematics must be conceived of as deducing consequences from given axioms, this means both (1) that machines are able to do mathematics, and so that machines can in a sense think, and (2) that doing mathematics really means to deduce consequences from given axioms, i.e. that humans actually do mathematics by deducing consequences from given axioms. Now, if humans actually do mathematics by deducing consequences from given axioms, then human brains can be regarded as Turing Machines, at least in the sense that what we do when we think is, in the ultimate analysis, a kind of computation. Since mathematics and science are subsets of our thinking, this means that mathematics and science are computational activities. Indeed, a Turing Machine is a mathematical model of computation that defines an abstract machine which is capable of simulating any algorithm's behaviour. Since computational activities can be performed by computer machines, there is no compelling theoretical reason why machines might not be able to do mathematics and science. Nor is there any *fundamental* difference between Turing Machines and our brains. This is just a sketch of how the so-called Computational Theory of Mind (CTM)[1] is usually argued for. In this article, for the sake of convenience, I will refer to those who are broadly sympathetic to this line of reasoning and the idea that science can be automated as to 'the computationalists'.

It might be objected that those limitative results, such as Gödel's and Turing's, which paved the way to the development of computer machines, showed also that Hilbert's formalist program was unfeasible, and that this should had been enough to refuse the axiomatic view of mathematics. Indeed, according to Gödel's first incompleteness theorem, any sufficiently strong, consistent formal system F is incomplete, since there are statements of the language of F which are undecidable, i.e. they can neither be proved nor disproved in F. This theorem proves that the first one of the two main goals of Hilbert's program, i.e. the formalization of all of mathematics in axiomatic form, is unattainable. Moreover, according to Gödel's second incompleteness theorem, for any sufficiently strong, consistent formal system F, the consistency of F cannot be proved in F itself. This theorem proves that the second one of the two main goals of Hilbert's program, i.e. to give a proof that the axiomatization of all of mathematics is consistent, is unattainable.[2] Finally, Turing showed the relation between undecidability and incomputability, since it proved that there is no algorithm for deciding the truth of statements in Peano arithmetic. So, this objection is in a sense correct: limitative results showed that Hilbert's program was unfeasible. But it has not to be overlooked that the 'mathematical machinery' developed by mathematicians such as Gödel and Turing to adequately deal with Hilbert's problems and program was developed *within* a shared axiomatic perspective on

[1] On CTM, see Rescorla (2017).

[2] For a survey on Gödel's theorems, see Raatikainen (2018).

mathematics and was perfectly suited to meet Hilbert's standard of rigor and formalization (Longo 2003). So, despite those results showed that Hilbert's program cannot be pursued, since they proved extremely useful and were developed within what can be called a Hilbertian theoretical framework, they contributed to perpetuate Hilbert's view of mathematics, according to which the method of mathematics is the axiomatic method.

Still today, many mathematicians and philosophers think that despite Hilbert's program cannot be entirely realized, the ideas conveyed by such program are appealing and not completely out of track. In this view, the impact of limitative results such as Gödel's and Turing's on the formalist approach to mathematics has not to be overstated. For example, Calude and Thompson recently wrote that although it is not possible "to formalise all mathematics, it is feasible to formalise essentially all the mathematics that 'anyone uses'. Zermelo-Fraenkel set theory combined with first-order logic gives a satisfactory and generally accepted formalism for essentially all current mathematics" (Calude and Thompson 2016, p. 139). In their view, Hilbert program was not completely refuted by limitative results, rather it "can be and was salvaged by changing its goals slightly:" although it is not possible "to prove completeness for systems at least as powerful as Peano arithmetic," it is nevertheless "feasible to prove completeness for many weaker but interesting systems, for example, first-order logic [...], Kleene algebras and the algebra of regular events and various logics used in computer science" (Ibidem).[3] Examples of this kind testify that the axiomatic view, despite its inadequacy, is still the received view in the philosophy of mathematics.

The Analytic View of the Method of Mathematics

I argue that since the axiomatic view is inadequate, we should prefer an alternative view, namely the analytic view of the method of mathematics (Cellucci 2013, 2017), according to which knowledge is increased through the analytic method. According to the analytic method, to solve a problem one looks for "some hypothesis that is a sufficient condition for solving it. The hypothesis is obtained from the

[3] This approach seems to forget that Hilbert did not seek formalization for the sake of formalization. Formalization was not an end, rather it was a means in Hilbert's view. His main aim was to give a secure foundation to mathematics through the formalization of a part of it. And this goal cannot be reached because of Gödel's results. So, it is difficult to overstate the relevance of those results for Hilbert's view. That some limited portion of mathematics can be formalized or shown to be complete, but it is not possible to formalize the all of mathematics or prove its completeness in general, it is not something that can salvage Hilbert's perspective "by changing its goals slightly," rather it is a complete defeat of Hilbert's view, since it shows the unfeasibility of its *main* goal. For example, according to Weyl, the relevance of Gödel's results cannot be overstated, since because of those results the "ultimate foundations and the ultimate meaning of mathematics remain an open problem [...]. The undecisive outcome of Hilbert's bold enterprise cannot fail to affect the philosophical interpretation" (Weyl 1949, p. 219).

problem […], by some non-deductive rule, and must be plausible […]. But the hypothesis is in its turn a problem that must be solved," and is solved in the same way (Cellucci 2013, p. 55).[4]

The assessment of the *plausibility* of any given hypothesis is crucial in this perspective. But how has plausibility to be understood? The interesting suggestion made by the analytic view is that the plausibility of a hypothesis is assessed by a careful examination of the arguments (or reasons) for and against it. According to this view, in order to judge over the plausibility of a hypothesis, the following 'plausibility test procedure' has to be performed: (1) "deduce conclusions from the hypothesis"; (2) "compare the conclusions with each other, in order to see that the hypothesis does not lead to contradictions"; (3) "compare the conclusions with other hypotheses already known to be plausible, and with results of observations or experiments, in order to see that the arguments for the hypothesis are stronger than those against it on the basis of experience" (Ibidem, p. 56). If a hypothesis passes the plausibility test procedure, it can be temporarily accepted. If, on the contrary, a hypothesis does not pass the plausibility test, it is put on a 'waiting list', since new data may always emerge, and a discarded hypothesis may successively be re-evaluated. Thus, according to the analytic view of method, what in the ultimate analysis we really do in the process of knowledge ampliation, is to produce hypotheses, assess the arguments/reasons for and against each hypothesis, and provisionally accept or refuse such hypotheses.

In the last century, the dominance of a foundationalist perspective on scientific and mathematical knowledge, the influence of Hilbert's thought, and the diffusion of the idea that a logic of discovery cannot exist, led to the widespread conviction that the method of mathematics and science is (or should be) the axiomatic method.[5] Analysis, i.e. the search for new hypotheses by means of which problems can be solved, has been overlooked or neglected (Schickore 2014). Philosophers rejected the goal of "traditional epistemology from Plato to Boole: a theory of discovery" (Glymour 1991, p. 75). Indeed, since Plato and Aristotle philosophers "thought the goal of philosophy, among other goals, was to provide methods for coming to have knowledge" (Ibidem). But in the twentieth century, "there was in philosophy almost nothing more of methods of discovery. A tradition that joined together much of the classical philosophical literature simply vanished" (Ibidem). For example, Popper famously stated that "there is no such thing as a logical method of having new ideas, or a logical reconstruction of this process" (Popper 2005, p. 8).

Contrary to this perspective, the analytic view maintains that there is a logic of discovery, and that one of the goals of philosophy is to provide methods for coming

[4] The origin of the analytic method may be traced back to the works of the mathematician Hippocrates of Chios and the physician Hippocrates of Cos, and was firstly explicitly formulated by Plato in *Meno*, *Phaedo* and the *Republic*. Here I can only give a sketch of the analytic method. For an extensive presentation of the analytic view, see Cellucci (2013, 2017).

[5] For a survey of the main conceptions of method that have been put forward so far, see Cellucci (2013, 2017). On the analytic method, see also Hintikka and Remes (1974), and Lakatos (1978, Vol. 2, Chap. 5). On the axiomatic method, see also Rodin (2014, part I).

to have knowledge. Indeed, the analytic method "is a logical method" and from the fact that "knowledge is the result of solving problems by the analytic method, it follows that logic provides means to acquire knowledge" (Cellucci 2013, p. 284). Since logic is a branch of philosophy, this means that philosophy does provide methods for coming to have knowledge. This also means that according to the analytic view, logic has not to be understood as an exclusively deductive enterprise. If logic is understood as an exclusively deductive enterprise, then there cannot be a logic of discovery, since knowledge ampliation requires non-deductive reasoning. Indeed, deductive rules are usually regarded as non-ampliative, because the conclusion is contained in the premises, while non-deductive rules are usually regarded as ampliative, because the conclusion is not contained in the premises.[6] However, that there cannot be a deductive logic of discovery does not mean that there cannot be any logic of discovery. Indeed, logic does not need to be an exclusively deductive enterprise.[7] According to the analytic view, there can be a logic of discovery, but such logic cannot be exclusively deductive.

The Analytic Method as a Heuristic Method

That according to the analytic view there can be a logic of discovery, but such logic cannot be exclusively deductive, implies that the analytic method is not an algorithmic method. Methods can indeed be divided into algorithmic and heuristic (Newell et al. 1957). An algorithmic method "is a method that guarantees to always produce a correct solution to a problem. Conversely, a heuristic method is a method that does not guarantee to always produce a correct solution to a problem" (Cellucci 2017, p. 142). Algorithmic methods are closely associated with deductive reasoning and can be mechanized. Heuristic methods are instead closely associated with ampliative reasoning and cannot be mechanized. Algorithmic methods are regarded as closely associated with deductive reasoning, because algorithms are deeply related to computation, and computation can in a sense be regarded as a special form of deduction. For example, Kripke states that "computation is a deductive argument from a finite number of instructions," namely a "special form of mathematical argument" where one "is given a set of instructions, and the steps in the computation are supposed to follow – follow deductively – from the instructions as given. *So a*

[6]The claim that in deduction the conclusion is contained in the premises has to be understood as meaning that the conclusion either is literally a part of the premises or implies nothing that is not already implied by the premises. The claim that deduction is non-ampliative has been disputed by some philosophers. For example, Dummett famously objects that, if deductive rules were non-ampliative, then, "as soon as we had acknowledged the truth of the axioms of a mathematical theory, we should thereby know all the theorems. Obviously, this is nonsense" (Dummett 1991, p. 195). On this issue, which cannot be treated here for reason of space, and for a possible rejoinder to Dummett's objection, see Cellucci (2017, Sect. 12.7), and Sterpetti (2018, Sect. 6).

[7]This view is controversial. For a defense of the claim that a logic of discovery has to be deductive, see e.g. Jantzen (2015).

computation is just another mathematical deduction, albeit one of a very specialized form" (Kripke 2013, p. 80). Although it is not *really* possible to equate computation with deduction in a strict sense, since despite deductions can be regarded as isomorphic to computable functions (see below, section "Proofs and programs"), there are computable functions that are not isomorphic to any deduction, and so it cannot be claimed that *every* algorithmic method is deductive in character and, therefore, that every algorithmic method is non-ampliative, there is in any case a close relation between the non-ampliativity of deductive reasoning and the mechanizability of algorithmic methods on the one hand, and the ampliativity of non-deductive reasoning and the non-mechanizability of heuristic methods on the other hand. Since according to the analytic method, discovery is pursued by forming hypotheses through non-deductive inferences, which are ampliative, the analytic method is a heuristic method, and so it cannot be mechanized.

Here some clarifications are in order. Indeed, both deductive and non-deductive inference rules can be formalized (Cellucci 2013), and so one might expect that both deductive and non-deductive reasoning could be mechanized. But things are more complicated. The crucial point is that there is no algorithmic method, i.e. no mechanizable method, to choose what inference rule to apply to what premise in order to produce the desired conclusion in a given context. For example, Gigerenzer states that although algorithms (i.e. formalized rules) for scientific inferences exist, "there is no 'second-order' algorithm for choosing among them," but, he continues, despite there is no algorithm for choosing among algorithms, "scientists nonetheless do somehow choose, and with considerable success" (Gigerenzer 1990, p. 663). How can scientists do that? They "argue with one another, offer reasons for" their choices, and "sometimes even persuade one another" (Ibidem). In other words, scientists choose which non-deductive inference rule to use in a given context by assessing the plausibility of such choice. This fits well with the analytic view, since which rule to use to find a hypothesis to solve a given problem is a hypothesis in its turn. So, also the process by which such hypothesis is evaluated has to be accounted for in terms of plausibility. And indeed, as Gigerenzer says, scientists provide and assess reasons for and against each hypothesis about which rule to use, i.e. they assess its plausibility. Gigerenzer says also that there is no algorithm for deciding what inference rule to use. This amounts to say that the process by which plausibility is assessed, i.e. the process by which reasons for and against a given hypothesis are evaluated, cannot be reduced to computation (more on this below). And this is true even for deductive reasoning. Indeed, even if a mathematical proof consists exclusively of deductions, there is no algorithm to automatically find out the 'correct' inferential path from a given set of premises to the desired conclusion, namely an algorithm to decide what deductive inference rule to apply to what premise and in what order. So, one should say that, in the strict sense, both deductive and non-deductive reasoning cannot be mechanized (Cellucci 2013).

But, it might nevertheless be objected that in the case of deductive reasoning there is at least an algorithm, i.e. a mechanizable procedure, for enumerating all deductions from given premises. By means of the so-called British Museum algorithm (Newell et al. 1957), one can deduce all possible consequences from a given set of premises. One can apply, for instance, Modus Ponens to all premises and derive all two-steps conclusions that are so derivable. Then, one can apply Modus Ponens to the set of two-steps conclusions previously derived and derive all three-steps conclusions that are so derivable, and so on. Thus, one might be confident that *if* the conclusion one wishes to reach is deductively derivable from a given set of premises, the desired conclusion will sooner or later be derived by such algorithm. So, one might claim that, despite there is no algorithmic method to prove that a given statement *C* can be deduced from a given set of premises *P*, deducing consequences from *P* and checking whether *C* is among such consequences is in principle a mechanizable enterprise. Thus, although deduction is not mechanizable in the strict sense, it can be conceded to computationalists that deduction can in a sense be mechanized, because an algorithm can, at least in principle, be developed to derive all deducible conclusions from given premises.

On the contrary, despite non-deductive inferences rules can be formalized as well as deductive rules, there is no algorithm for enumerating all the consequences that can be derived by means of non-deductive inference rules from a given set of premises. Non-deductive inference rules are ampliative, and it is ampliativity that makes a difference with regard to mechanizability. Indeed, if one has to mechanize an inferential process, the conclusion needs to be uniquely determined by its premises (Cellucci 2011). If it were otherwise, a mechanical procedure would be unable to decide what conclusion should be derived from a given premise. Analogously, when one deals with algorithms, each step needs to be determined by the previous one. In deductive inferences, conclusions are uniquely determined by premises. If, for instance, one applies the rule 'Modus Ponens' to premises 'A → B' and 'A', conclusion cannot but be 'B'. This is one of the reasons why many computer scientists regard computation as deeply related to deduction. Hayes and Kowalski, for example, define computation as 'controlled deduction' (Hayes 1973; Kowalski 1979). Computer programs are indeed algorithms, and algorithms are deeply related to computation. In order to develop algorithms that perform in *predictable* ways, computer programs need to be made of predictable elementary computational steps, i.e. steps whose effects are predictable. For example, Pereira states that a computation is a sequence of "atomic computing actions [...]. Without predictability of the effects of sequences of elementary actions, the art of programming as we know it would be impossible, as would the art of painting if pigments changed their colors randomly under the brush" (Wos et al. 1985, p. 9). If elementary computational steps are made of deductions, programs are predictable, since, as noted, in deduction conclusion is uniquely determined by premises. This explains why algorithmic methods, although they are not necessarily deductive in character (more on this below), are usually regarded as closely associated with deductive reasoning.

But when one deals with non-deductive inference rules, which are ampliative rules, conclusions are generally not uniquely determined by premises. From the very same premise, different (and possibly incompatible) conclusions can be inferred by means of the same non-deductive inference rule. Consider induction. From the (true) premise 'all emeralds examined so far are green', at least two different and incompatible conclusions can be inductively inferred, namely 'all emeralds are green' and 'all emeralds so far examined are green, but all emeralds that will subsequently be examined will be blue' (Goodman 1983, Chap. 3; Cellucci 2011). The fact that when one deals with non-deductive inferences, conclusions are generally not uniquely determined by premises, makes heuristic methods, which rest on non-deductive inferences, non-mechanizable. Since conclusion is not uniquely determined by premises, in order to decide which is the conclusion that one has to draw from a given set of premises, one has to assess the plausibility of each possible conclusion. As stated, the process by which plausibility is assessed, i.e. the process by which reasons for and against a given hypothesis are evaluated, cannot be reduced to computation (more on this below). Therefore, the process of plausibility assessment cannot be made algorithmic. Since heuristic methods are based on non-deductive inferences, which in their turns are based on the process of plausibility assessment; and since the process of plausibility assessment cannot be made algorithmic, heuristic methods cannot be made algorithmic, i.e. they cannot *really* be mechanized.

Moreover, as research develops, new inference rules might always be added to the set of non-deductive inference rules that we use to find hypotheses, and so new conclusions, which previously were underivable from a given set of premises, might eventually become derivable from that very same set of premises by means of such new rules. Again, there is no algorithm to foresee what non-deductive inference rules will be added. As Bacon states, "the art of discovery may grow with discoveries" (Bacon 1961–1986, I, p. 223).

Since, contrary to deductive inferences, in the case of non-deductive inferences conclusions are not uniquely determined by premises and new rules might always be added, when one is dealing with non-deductive reasoning, there is no algorithm, not even in principle, to derive all possible consequences from given premises and check whether the desired conclusion is among them. Thus, in a sense, it can be said that non-deductive reasoning is non-mechanizable in an even stricter sense than deductive reasoning.[8]

Finally, it must be taken into consideration that deduction is usually regarded as truth-preserving, so that if premises are true, conclusion needs to be true. On the contrary, non-deductive inferences are usually regarded as non-truth-preserving, so that even if premises are true, conclusion might be false. So, when it is said that *there is* an algorithmic method to solve a given problem, it can safely be inferred that there is a solution to that problem and that that solution is 'correct', at least in the sense that it is true, provided that premises are true. On the contrary, when it is said that there is a heuristic method to solve a given problem, it cannot safely be inferred that the solution will be the 'correct' one, since non-deductive inference rules are not truth-preserving, and so conclusion might be false despite premises are true.

---

[8] On this sort of asymmetry between deductive and non-deductive reasoning with respect to mechanizability, cf. Cellucci (2017, p. 306): "there is no algorithm for discovering hypotheses, and hence for obtaining the solution [of a given mathematical problem] by analysis, while [...] there is an algorithm for enumerating all deductions from given axioms, and hence for obtaining the solution [of that problem] by synthesis."

It should now be clearer why when it comes to discovery, one usually deals with heuristic methods. As already said, while deduction is non-ampliative, non-deductive inferences are ampliative. Since discovery is related to knowledge ampliation, it is an ampliative enterprise. So, usually it cannot be carried on by means of algorithmic methods, which usually rest on deduction, which is non-ampliative. Rather, discovery must be carried on by means of heuristic methods, since they rest on non-deductive inferences, to extend our knowledge. And since non-deductive reasoning is ampliative but it cannot be mechanized, heuristic methods are ampliative, but they cannot be mechanized either.

To recapitulate, algorithmic methods are methods by means of which it is possible to uniquely and automatically determine the solution of a problem. They are closely associated with deductive reasoning, because in deduction conclusion is uniquely determined by premises, and so deduction can be automated. Nevertheless, algorithmic methods need not to be deductive in character. There can well be algorithmic methods that are non-deductive in character. Consider the British Museum algorithm illustrated above. This algorithm is not deductive in character, despite it generates nothing but deductions. Nevertheless, it is able to uniquely and automatically determine the solution of a problem, i.e. it is an algorithmic method. Indeed, the British Museum algorithm can be mechanized, since it rests on deduction in order to solve a problem, and in deduction conclusion is uniquely determined by premises. Deduction can thus be automated. So, the British Museum algorithm, although it is not deductive in character, can be automated as well. Consider also "Baconian" induction, which is a kind of induction in which conclusion can be uniquely determined by given premises, and which can therefore be mechanized. This kind of inference is studied, for instance, by Inductive Logic Programming (Muggleton and De Raedt 1994). Nevertheless, this kind of "mechanical induction is an extremely limited one and in a sense can be reduced to deduction, so it is not really ampliative and hence is generally inadequate for discovery" (Cellucci 2013, p. 160). So, despite it cannot be claimed that algorithmic methods are deductive in character, it is fair to say that usually in non-deductive inferences conclusion is not uniquely determined by premises, and this makes those inferences non-mechanizable. So, those methods that rely on non-deductive inferences to solve a problem are usually non-mechanizable methods, i.e. they are heuristic methods. This is why algorithmic methods are usually closely associated with deductive reasoning, while heuristic methods are usually closely associated with ampliative reasoning.

According to the analytic view, the axiomatic method is inadequate to explain advancement in mathematics and natural sciences precisely because, since it conceives of logic as an exclusively deductive enterprise, and deduction is non-ampliative, the axiomatic method is unable to account for the process of *hypotheses production*.[9] This means that the axiomatic method cannot improve our understanding of how we acquire *new* mathematical knowledge. From an epistemological point of

[9]This view can, at least in part, be traced back to Lakatos (1976), where Lakatos, by relying on the work of Pólya, strongly criticized the occultation of the heuristic steps that are crucial to the development of mathematics.

view, it is disappointing that the alleged *method* of mathematics is unable to say anything relevant about *how* new mathematical knowledge is acquired. In the light of the axiomatic method, knowledge ampliation remains a mystery. On the contrary, in the analytic view the path that has been followed to reach a result and solve a problem is not occulted, since in this view the context of discovery is not divorced from the context of justification. For instance, an analytic demonstration consists in a non-deductive derivation of a hypothesis from a problem and possibly other data, where the hypothesis is a sufficient condition for the solution of the problem, and is plausible (Cellucci 2017, Chap. 21).

It is important to underline that the analytic method involves both deductive and non-deductive reasoning. Indeed, to find a hypothesis we proceed from the problem by performing ampliative inferences, and then in order to assess the plausibility of such hypothesis we deduce conclusions from it. But the role that deduction plays in the analytic view is not the exclusive role that deduction is supposed to play in the axiomatic view. According to the analytic view, axioms are not the source of mathematical knowledge, and we shouldn't overestimate their role, which is limited to give us the possibility of presenting, for didactic or rhetorical purposes, some body of already acquired knowledge in deductive form. Axioms do not enjoy any temporal or conceptual priority in the development of mathematical knowledge, nor do they play any special epistemological role. As Hamming states, if "the Pythagorean theorem were found to not follow from postulates, we would again search for a way to alter the postulates until it was true. Euclid's postulates came from the Pythagorean theorem, not the other way" (Hamming 1980, p. 87).

Finally, it is worth noting that the concept of plausibility has not to be confused with the concept of probability. As Kant points out, "plausibility is concerned with whether, in the cognition, there are more grounds for the thing than against it" (Kant 1992, p. 331), while probability measures the relation between the winning cases and possible cases. Plausibility involves a comparison between the arguments for and the arguments against, so it is not a mathematical concept. Conversely, probability is a mathematical concept (see Cellucci 2013, Sect. 4.4). It may be objected that, although probability and plausibility appear to be distinct concepts, we may account for plausibility-based considerations in terms of probability, because plausibility obeys the law of probability (Pólya 1941). But this objection is inadequate. To see that plausibility is not equivalent to probability, consider that, since probability is "a fraction whose numerator is the number of favorable cases and whose denominator is the number of all the cases possible" (Laplace 1951, p. 7), in order to effectively calculate the probability of a given hypothesis h, we have to know the denominator, i.e. the number of all the cases possible. But in many cases, we do not know (and perhaps we cannot even know) the number of all the cases possible. Thus, if plausibility were to be understood in terms of probability, we should not be able to evaluate the plausibility of all those hypotheses for which we are unable to determine in advance the set of all the possible rival alternatives. But, scientist routinely evaluate the plausibility of that kind of hypotheses, so it cannot be the case that probability is equivalent to plausibility.[10]

That plausibility is not a mathematical concept and cannot be reduced to probability is crucial. Indeed, this implies that plausibility assessment cannot be reduced to computation and so made algorithmic. Thus, computer machines cannot perform plausibility assessment. Since plausibility assessment is central to the analytic method, that computer machines cannot perform plausibility assessment means that *if* the analytic method is the method by which scientific discovery is pursued, contrary to what computationalists claim, scientific discovery cannot be automated.

The Analytic View and the Automation of Science

There are three main arguments against the claim that science can be automated. The first one is that since there is no logic of discovery, machines cannot be programmed to make genuine discoveries. To face this argument, computationalists should prove that there is in fact an algorithmic method for discovery. Many computationalists attempt to do this, i.e. to provide evidences that programs do make scientific discoveries (see e.g. King et al. 2009; Colton 2002). But to prove that machines made *genuine* scientific discoveries that are able to extend our knowledge is not an easy task, since, as this chapter tries to clarify, knowledge cannot *really* be extended by exclusively computational means.[11] The second argument against the claim that science can be automated is that, since human minds outstrip Turing Machines, machines cannot equate human computational performances, and so discovery will remain a human enterprise. To face this argument, computationalists need to show that some of its assumptions is unjustified. In section "Mathematical knowledge", I deal with this argument and illustrate how difficult might be to defend some of its assumptions. The third argument against the claim that science can be automated is that there is a logic of discovery, but this logic is not exclusively deductive. According to this line of reasoning, the method of mathematics and science is the analytic method, and this implies that machines cannot implement it, because it is not an algorithmic method, i.e. it is not mechanizable. Thus, neither mathematics nor natural sciences can be automated. My claim is that this last argument is the

[10] For an extensive treatment of this issue, see Cellucci (2013, Sect. 20.4); Sterpetti and Bertolaso (2018).

[11] I cannot analyse here the discoveries allegedly made by computer programs. This is a topic for future research. Briefly, the main questions one has to address when dealing with this issue are: (1) whether hypotheses are really produced by programs, since often either a set of hypotheses or a set of heuristic strategies to routinely produce hypotheses from given inputs are already present in the so-called background knowledge of programs (see e.g. Marcus 2018); (2) whether programs can only produce results that can be obtained through a merely exploratory search of a well-defined space of possibilities, or they are also able to make innovative discoveries, i.e. discoveries that originate from the formulation of new concepts, i.e. concepts that cannot easily be derived from current ones, and modify the very space of possibilities (see e.g. Wiggins 2006).

most difficult one for the computationalists to refute, since it represents the most serious theoretical objection to the claim that science can be automated. Indeed, if one maintains the axiomatic view, as the supporters of the first two arguments against the claim that science can be automated usually do, it is likely that computationalists will find a way to defend their position, because the axiomatic method is deductive in character, and thus every instance of it can be mechanized, at least in principle. So, if one adopts the axiomatic view, one can hardly find out a truly compelling theoretical reason to support the claim that science cannot be automated.

Proofs and Programs

To allow my argument to take off the ground, it is important to clarify the connection between computer science and mathematics, better the connection between the claim that scientific discovery can be automated and the idea that mathematics is theorem proving.[12] Indeed, those who think that the method of mathematics is the axiomatic method, usually also think that mathematics is theorem proving (Cellucci 2017, Chap. 20). According to them, mathematicians start from a set of axioms and deduce theorems from them. On the contrary, according to those who think that mathematics is problem solving, mathematicians analyze a problem and then try to infer, from the problem and other relevant available knowledge, a hypothesis that is able to solve it. As Mäenpää states, with his *Grundlagen der Geometrie*, Hilbert "reduced geometry to theorem proving" (Mäenpää 1997, p. 210). Then, "Hilbert's model has spread throughout mathematics" in the twentieth century, "reducing it to theorem proving. Problem solving, which was the primary concern of Greek mathematicians, has been ruled out" (Ibidem). I argue that mathematics is not theorem proving, because this view is unable to account for how mathematical knowledge is ampliated. For example, when Cantor demonstrated that to every transfinite cardinal there exist still greater cardinals, "he did not deduce this result from truths already known […] because it could not be demonstrated within the bounds of traditional mathematics. Demonstrating it required formulating new concepts and new hypotheses about them" (Cellucci 2017, p. 310). So, mathematical knowledge cannot *really* be extended by exclusively deductive means. Mathematics is thus better conceived of as problem solving, since this view allows us to appreciate the role played by ampliative reasoning in finding hypotheses to solve problems. But if mathematics is not theorem proving, the axiomatic view is inadequate as a view of the method of mathematics, and so it cannot be claimed that science can be automated. To see why this may be the case, we need firstly to consider the relation between computer programs and mathematical proofs. Indeed, by the Curry-Howard isomorphism, we know that there is a deep relation between computer programs and mathematical proofs.

[12] On whether mathematics is theorem proving or problem solving, see Cellucci (2017, Chap. 20).

The Curry-Howard isomorphism establishes a correspondence between systems of formal logic as encountered in proof theory and computational calculi as found in type theory (Sørensen and Urzyczyn 2006, p. v; Prawitz 2008). Proof theory is focused on formal proof systems. It was developed in order to turn "the concept of specifically mathematical proof itself into an object of investigation" (Hilbert 1970, p. 12). The λ-calculus was originally proposed as a foundation of mathematics around 1930 by Church and Curry, but it was a "somewhat obscure formalism until the 1960s," when its "relation to programming languages was [...] clarified" (Alama and Korbmacher 2018, Sect. 3). The λ-calculus is a model of computation. It was introduced few years before another famous model of computation was introduced, namely Turing Machines. In the latter model, "computation is expressed by reading from and writing to a tape, and performing actions depending on the contents of the tape" (Sørensen and Urzyczyn 2006, p. 1). Turing Machines resemble "programs in imperative programming languages, like Java or C" (Ibidem). In contrast, in λ-calculus one is concerned with functions, and "these may both take other functions as arguments, and return functions as results. In programming terms, λ-calculus is an extremely simple higher-order, functional programming language" (Ibidem). So, with the invention of computer machines, the λ-calculus proved to be a useful tool in designing and implementing programming languages. For instance, the λ-calculus can be regarded as an idealized sublanguage of some programming languages like LISP.

Proof theory and the λ-calculus finally met, since Curry and Howard realized that the programming language was a logic, and that the logic was a programming language. In 1934, Curry firstly observed that every type of a mathematical function ($A \rightarrow B$) can be read as a logical proposition ($A \supset B$), and that under this reading the type of any given function always corresponds to a provable proposition. Conversely, for every provable proposition there is a function with the corresponding type (Curry 1934). In later years, Curry extended the correspondence to include terms and proofs (Wadler 2015). In 1969, Howard circulated a manuscript, unpublished until 1980, where he showed that an analogous correspondence obtains between Gentzen's natural deduction and simply-typed λ-calculus (Howard 1980). In that paper, Howard made also explicit the deepest level of the correspondence between logic and programs, namely that simplification of proofs corresponds to evaluation of programs. This means that for each way to simplify a proof there is a corresponding way to evaluate a program, and vice versa (Wadler 2015).

The Curry-Howard isomorphism is based on the so-called 'propositions-as-sets' principle. In this perspective, a proposition is thought of as its set of proofs. Truth of a proposition corresponds to the non-emptiness of the set. To illustrate this point, we will stick to an example made by Dybjer and Palmgren (2016). Consider a set $\mathrm{E}_{m,n}$, depending on $m, n \in \mathbb{N}$, which is defined by:

$$\mathrm{E}_{m,n} = \begin{cases} \{0\} & if\ m = n \\ \varnothing & if\ m \neq n. \end{cases}$$

$E_{m,n}$ is nonempty when $m = n$. The set $E_{m,n}$ corresponds to the proposition $m = n$, and the number 0 is a proof-object inhabiting the sets $E_{m,n}$. Consider now the proposition: *m is an even number*, expressed as the formula $\exists n \in \mathbb{N}.\ m = 2n$. A set of proof-objects can be built which corresponds to this formula by using the general set-theoretic sum operation. Suppose that A_n ($n \in \mathbb{N}$) is a family of sets. Then its disjoint sum is given by the set of pairs

$$(\Sigma n \in \mathbb{N})A_n = \{(n, a) : n \in \mathbb{N},\ a \in A_n\}.$$

If we apply this construction to the family $A_n = E_{m,2n}$ we can see that $(\Sigma n \in \mathbb{N})E_{m,2}$ is nonempty when there is an $n \in \mathbb{N}$ with $m = 2n$. By using the general set-theoretic product operation $(\Pi n \in \mathbb{N})A_n$ we can similarly obtain a set corresponding to a universally quantified proposition.

So, in this context proofs of $A \supset B$ are understood as "functions from (proofs of) A to (proofs of) B and $A \supset B$ itself [as] the set of such functions" (von Plato 2018, Sect. 6). If we take, for instance, "$f : A \supset B$ and $a : A$, then functional application gives $f(a) : B$. The reverse, corresponding to the introduction of an implication, is captured by the principle of functional abstraction of [...] Church's λ-calculus" (Ibidem). As von Plato states, the Curry-Howard isomorphism made intuitionistic natural deduction crucial to computer science. Indeed, the Curry-Howard isomorphism "gives a computational semantics for intuitionistic logic in which computations, and the executions of programs more generally, are effected through normalization" (Ibidem). A proof of an implication $A \supset B$, for instance, is a "program that converts data of type A into an output of type B. The construction of an object (proof, function, program) f of the type $A \supset B$ ends with an abstraction" (Ibidem). When an object a of type A is put into f as an argument, the resulting expression is not normal, "but has a form that corresponds to an introduction followed by an elimination. Normalization now is the same as the execution of the program f" (Ibidem).

For the aim of this chapter, the relevance of the Curry-Howard isomorphism lies in that it shows that computer programs are strictly equivalent to formalized mathematical proofs. Indeed, we have that for each proof of a given proposition, there is a program of the corresponding type, and vice versa. But the correspondence is even deeper, in that for each way to simplify a proof there is a corresponding way to evaluate a program, and vice versa. This means that "we have not merely a shallow bijection between propositions and types, but a true isomorphism preserving the deep structure of Proofs and Programs" (Wadler 2015, p. 75). In other words, we can understand programs as proofs and proofs as programs.

Why the fact that programs are proofs is relevant to the discussion of the claim that science can be automated? Because to claim that science can be automated amounts to claim that computer machines are able to contribute to knowledge ampliation. Now, if one wishes to claim that machines are able to contribute to knowledge ampliation, and programs are equivalent to mathematical proofs, and mathematical proofs are chains of deductions from given axioms, then one needs to claim that mathematical knowledge can be ampliated by means of deductions from given axioms, or by means that can be shown to be equivalent to deduction. And this amounts to claim that the method by which mathematical knowledge is ampliated is the axiomatic method. More generally, to claim that machines are able to contribute to knowledge ampliation amounts to claim that the process of knowledge ampliation can be entirely reduced to computation, since deductions can be regarded as isomorphic to computable functions. Indeed, if according to the axiomatic view, mathematical proofs are crucial to the ampliation of mathematical knowledge, and mathematical proofs are chains of deductions from given axioms, and mathematical proofs are equivalent to programs, so that proofs can be mechanized, then the process of knowledge ampliation can be accounted for in computational terms, i.e. it can be reduced to computation. In other words, if one wishes to claim that machines are able to contribute to knowledge ampliation, one commits oneself to the claim that the axiomatic view is an adequate view of how knowledge is extended. So, the claim that scientific knowledge can be extended by computer machines is equivalent to the claim that the method of mathematics is the axiomatic method and mathematical knowledge can be extended by means of that method. This implies that confuting the claim that mathematical knowledge can be extended by the axiomatic method is equivalent to confuting the claim that science can be automated. This is why I focus here on the issue of the method of mathematics in order to assess the claim that scientific discovery can be automated.

Mathematical Knowledge

In order to support the claim that, since the analytic method is the method of mathematics, neither mathematics nor natural sciences can be automated, in this section I illustrate one of the main reasons that can be provided to show that the axiomatic view is inadequate as a view of the method of mathematics.

According to the traditional view of mathematics, mathematical knowledge is acquired by exclusively deductive means, namely by deductive proofs from previously acquired mathematical truths. For example, Prawitz states that "mathematics [...] is essentially a deductive science, which is to say that it is by deductive proofs that mathematical knowledge is obtained" (Prawitz 2014, p. 78). This view gives raise to several problems, but here I focus on the problem of accounting for how we acquired the initial body of mathematical truths from which mathematics originated. More precisely, I argue that since the axiomatic method is unable to account for how we acquired the initial body of mathematical truths from which mathematics originated, the axiomatic view is unable to secure the epistemic superiority of mathematical knowledge over scientific knowledge and provide a secure foundation to mathematics, and that this fact undermines one of the main reasons for why the axiomatic view was so appealing to many philosophers in the first place.

Mathematical Starting Points

The problem of accounting for how we acquired the initial body of mathematical truths from which mathematics originated is deeply related to the issue of whether mathematics is distinct from other sciences. Indeed, the degree of certainty of mathematical knowledge is usually thought to be higher than that of scientific knowledge. Still today, mathematics is regarded as "the paradigm of certain and final knowledge" (Feferman 1998, p. 77) by most mathematicians and philosophers. According to many mathematicians and philosophers, the degree of certainty that mathematics is able to provide is one of its qualifying features. For example, Byers states that the certainty of mathematics is "different from the certainty one finds in other fields [...]. Mathematical truth has [...] [the] quality of inexorability. This is its essence" (Byers 2007, p. 328). The higher degree of certainty and justification displayed by mathematical knowledge is usually supposed to be due to the method of mathematics, which is commonly taken to be the axiomatic method. In this view, the method of mathematics differs from the method of investigation in the natural sciences: whereas "the latter acquire general knowledge using inductive methods, mathematical knowledge appears to be acquired [...] by deduction from basic principles" (Horsten 2015). So, it is the deductive character of mathematical demonstrations that confers its characteristic certainty to mathematical knowledge, since demonstrative "reasoning is safe, beyond controversy, and final" (Pólya 1954, I, p. v), precisely because it is deductive in character.

Now, a deductive proof "yields categorical knowledge [i.e. knowledge which is independent of any particular assumptions] only if it proceeds from a secure starting point and if the rules of inference are truth-preserving" (Baker 2016, Sect. 2.2). Let us concede, for the sake of the argument, that it can safely be asserted that deduction is truth-preserving, and so that if premises are true, conclusion needs to be true.[13] Then, if one embraces the axiomatic view, to prove that it is true that mathematical knowledge displays a higher degree of justification because of its deductive character, one has to prove that the mathematical starting points of mathematical reasoning, i.e. axioms, are known by some means that guarantees a degree of justification higher than the degree of justification provided by the means by which in the natural sciences the non-mathematical starting points of inductive inferences are known. Otherwise, the claim that mathematics is epistemically superior to natural sciences would be ungrounded. Indeed, according to the axiomatic method, mathematical knowledge is extended by deductions. So, the certainty of mathematical results depends on whether the axioms from which they are derived are known to be true with certainty. Mathematical results derived from axioms can in their turn become starting points for other deductions, and so on. At any stage of this process of ampliation of mathematical knowledge, the newly obtained results will be as certain as

[13] For a defense of the claim that the axiomatic view is inadequate also because there is no non-circular way of proving that deduction is truth-preserving, see Cellucci (2006).

the initial starting points, since deduction is truth-preserving. This point has been clearly illustrated by Williamson:

> At any given time, the mathematical community has a body of knowledge, including both theorems and methods of proof. Mathematicians expand mathematical knowledge by recursively applying it to itself [...]. Of course, present mathematical knowledge itself grew out of a smaller body of past mathematical knowledge by the same process. Since present mathematical knowledge is presumably finite, if one traces the process back far enough, one eventually reaches 'first principles' of some sort that did not become mathematical knowledge in that way. (Williamson 2016, p. 243).

The difficult question is: How such 'first principles' became mathematical knowledge? There is no clear and undisputed answer to this question. And yet answering to this question it is crucial, since the epistemic status of mathematical knowledge depends on the epistemic status of those first principles.

The problem is that the axiomatic view is unable to account for *how* such 'first principles' became mathematical knowledge, nor is it able to justify their alleged epistemic superiority. Indeed, if mathematical knowledge is knowledge of the most certain kind, and the method of mathematics is the axiomatic method, in order to claim that knowledge produced by that method is certain, the starting points by which such knowledge is derived have to be known to be true with certainty. In an axiomatic context, this amounts to claim that one can know with certainty that the axioms that constitute our mathematical starting points are consistent. But, it is almost uncontroversial that it is generally impossible to mathematically prove that axioms are consistent, because of Gödel's results. Indeed, as already noted, by Gödel's second incompleteness theorem, for any consistent, sufficiently strong deductive theory *T*, the sentence expressing the consistency of *T* is undemonstrable in *T*. So, according to many philosophers, there must be some *other* way to know with certainty that those axioms which constitute our mathematical starting points are consistent, otherwise the claim that mathematical knowledge displays an epistemic status which is superior to that of scientific knowledge cannot be justified.

This issue deserves a careful examination, since it is related to an important discussion on the philosophical consequences of Gödel's results which took place in the last decades, namely the discussion on whether Gödel's results imply that CTM is untenable (Horsten and Welch 2016; Raatikainen 2005). That discussion is useful to illustrate how even those philosophers who reject the idea that mathematical reasoning can be automated, failed to recognize the inadequacy of the axiomatic view, and so were led astray in their attempts to argue against the computationalist perspective.

Gödel's Disjunction

Gödel regarded it as clear that the incompleteness of mathematics demonstrated that mathematical reasoning cannot be mechanized. In his view, it "is not possible to mechanize mathematical reasoning, i.e., it will never be possible to replace the mathematician by a machine, even if you confine yourself to number-theoretic problems" (Gödel *193?, p. 164). In order to support his view, Gödel famously formulated the so-called Gödel's Disjunction (GD), according to which either the human mathematical mind cannot be captured by an algorithm, or there are absolutely undecidable problems of a certain kind. More precisely, he states that "either [...] the human mind [...] infinitely surpasses the powers of any finite machine, or else there exist absolutely unsolvable diophantine problems" (Gödel 1951, p. 310), where 'absolutely' means that those problems would be undecidable, "not just within some particular axiomatic system, but by any mathematical proof that the human mind can conceive" (Ibidem). In other words, either absolute provability outstrips all forms of relative provability or there are absolutely undecidable sentences of arithmetic. That there are absolutely undecidable sentences implies that the mathematical world "is independent of human reason, insofar as there are mathematical truths that lie outside the scope of human reason" (Koellner 2016, p. 148). So, GD can be rephrased as follows. At least one of these claims must hold: either minds outstrip machines or mathematical truth outstrips human reason.

A crucial notion in Gödel's argument against the claim that minds are equivalent to machines is the notion of 'absolute provability'. Relative provability is mechanizable, since it is provability within some particular axiomatic (i.e. formal) system, and any axiomatic system can be represented by a Turing Machine, i.e. it can be represented by an algorithm. The problem is that by Gödel's first incompleteness theorem, for any sufficiently strong axiomatic system F, there are statements which are undecidable within F, i.e. they can neither be proved nor disproved in F. Gödel shows how the issue of the decidability of statements in an axiomatic system F, is equivalent to the issue of the solvability of a certain kind of problems in arithmetic, namely diophantine problems. So, the argument goes, if there are absolutely undecidable sentences in mathematics, i.e. sentences that cannot be decided algorithmically, and if the human mind is equivalent to a Turing Machine, i.e. it is able to decide only algorithmically decidable sentences, then there are mathematical truths that cannot be known by human minds (Horsten and Welch 2016).

Why did Gödel focus on diophantine problems? Because in order to support his idea that mathematical reasoning cannot be mechanized, Gödel aimed to show that, despite some portions of mathematics can be completely formalized, the all of mathematics cannot be formalized. He was thus interested in finding the smallest portion of mathematics which cannot be formalized (Gödel *193?), and diophantine problems are among core problems in number theory. Moreover, the tenth problem in the famous list of Hilbert asked for "a procedure which in a finite number of steps could test a given (polynomial) diophantine equation for solvability in integers," and it "is easy to see that it is equivalent to ask for a test for solvability in natural numbers" (Davis 1995, p. 159). Gödel's result on the unsolvability of certain diophantine problems was not able to prove that Hilbert's tenth problem is unsolvable. Nevertheless, the relation that Gödel highlighted between a set being computable and the solvability of a diophantine equation is deeply related to the hypothesis that Hilbert's tenth problem is unsolvable, since to prove that Hilbert's tenth problem is unsolvable amounts to show that "every recursively enumerable [...] relation is diophantine, or, equivalently, that every primitive recursive relation is diophantine" (Ibidem), as Matiyasevič proved in 1970 (Matiyasevič 2003).

To better see the relationship between absolutely unsolvable diophantine problems and the issue of whether human minds are equivalent to Turing Machines, consider that Gödel subscribes to the iterative conception of sets. According to this conception, in order to construct ever larger sets, one begins with the integers and iterate the power-set operation through the finite ordinals. This iteration "is an instance of a general procedure for obtaining sets from a set *A* and well-ordering R" (Boolos 1995, p. 291). Axioms can be formulated to describe the sets formed at various stages of this process, but "as there is no end to the sequence of operations to which this iterative procedure can be applied, there is none to the formation of axioms" (Ibidem). Gödel observes that higher-level set-theoretic axioms will entail the solution of certain problems of inferior level left undecided by the preceding axioms; those problems take a particularly simple form, namely to determine the truth or falsity of some diophantine propositions. Diophantine propositions are sentences of the form:

$$\left(\forall x_1,\ldots,x_n \in \mathbb{Z}\right)\left(\exists y_1,\ldots,y_m \in \mathbb{Z}\right) p\left(x_1,\ldots,x_n,y_1,\ldots,y_m\right) = 0$$

where p is a diophantine polynomial, i.e. a polynomial with integer coefficients.

Now, it can be proved that the question of whether a given Turing Machine produces a certain string as an output is equivalent to the question of whether a certain diophantine proposition P is true: the decision problem for diophantine propositions is essentially the decision problem for Turing Machines, under another description (Leach-Krouse 2016). Along this line of reasoning, Gödel proves that Gödel's first theorem "is equivalent to the fact that there exists no finite procedure for the systematic decision of all diophantine problems" (Gödel 1951, p. 308). Thus, the decision problem for diophantine propositions is absolutely unsolvable, i.e. it is impossible to find a mechanical procedure for deciding every diophantine proposition.

Now, recall that according to GD, either the human mind surpasses the powers of any finite machine, or else there exist absolutely unsolvable diophantine problems. We have seen that there are absolutely unsolvable diophantine problems. Does this mean that Gödel endorses the second disjunct of GD? The answer is in the negative.

In Gödel's view, that there is an *algorithmically* unsolvable decision problem for diophantine propositions does not mean that the answer to the question about whether there are diophantine unsolvable problems has to be in the affirmative. According to Gödel, it is true that we have "found a problem which is absolutely unsolvable [...]. But this is not a problem in the form of a question with an answer Yes or No, but rather something similar to squaring the circle with compass and ruler" (Gödel *193?, p. 175). Gödel's idea is that the decision problem for diophantine propositions is undecidable because "the problem of finding a mechanical procedure restricts the types of possible solutions, just as the problem of squaring the circle with compass and straightedge restricts possible solutions" (Leach-Krouse 2016, p. 224). In Gödel's view, mechanical solutions aren't the only intelligible solutions that can be offered to mathematical problems. In other words, Gödel believes that it is possible to decide diophantine propositions in some non-mechanical way. Indeed, Gödel credits Turing with having established beyond any doubt that the recursive functions are exactly the functions that can actually be computed. So, Gödel accepted the Church-Turing thesis. But Gödel understood the thesis as stating that the "recursive functions are exactly the *mechanically* computable functions, *not* the functions computable by a humanly executable method" (Ibidem).

Why Gödel thinks that some mathematical problems are solvable even if there are no mechanical solutions to such problems? One of the reasons is that Gödel thinks "in a somewhat Kantian way that human reason would be fatally irrational if it would ask questions it could not answer" (Raatikainen 2005, p. 525). It is because he subscribes to this kind of 'rational optimism' (Shapiro 2016) that Gödel rejects the second disjunct of GD, i.e. that there are absolutely unsolvable problems in mathematics. Gödel shares Hilbert's belief that "for any precisely formulated mathematical question a unique answer can be found" (Gödel *193?, p. 164), and so that there cannot be absolutely unsolvable problems in mathematics. Hilbert was "so firm in this belief that he even thought a mathematical proof could be given for it, at least in the domain of number theory" (Ibidem). A mathematical proof of Hilbert's idea that there aren't absolutely unsolvable problems in mathematics can be understood as the proof of the following statement (H): given "an arbitrary mathematical proposition *A* there exists a proof either for *A* or for not-*A*, where by 'proof' is meant something which starts from evident axioms and proceeds by evident inferences" (Ibidem). But "formulated in this way the problem is not accessible for mathematical treatment because it involves the non-mathematical notion of evidence" (Ibidem), where 'evidence' can be understood as 'justification'. But if one tries to reduce this informal notion of proof to a formal one, so that it can be accessible for mathematical treatment, then it comes out that it is not possible to prove (H), because, as Gödel himself proved, it is impossible to prove that there are not unsolvable mathematical problems, i.e. it is impossible to prove that for any mathematical proposition *A* there exists a proof either for *A* or for not-*A*. According to Gödel, this negative result can have two different meanings: "(1) it may mean that the problem in its original formulation has a negative answer, or (2) it may mean that through the transition from evidence to formalism something was lost" (Ibidem). In his view, it "is easily seen that actually the second is the case" (Ibidem).

Why Gödel thinks that (2) is more plausible than (1)? Because according to him, "the number-theoretic questions which are undecidable in a given formalism are always decidable by evident inferences not expressible in the given formalism" (Ibidem). In other words, undecidability results are a reflection of the inadequacy of our *current* axioms. But new and better axioms can always be produced in order to answer questions left unanswered, because, as noted above, there is no limit to the process of axioms formation. It is for this reason that in Gödel's view, questions which are undecidable in a given formalism are always decidable by evident inferences not expressible in the given formalism. For example, consider Peano Arithmetic (PA). According to Gödel, if one climbs the hierarchy of types, the axiom system for second order arithmetic PA^2 decides the statement left undecided at the lower level, namely the consistency of PA, i.e. *Con*(PA). Now one has to prove *Con*(PA^2). But, for Gödel's second incompleteness theorem, PA^2 does not decide *Con*(PA^2). However, the axiom system for third-order arithmetic PA^3 settles the statement left undecided at the lower level, namely *Con*(PA^2). And so on. Thus, in Gödel's view, questions which are undecidable in a given formalism are always decidable by evident inferences not expressible in the given formalism. This can be done by introducing new and more powerful axioms and by working with such a more powerful formalism.[14] What about the 'evidence' of those new 'evident inferences'? According to Gödel, those new evident inferences "turn out to be exactly as evident as those of the given formalism" (Gödel *193?, p. 164).

But this view is unsatisfactory, since it does not answer the very question we started with: How is it that the axioms of a given formalism are justified? If more powerful axioms can always be formed from inferior axioms, and if those more powerful axioms are as justified as the less powerful axioms from which they are derived, it is crucial to know to what extent initial axioms are justified. It cannot simply be answered that to prove what is not provable in a given formalism, one can always resort to some stronger axioms, which are as justified as the weaker axioms, because unless initial weaker axioms are known to be true with certainty, the fact that stronger axioms are as justified as weaker ones does not tell anything about the degree of certainty to which stronger axioms are justified. To prove something is to provide justification for something. If to prove something in system A, one has to rely on axioms of system A', which are stronger than axioms of system A, and so on, then a regression is lurking.

Intrinsic and Extrinsic Justification

So, we need to focus again on the following question: How can mathematical starting points be justified? According to Gödel, axioms can be justified either intrinsically or extrinsically. Axioms that are intrinsically justified are those "new axioms which only unfold the content of the [iterative] concept of set" (Gödel 1964, p. 261). On the contrary, axioms are extrinsically justified if even "disregarding the intrinsic necessity of some new axiom, and even in case it has no intrinsic necessity at all, a probable decision about its truth is possible also in another way, namely, inductively by studying its 'success'" (Ibidem), where by 'success' Gödel means 'fruitfulness in consequences' (Koellner 2014, Sect. 1.4.2). Both these accounts of how axioms can be justified are unsatisfactory.

[14] In fact, things are more complicated. When one climbs the hierarchy of sets, the stronger axioms that become available lead to "more intractable instances of undecidable sentences" (Koellner 2011, Sect. 1). For example, at the third infinite level one can formulate Cantor's Continuum Hypothesis. These instances of independence "are more intractable in that no simple iteration of the hierarchy of types leads to their resolution" (Ibidem). I will not address this issue here.

As regard intrinsic justification, according to Gödel by "focusing more sharply on the concepts concerned" (Gödel *1961/?, p. 383) one clarifies the meaning of those concepts. By such procedure, "new axioms, which do not follow by formal logic from those previously established, again and again become evident" (Ibidem, p. 385). In Gödel's view, this explains why minds and machines are not equivalent, since "it is just this becoming evident of more and more new axioms on the basis of the meaning of the primitive notions that a machine cannot imitate" (Ibidem). But it is not easy to determine with certainty whether a new axiom is merely the unfolding of the content of the iterative concept of set, which is supposed to be sufficiently evident and unambiguous to be easily graspable by human minds. For example, against the widespread idea that the more familiar axioms of Zermelo-Fraenkel set theory with the axiom of Choice (ZFC) follow directly from the iterative conception of set, i.e. that they are intrinsically justified, while other stronger axioms, such as e.g. large cardinal axioms, are only supported by extrinsic justification, Maddy writes that even "the most cursory look at the particular axioms of ZFC will reveal that the line between intrinsic and extrinsic justification, vague as it might be, does not fall neatly between ZFC and the rest" (Maddy 1988, p. 483). According to Maddy, that the more familiar axioms of ZFC are "commonly enshrined in the opening pages of mathematics texts should be viewed as an historical accident, not a sign of their privileged epistemological or metaphysical status" (Ibidem).

Besides the difficulty of demarcating intrinsic justification from extrinsic justification, there is also the difficulty of adjudicating between different possible but incompatible intrinsic justifications. For example, Cellucci criticizes Gödel's idea that we can extend our knowledge of the concepts of set theory by focusing more sharply on the concept of set as follows:

> Suppose that, by focusing more sharply on the concept of set Σ, we get an intuition of that concept. Let S be a formal system for set theory, whose axioms this intuition ensures us to be true of Σ. So Σ is a model of S, hence S is consistent. Then, by Gödel's first incompleteness theorem, there is a sentence A of S which is true of Σ but is unprovable in S. Since A is unprovable in S, the formal system $S' = S \cup \{\neg A\}$ is consistent, and hence has a model, say Σ'. Then $\neg A$ is true of Σ' and hence A is false of Σ'. Now, Σ and Σ' are both models of S, but A is true of Σ and false of Σ', so Σ and Σ' are not equivalent. Suppose next that, by focusing more sharply on the concept of set Σ', we get an intuition of this concept. Then we have two different intuitions, one ensuring us that Σ is the concept of set, and the other ensuring us that Σ' is the concept of set, where the sentence A is true of Σ and false of Σ'. This raises the question: Which of Σ and Σ' is the genuine concept of set? Gödel's procedure gives no answer to this question. (Cellucci 2017, p. 255).

This scenario cannot be easily dismissed, because Gödel does not require intuition to be infallible (Williamson 2016), so we cannot exclude that we can find ourselves in the situation described above, where we have two different and incompatible intuitions of the concept of set, one ensuring us that Σ is the concept of set, and the other ensuring us that Σ' is the concept of set. More generally, although it is often claimed that "axioms do not admit further justification since they are *self-evident*" (Koellner 2011, Sect. 1), it is very difficult to neatly distinguish what is self-evident and what is not. Indeed, "there is wide disagreement in the foundations of mathematics as to which statements are self-evident" (Koellner 2014, Sect.

1.4.1).[15] As Hellman and Bell write, contrary to the "popular (mis)conception of mathematics as a cut-and-dried body of universally agreed-on truths and methods, as soon as one examines the foundations of mathematics, one encounters divergences of viewpoint [. . .] that can easily remind one of religious, schismatic controversy" (Bell and Hellman 2006, p. 64).

Moreover, one should also keep in mind that "even such distinguished logicians as Frege, Curry, Church, Quine, Rosser and Martin-Löf have seriously proposed mathematical theories that have later turned out to be inconsistent" (Raatikainen 2005, p. 523). As Davis states, in all those cases insight "didn't help" (Davis 1990, p. 660).

Finally, by Gödel's results we know that we cannot define once and for all a set of axioms, and then try to justify those axioms by claiming that they are self-evident because they are so simple and elementary that they cannot fail to appear as self-evident to anyone. Rather, we know that we will always need to introduce new ever stronger axioms. And those axioms are increasingly less simple than the simple ones we started with. So, even if one concedes, for the sake of the argument, that simplest axioms might appear as self-evident to the majority of mathematicians, the fact that we need to introduce new stronger axioms raises the question of how one is to justify these new axioms, for "as one continues to add stronger and stronger axioms the claim that they are [...] self-evident [...] will grow increasingly more difficult to defend" (Koellner 2011, Sect. 1), since it is usually perceived that the more one moves along the hierarchy of sets, the less axioms are self-evident, and rather they become increasingly disputable.

As regard extrinsic justification, Gödel famously writes that:

> There might exist axioms so abundant in their verifiable consequences, shedding so much light upon a whole field, and yielding such powerful methods for solving problems [...] that, no matter whether or not they are intrinsically necessary, they would have to be accepted at least in the same sense as any well-established physical theory. (Gödel 1964, p. 261).

Now, the problem is that this kind of justification makes mathematical knowledge on a par with scientific knowledge with respect to epistemic justification. In other words, this kind of justification of the axioms is unable to support the claim that mathematical knowledge is epistemically superior to scientific knowledge because it is deductively derived from premises which are certain. For example, Davis states that in this perspective new "axioms are just as problematical as new physical theories, and their eventual acceptance is on similar grounds" (Davis 1990, p. 660). But if mathematical starting points are justified because of their consequences, things do not go in the way predicted by the axiomatic view. Rather, things go the other way around. It is no more the case that conclusions are justified because they are deductively inferred from premises which are certain, rather premises are regarded as justified because it is possible to derive from them consequences which have so far proved interesting or useful. This kind of reasoning is not deductive in character, rather it is inductive, and so it seems inadequate to support the axiomatic

[15] On disagreement in mathematics, see Sterpetti (2018).

view of the method of mathematics and the alleged epistemic superiority of mathematical knowledge over scientific knowledge.

Moreover, extrinsic justification makes mathematics susceptible to those criticisms that are usually reserved to natural sciences. For example, it might be objected that even if consequences derived from some new axiom *A* are useful and interesting, and no contradiction has so far been derived from *A*, one cannot know that things will continue this way. A contradiction may always emerge, unless *A* is known to be true with certainty. If we do not know *A* to be true with certainty in advance, how can we know that a contradiction will not be derived from *A* within one hundred years? So, if mathematics is extrinsically justified, i.e. it is inductively justified, the justification of mathematical knowledge is prone to the same challenges to which the justification of scientific knowledge is prone.

Lucas' and Penrose's Arguments

The fact that there is no way to show that the all of mathematics can be derived from some self-evident axioms or by merely elaborating on the concept of set, and so that mathematics can be intrinsically justified, implies that the axiomatic view is unable to support the claim that mathematical knowledge is epistemically superior to scientific knowledge. It also implies that the 'inductive challenge' just illustrated can be moved against those who wish to support the claim that scientific discovery cannot be mechanized because humans outstrip machines, if they maintain the axiomatic view of mathematics.

To see this point, consider Lucas' (1961) and Penrose's (1989) arguments in support of the first disjunct of GD, i.e. the claim that minds outstrip machines, and so mathematics cannot be mechanized. It is worth noting that Gödel did not think that either one of the disjuncts of GD could be established solely by appeal to the incompleteness theorems. He thought instead that the disjunction as a whole, i.e. GD, was a "mathematically established fact" (Gödel 1951, p. 310), and that it was implied by his incompleteness theorems. In contrast, Lucas and Penrose argued that the incompleteness theorems imply the first disjunct of GD, i.e. the claim that minds outstrip machines (Koellner 2016).

According to Lucas (1961), while any recursively enumerable system *F* can never prove its Gödel sentence *G*, i.e. the sentence saying of itself that it is not provable in *F*, human minds can know that *G* is true. In this view, *G* is absolutely provable, although it is not provable in *F*. According to Lucas this shows that minds outstrip machines. However, in order to draw such conclusion, one needs to assume that human minds can know that *F* is consistent, i.e. that it is absolutely provable that *F* is consistent. Indeed, to claim that *G* is absolutely provable amounts to claim that the consistency of the axioms of *F* is absolutely provable. It is worth recalling that we refer here to 'absolute provability', because it is instead algorithmically provable that the consistency of the axioms of *F* is not provable in *F*, because of Gödel's second incompleteness theorem.

To see why for Lucas' argument to work one needs to assume that it is absolutely provable that *F* is consistent, recall that Gödel's second incompleteness theorem has a conditional form (Raatikainen 2005). Indeed, Gödel showed that for any sufficiently strong formal theory *F*, if *F* is consistent, a sentence *G* in the language of *F*, which is equivalent in *F* to the sentence expressing the consistency of *F*, cannot be proved in *F*. Thus, if *F* proves only true sentences, i.e. it is consistent, then *G* cannot be proved in *F*.

But how can one support the claim that it is absolutely provable that *F* is consistent? Penrose, following Lucas, claims that although *G* is unprovable in *F*, we can always 'see' that *G* is true by means of the following argument. If *G* is provable in *F*, then *G* is false, but that is impossible, because our formal system "should not be so badly constructed that it actually allows false propositions to be proved!" (Penrose 1989, p. 140). If it is impossible to prove *G* in *F*, then *G* is unprovable, and therefore it is true.

So, again, in order to 'see' the truth of *G* one has to be able to 'see' the consistency of the axioms of *F*, i.e. that it is not possible to derive contradictions from the axioms of our formal system. As Davis states, in this view if some form of "*insight* is involved, it must be in convincing oneself that the given axioms are indeed consistent, since otherwise we will have no reason to believe that the Gödel sentence [i.e. *G*] is true." (Davis 1990, p. 660). In this line of reasoning, to 'see' the consistency of the axioms of *F* is thus precisely what humans can do, but machines cannot. Thus, humans must be able to see the consistency of axioms in some non-algorithmic way. This means that for Lucas' and Penrose's arguments to work, one should be able account for how it is possible to see the consistency of axioms in some non-algorithmic way. If it cannot be proved that human minds are able to know with certainty that the axioms of *F* are consistent in some non-algorithmic way, Lucas' and Penrose's argument would amount to a merely conditional statement asserting that 'the Gödel sentence of *F*, i.e. *G*, is true, if *F* is consistent'. But this conditional statement is provable in *F*, and therefore to claim that humans can see the truth of such conditional statement is not sufficient for establishing the first disjunct of GD, i.e. the claim that minds outstrip machines.

Now, neither Lucas nor Penrose give us an adequate and detailed account of what it is a non-algorithmic way of proving the consistency of the axioms of *F*. For instance, they do not provide any account of how it is that a human brain can 'see' the consistency of the axioms of *F* which may be compatible with what we know about human cognitive abilities.[16] They simply assume that it is evident that (at least some portion of) mathematics *is* consistent, since no falsities can be derived in it. But, as we already noted, even if one concedes, for the sake of the argument, that there are certain limited formal theories of which the set of provable sentences can be seen to contain no falsehoods, such as e.g. Peano Arithmetic (PA); and even if one concedes, for the sake of the argument, that the Gödel sentence for PA is true and unprovable in PA, this does not amount to concede "that we can see the truth of Gödel sentences for more powerful theories such as ZF set theory, in which almost

[16] On this issue, see Sterpetti (2018).

the whole of mathematics can be represented" (Boolos 1990, p. 655). Such an 'extrapolation' is ungrounded, and neither Lucas nor Penrose give us compelling reason to think that instead we should rely on it. Rather we have reason to be skeptical about such an extrapolation. Indeed, in the absence of an adequate justification of the certainty by which we are supposed to know the consistency of the axioms of *F*, a sort of pessimistic meta-induction over the history of mathematics can be raised, which is analogous to the one raised by Laudan (1981) over the history of science. If, as already noted, distinguished logicians as Frege, Curry, Church, Quine, Rosser and Martin-Löf have proposed mathematical theories that have later turned out to be inconsistent, how can we be sure that our 'seeing' that *F* is consistent is reliable? Consider Zermelo-Fraenkel set theory, ZF. According to Boolos, there is no way to be certain that "we are not in the same situation vis-a-vis ZF that Frege was in with respect to naive set theory [...] before receiving, in June 1902, the famous letter from Russell, showing the derivability in his system of Russell's paradox" (Boolos 1990, p. 655).

It should now be clearer why if one does not have a compelling justification for the claim that the axioms of *F* are consistent, then mathematical knowledge cannot be claimed to be epistemically superior to scientific knowledge. As Boolos says, are "we really so certain that there isn't some million-page derivation of '0 = 1' that will be discovered some two hundred years from now? Do we know that we are really better off than Frege in May 1902?" (Ibidem).

Lucas's and Penrose's Arguments and the Axiomatic View

The discussion above shows that those who support the idea that minds are not equivalent to machines along the lines of Lucas and Penrose fail because, even if they wish to show that mathematical reasoning, and scientific discovery more generally, cannot be reduced to mechanical computation, their view of the method of mathematics is basically equivalent to the axiomatic view, according to which to do mathematics is to provide axiomatic proofs, i.e. to deductively prove theorems. This is the reason why they are committed to show that human minds can somehow 'see' something which is uncomputable by machines, i.e. something that cannot be deductively derived by machines. Their formalist and foundationalist approach to knowledge and mathematics leads them to think that in order to show that mathematical knowledge cannot be mechanized, it should be proved that humans can realize the axiomatic ideal that it has been proved machines cannot realize. But they do not really put into question the axiomatic view of mathematics. The problem is that if one claims that the method of mathematics is the axiomatic method, which relies exclusively on deductive inferences, since, as noted above in section "The analytic method as a heuristic method", in deduction conclusions are uniquely determined by premises, so that deduction can be made algorithmic and thus mechanized, it is then difficult for one to justify the claim that mathematical knowledge cannot be mechanized. So, one of the reasons why Lucas' and Penrose's arguments fail is that they both remain within the boundary of the traditional view of mathematics, according to which we "begin with self-evident truths, called 'axioms', and proceed by self-evidently valid steps, to all of our mathematical knowledge" (Shapiro 2016, p. 197).

I think that the failure of Lucas' and Penrose's attempts invites us to rethink such a conception of the method of mathematics. Their failure does not show that minds are equivalent to machines, because it cannot *really* be proved that minds make computations that machines cannot make. Rather, their failure suggests that their view of the method of mathematics is inadequate, since it is unable to account for why it is that minds and machines are not equivalent. Since we have good reason to think that it is not so easy to prove that minds are equivalent to machines, if assuming the traditional view of mathematics is an impediment in showing why this is the case, we should reject such view of the method of mathematics and search for another view of mathematics that allows us to adequately account for that fact. In my view, such a different view of mathematics is already available, and it is the analytic view.

What good reason do we have to think that it is not so easy to prove that minds are equivalent to machines? If minds were equivalent to machines, we would expect mathematical knowledge be advanced by deductions from given axioms. But the axiomatic view is challenged by the *actual* development of mathematics. Basically, mathematical knowledge is not advanced exclusively by means of deductions from given axioms. For example, Portoraro states that mathematicians "combine induction heuristics with deductive techniques when attacking a problem. The former helps them guide the proof-finding effort while the latter allows them to close proof gaps" (Portoraro 2019, Sect. 4.8). Shapiro writes that the standard way to establish new theorems is "to embed them in a much richer structure, and take advantage of the newer, more powerful expressive [...] resources. The spectacular resolution of Fermat's theorem, via elliptical function theory, is not atypical in this respect" (Shapiro 2016, pp. 197–198). This 'holistic' epistemology of mathematics is compatible with the analytic view of the method of mathematics.[17] Indeed, in this view, we do not solve problems because we are able to derive the solution from a fixed set of axioms known to be consistent from the start, rather mathematical knowledge is ampliated by formulating new and richer hypotheses, which need not necessarily be new *axioms*, and which can lead to the solution of the problems we are dealing with. So, it is not the case that complex mathematical problems are solved by starting from simple and consistent axioms. Rather, it is the case that some aspects of mathematics are illuminated by expanding our theoretical resources, i.e. by creating new and more complex mathematics, which can shed light on the mathematical problems that arise at a 'lower' level of complexity and give raise to new problems to be solved.

It is important to stress that if one conceives of how mathematics advances in this way, one can more clearly see why mathematical reasoning cannot be mechanized.

[17] For an account of the resolution of Fermat's Last Theorem inspired by the analytic view, see (Cellucci 2017, Sect. 12.13).

Basically, machines are fixed devices, which need to be built to produce desired outputs. In order to do that, one needs to know in advance almost everything about the problem the machine will be asked to solve. On the contrary, when we humans try to extend our knowledge, we do not know in advance what we will have to face, what shape the space of possibilities may have (Sterpetti and Bertolaso 2018). So, it is quite difficult to program a machine to make discoveries in the same way it happens to us humans to make discoveries. Since we do not know in advance how mathematical knowledge will be developed, we do not know how to program a machine that can really advance mathematical knowledge on its own. For example, Shapiro states that since "we do not know, in advance, just what rich theories we will need to prove future theorems about the natural numbers, we are not in a position to say what" the available theoretical resources are if we try to deal with issues such as whether "there are arithmetic sentences unknowable in principle" (Shapiro 2016, p. 198). The point is that while a Turing Machine has to have a fixed alphabet, a single language, and a fixed program for operating with that language, real "mathematicians do not have that. There is no fixed language, no fixed set of expressive resources, and there is no fixed set of axioms, once and for all, that we operate from" (Ibidem).

It is because the axiomatic view is unable to satisfactorily account for how knowledge is ampliated that I think that there are good reasons to think that it is not easy to prove that machines are equivalent to minds. I claimed that the analytic view is instead able to provide a more satisfactory account of how knowledge is ampliated. But the analytic view does that at a cost. According to the analytic view, knowledge is advanced by non-deductive means. Non-deductive inferences rest on plausibility assessment. Plausibility is not a mathematical concept, so plausibility assessment cannot be reduced to computation, i.e. it cannot be mechanized. This implies that discovery cannot be mechanized and that there is no epistemic difference between scientific knowledge and mathematical knowledge. So, according to the analytic view minds are not equivalent to machines independently from whether machines and minds can make the same computations. The axiomatic view and computationalism fall together. And the alleged epistemic superiority of mathematics falls as well. In this perspective, "there is no difference in kind between a mathematical proof, an entrenched scientific thesis, and a well-confirmed working hypothesis. Those are more matters of degree. In principle, nothing is unassailable in principle" (Ibidem).

Absolute Provability and the Axiomatic View

We have seen that it is not easy to claim that minds outstrip machines and that mathematics is epistemically superior to natural sciences, because it is not easy to adopt the axiomatic view *and* defend the claim that axioms are known to be consistent. Someone might try to rescue the axiomatic view and Gödel's and Hilbert's belief that "for any precisely formulated mathematical question a unique answer can be found," by both (1) defending the claim that the all of mathematical knowledge is absolutely provable and (2) weakening the claim that mathematics is epistemically superior to natural sciences. The problem is that this strategy to rescue the axiomatic view leads to the explicit acceptance of the claim that minds are equivalent to machines. So, there is a sort of dilemma here: either one accepts CTM in order to rescue the axiomatic view, or one rejects the axiomatic view in order to defend the claim that minds are not equivalent to machines.

For example, Williamson (2016) elaborates on Gödel's view in order to shed light on the concept of *absolute provability*. He suggests applying the term 'normal mathematical process' to all "those ways in which our mathematical knowledge can grow. Normal mathematical processes include both the recursive self-application of pre-existing mathematical knowledge and the means, whatever they were, by which first principles of [...] mathematics originally became mathematical knowledge" (Williamson 2016, p. 243). In this view, a mathematical hypothesis is "*absolutely provable* if and only if it can in principle be known by a normal mathematical process" (Ibidem). So far, so good. If one follows Williamson's argument, one can reach the conclusion that "every true formula of mathematics is absolutely provable, and every false formula is absolutely refutable" (Ibidem, p. 248). The problem is precisely that there is no satisfactory account of the means by which first principles of mathematics originally became mathematical knowledge. So, Williamson is not able to provide a justification of the claim that mathematical starting points became mathematical knowledge in a way which makes them epistemically superior to scientific knowledge, nor is he interested in providing such a justification, since he rejects the thesis that mathematical knowledge is certain in a way which is intrinsically different from the way in which scientific knowledge is certain. According to Williamson, that all arithmetical truths are absolute provable, "does not imply that human [...] minds can somehow do more than machines" (Ibidem). It merely means that for "every arithmetical truth A, it is possible for a finitely minded creature [...] to prove A" (Ibidem). But there is nothing in this view that excludes the hypothesis that "for every arithmetical truth A, some implementation of a Turing machine can prove A" (Ibidem). Rather, in this view "the implementation of the Turing machine is required to come to know A by a normal mathematical process, and so to be minded" (Ibidem), precisely because Williamson conceives of mathematical reasoning exclusively in terms of computability. As we noted, if all of mathematical reasoning is accounted for in terms of computational activity, no room is left in this view for plausibility assessment, i.e. for non-deductive inferences and for those heuristic methods that rest on such inferences, since plausibility assessment cannot be reduced to computation. This means that, in this view, the method of mathematics is the axiomatic method, which rests exclusively on deduction. As already noted, deduction can be regarded as reducible to computational activity. This seems to imply that in this view, the axiomatic method is an algorithmic method and can therefore be mechanized. So, in Williamson's view, there is no radical difference between machines and minds. If it is true that we know that no "possible implementation of a Turing machine can prove all arithmetical truths" (Ibidem), we got no reason to support the claim that finitely minded creatures like us can instead prove all arithmetical truths. Thus, minds do not outstrip machines.

Why Williamson rejects the idea that mathematical knowledge is more certain than scientific knowledge? Precisely because there is no satisfactory way to support such a claim of superiority, and so he thinks that such claim should be rejected. The point is that there is no way to convincingly argue for the epistemic superiority of mathematical starting points. According to Williamson, for anyone who objects that his approach is unable to guarantee the mathematical certainty of the new stronger axioms that we need to introduce to prove statements left unsettled by weaker axioms, the "challenge is to explain the nature of this 'mathematical certainty' we are supposed actually to have for the current axioms, but lack for the new one" (Ibidem, p. 251).

To recapitulate, Williamson's view is unable to support the claim that mathematical knowledge is epistemically superior to scientific knowledge, because although Williamson conceives of mathematical reasoning in terms of computability, i.e. almost exclusively in deductive terms, he, unlike Gödel, does not believe that there is a way to prove that mathematical starting points are true which is distinct from the way by which scientific hypotheses are known to be true. And it is because of the supposed computational character of mathematical reasoning that in this view machines and minds are equivalent. Williamson's view is thus unable to support the claim that machines and minds are not equivalent because, as Lucas' and Penrose's, it endorses the axiomatic view of mathematics, according to which to do mathematics is to make deductions from given axioms. Since deductive reasoning can, at least in a non-strict sense, be mechanized, in this view it cannot be adequately supported the claim that machines are not equivalent to human minds.

The Debate on Gödel's Disjunction and the Axiomatic View

The debate just illustrated can be summarized by describing it in a dilemmatic form: either (1) mathematics is epistemically superior to natural sciences, and so there must be a way to prove the consistency of axioms by some non-mechanical means, and this implies that human minds outstrip machines (Lucas' and Penrose's arguments), so that discovery cannot be automated; or (2) mathematics is not epistemically superior to natural sciences, and so there is no way to prove the consistency of axioms by some non-mechanical means, and this implies that human minds do not outstrip machines (Williamson's argument), so that it cannot be excluded that discovery can be automated.

I wish to stress again that what is taken for granted by both those who argue for (1) and those who argue for (2) is the axiomatic view, according to which mathematical knowledge is advanced by purely *deductive* means. In this perspective, whether minds and machines are equivalent depends on whether they are able to make the same computations and solve the same problems. But this means that it has already been accepted that mathematical reasoning, with the possible exception of the ability to 'see' the consistency of axioms, can really be reduced to computation.

On the contrary, the analytic view does not have to face this dilemma. Indeed, according to the analytic view, mathematics and other sciences share the very same method, namely the analytic method, so there is no difference between mathematical knowledge and scientific knowledge with respect to epistemic justification. Both mathematics and sciences are advanced in the same way, i.e. by forming hypotheses through non-deductive inferences that are able (at least provisionally) to solve some given problems. In this view, deductive and non-deductive inferences are on a par with respect to their justification. So, the analytic view is able to account for the continuity between mathematics and sciences that it asserts we should expect, while the axiomatic view is not really able to justify its claim that mathematics and sciences display different degrees of epistemic justification. According to Cellucci, "the fact that generally there is no rational way of knowing whether primitive premises" of axiomatic proofs "are true [...] entails that primitive premises of axiomatic proofs [...] have the same status as hypotheses" (Cellucci 2008, p. 12) in both mathematics and natural sciences.

According to the analytic view, that there is no difference in the epistemic status of mathematical and scientific knowledge does not mean that machines are equivalent to minds, since in this view mathematical reasoning cannot be reduced to computation. Rather, in this view, non-deductive reasoning is essential to the ampliation of mathematical knowledge. So, in this view, that machines are not equivalent to minds does not imply that minds perform *computations* that machines cannot perform. In this view, machines are not equivalent to minds because non-deductive reasoning is crucial to the ampliation of knowledge. Since non-deductive reasoning cannot really be mechanized,[18] because it rests on the process of plausibility assessment, which cannot be mechanized, machines are unable to perform non-deductive reasoning. And it is *this* fact that implies that mathematical knowledge cannot be ampliated by machines alone.

Conclusions

In this chapter, I showed how the idea that science can be automated is deeply related to the idea that the method of mathematics is the axiomatic method, so that confuting the claim that mathematical knowledge can be extended by the axiomatic method is almost equivalent to confuting the claim that science can be automated. I defended the thesis that, since the axiomatic view is inadequate to account for how mathematical knowledge is extended, the analytic view should be preferred. To do that, I analysed whether the axiomatic view can adequately account for two aspects of its own conception of mathematical knowledge, namely (1) how we acquired the initial body of mathematical truths from which the all of mathematics is supposed

[18] This claim might appear disputable to those who claim that ampliative reasoning is actually performed by machines. As already said, for reason of space, I have to leave the analysis of the claim that machines do autonomously perform ampliative reasoning for future work.

to be originated and (2) the alleged epistemic superiority of mathematics over natural sciences. Then, I developed an argument that can be summarized as follows. If the method of mathematics and science is the analytic method, the advancement of knowledge cannot be mechanized, since ampliative reasoning, i.e. non-deductive reasoning, plays a crucial role in the analytic method, and non-deductive reasoning cannot be fully mechanized.

Acknowledgements I wish to thank Carlo Cellucci for careful reading and commenting on an earlier draft of this chapter.

References

Alama, J., and J. Korbmacher. 2018. The Lambda Calculus. In: *The Stanford Encyclopedia of Philosophy*, ed. E.N. Zalta. https://plato.stanford.edu/archives/fall2018/entries/lambda-calculus/.

Allen, J.F. 2001. In Silico Veritas. Data-Mining and Automated Discovery: The Truth Is in There. *EMBO Reports* 2: 542–544.

Anderson, C. 2008. The End of Theory: The Data Deluge Makes the Scientific Method Obsolete. *Wired Magazine*, 23 June.

Bacon, F. 1961–1986. *Works*. Stuttgart Bad Cannstatt: Frommann Holzboog.

Baker, A. 2016. Non-Deductive Methods in Mathematics. In: *The Stanford Encyclopedia of Philosophy*, ed. E.N. Zalta. https://plato.stanford.edu/archives/win2016/entries/mathematics-nondeductive/.

Bell, J., and G. Hellman. 2006. Pluralism and the Foundations of Mathematics. In: *Scientific Pluralism*, ed. S.H. Kellert, H.E. Longino, and C.K. Waters, 64–79. Minneapolis: University of Minnesota Press.

Boolos, G. 1990. On "Seeing" the Truth of the Gödel Sentence. *Behavioral and Brain Sciences* 13: 655–656.

———. 1995. Introductory Note to *1951. In: *Kurt Gödel. Collected Works. Volume III*, ed. S. Feferman et al., 290–304. Oxford: Oxford University Press.

Byers, W. 2007. *How Mathematicians Think*. Princeton: Princeton University Press.

Calude, C.S., and D. Thompson. 2016. Incompleteness, Undecidability and Automated Proofs. In: *Computer Algebra in Scientific Computing. CASC 2016*, ed. V. Gerdt et al., 134–155. Cham: Springer.

Cellucci, C. 2006. The Question Hume Didn't Ask: Why Should We Accept Deductive Inferences? In: *Demonstrative and Non-Demonstrative Reasoning*, ed. C. Cellucci and P. Pecere, 207–235. Cassino: Edizioni dell'Università degli Studi di Cassino.

———. 2008. Why Proof? What is a Proof? In: *Deduction, Computation, Experiment. Exploring the Effectiveness of Proof*, ed. R. Lupacchini and G. Corsi, 1–27. Berlin: Springer.

———. 2011. Si può meccanizzare l'induzione? In: *Vittorio Somenzi. Antologia e Testimonianze 1918-2003*, B. Continenza et al. (a cura di), 362–364. Mantova: Fondazione Banca Agricola Mantovana.

———. 2013. *Rethinking Logic*. Dordrecht: Springer.

———. 2017. *Rethinking Knowledge*. Dordrecht: Springer.

Colton, S. 2002. *Automated Theory Formation in Pure Mathematics*. London: Springer.

Curry, H.B. 1934. Functionality in Combinatory Logic. *Proceedings of the National Academy of Science* 20: 584–590.

Davis, M. 1990. Is Mathematical Insight Algorithmic? *Behavioral and Brain Sciences* 13: 659–660.

———. 1995. Introductory Note to *193? In: *Kurt Gödel. Collected Works. Volume III*, ed. S. Feferman et al., 156–163. Oxford: Oxford University Press.
Dummett, M. 1991. *The Logical Basis of Metaphysics*. Cambridge, MA: Harvard University Press.
Dybjer, P., and E. Palmgren. 2016. Intuitionistic Type Theory. In: *The Stanford Encyclopedia of Philosophy*, ed. E.N. Zalta. https://plato.stanford.edu/archives/win2016/entries/type-theory-intuitionistic/.
Feferman, S. 1998. *In the Light of Logic*. Oxford: Oxford University Press.
Gigerenzer, G. 1990. Strong AI and the Problem of "Second-Order" Algorithms. *Behavioral and Brain Sciences* 13: 663–664.
Glymour, C. 1991. The Hierarchies of Knowledge and the Mathematics of Discovery. *Minds and Machines* 1: 75–95.
Gödel, K. *193?. Undecidable Diophantine Propositions. In: *Kurt Gödel. Collected Works. Volume III* (1995), ed. S. Feferman et al., 164–175. Oxford: Oxford University Press.
———. 1951. Some Basic Theorems on the Foundations of Mathematics and Their Implications. In: *Kurt Gödel. Collected Works. Volume III* (1995), ed. S. Feferman et al., 304–323. Oxford: Oxford University Press.
———. *1961/?. The Modern Development of the Foundations of Mathematics in the Light of Philosophy. In: *Kurt Gödel. Collected Works. Volume III* (1995), ed. S. Feferman et al., 374–387. Oxford: Oxford University Press.
———. 1964. What Is Cantor's Continuum Problem? In: *Kurt Gödel. Collected Works. Volume II* (1990), ed. S. Feferman et al., 254–270. Oxford: Oxford University Press.
Goodman, N. 1983[4]. *Fact, Fiction, and Forecast*. Cambridge, MA: Harvard University Press.
Hamming, R.W. 1980. The Unreasonable Effectiveness of Mathematics. *The American Mathematical Monthly* 87: 81–90.
Hayes, P.J. 1973. Computation and Deduction. In: *Proceedings of the 2nd Mathematical Foundations of Computer Science Symposium*, 105–118. Prague: Czechoslovak Academy of Sciences.
Hilbert, D. 1970. Axiomatic Thinking. *Philosophia Mathematica*, ser. 1, 7: 1–12, 1st ed., 1918.
Hintikka, J., and U. Remes. 1974. *The Method of Analysis*. Dordrecht: Reidel.
Horsten, L. 2015. Philosophy of Mathematics. In: *The Stanford Encyclopedia of Philosophy*, ed. E.N. Zalta. http://plato.stanford.edu/archives/spr2015/entries/philosophy-mathematics/.
Horsten, L., and P. Welch. 2016. Introduction. In: *Gödel's Disjunction*, ed. L. Horsten and P. Welch, 1–15. Oxford: Oxford University Press.
Howard, W.A. 1980. The Formulae-as-Types Notion of Construction. In: *To H.B. Curry. Essays on Combinatory Logic, Lambda Calculus and Formalism*, ed. J.R. Hindley and J.P. Seldin, 479–490. New York: Academic Press.
Jantzen, B.C. 2015. Discovery Without a 'Logic' Would Be a Miracle. *Synthese*. https://doi.org/10.1007/s11229-015-0926-7.
Kant, I. 1992. *Lectures on Logic*. Cambridge: Cambridge University Press.
King, R.D., et al. 2009. The Automation of Science. *Science* 324: 85–89.
Koellner, P. 2011. Independence and Large Cardinals. In: *The Stanford Encyclopedia of Philosophy*, ed. E.N. Zalta. https://plato.stanford.edu/archives/sum2011/entries/independence-large-cardinals/.
———. 2014. Large Cardinals and Determinacy. In: *The Stanford Encyclopedia of Philosophy*, ed. E.N. Zalta. https://plato.stanford.edu/archives/spr2014/entries/large-cardinals-determinacy/.
———. 2016. Gödel's Disjunction. In: *Gödel's Disjunction*, ed. L. Horsten and P. Welch, 148–188. Oxford: Oxford University Press.
Kowalski, R.A. 1979. Algorithm = Logic + Control. *Communications of the ACM* 22: 424–436.
Kripke, S.A. 2013. The Church-Turing 'Thesis' as a Special Corollary of Gödel's Completeness Theorem. In: *Computability*, ed. B.J. Copeland, C.J. Posy, and O. Shagrir, 77–104. Cambridge, MA: MIT Press.
Lakatos, I. 1976. *Proofs and Refutations*. Cambridge: Cambridge University Press.
———. 1978. *Philosophical Papers*. In *2 Vol*. Cambridge: Cambridge University Press.

Laplace, P.S. 1951. *A Philosophical Essay on Probabilities,* 1st French edition, 1814. New York: Dover Publications
Laudan, L. 1981. A Confutation of Convergent Realism. *Philosophy of Science* 48: 19–49.
Leach-Krouse, G. 2016. Provability, Mechanism, and the Diagonal Problem. In: *Gödel's Disjunction*, ed. L. Horsten and P. Welch, 211–242. Oxford: Oxford University Press.
Longo, G. 2003. Proofs and Programs. *Synthese* 134: 85–117.
———. 2011. Reflections on Concrete Incompleteness. *Philosophia Mathematica* 19: 255–280.
Lucas, J.R. 1961. Minds, Machines, and Gödel. *Philosophy* 36: 112–127.
Maddy, P. 1988. Believing the Axioms I. *The Journal of Symbolic Logic* 53: 481–511.
Mäenpää, P. 1997. From Backward Reduction to Configurational Analysis. In *Analysis and Synthesis in Mathematics: History and Philosophy*, ed. M. Otte and M. Panza, 201–226. Dordrecht: Springer.
Marcus, G. 2018. Innateness, AlphaZero, and Artificial Intelligence. arXiv:1801.05667v1.
Matiyasevič, Y. 2003. Enumerable Sets Are Diophantine. In: *Mathematical Logic in the 20th Century*, ed. G.E. Sacks, 269–273. Singapore: Singapore University Press.
Mazzocchi, F. 2015. Could Big Data Be the End of Theory in Science? A Few Remarks on the Epistemology of Data-Driven Science. *EMBO Reports* 16: 1250–1255.
Muggleton, S., and L. De Raedt. 1994. Inductive Logic Programming. Theory and Methods. *Journal of Logic Programming* 19–20: 629–679.
Newell, A., J.C. Shaw, and H.A. Simon. 1957. Empirical Explorations of the Logic Theory Machine: A Case Study in Heuristic. In: *Proceedings of the 1957 Western Joint Computer Conference*, 218–230. New York: ACM.
Penrose, R. 1989. *The Emperor's New Mind*. Oxford: Oxford University Press.
Pólya, G. 1941. Heuristic Reasoning and the Theory of Probability. *The American Mathematical Monthly* 48: 450–465.
———. 1954. *Mathematics and Plausible Reasoning*. Princeton: Princeton University Press.
Popper, K.R. 2005. *The Logic of Scientific Discovery*. London: Routledge.
Portoraro, F. 2019. Automated Reasoning. In: *The Stanford Encyclopedia of Philosophy*, ed. E.N. Zalta. https://plato.stanford.edu/archives/spr2019/entries/reasoning-automated/.
Prawitz, D. 2008. Proofs Verifying Programs and Programs Producing Proofs: A Conceptual Analysis. In: *Deduction, Computation, Experiment. Exploring the Effectiveness of Proof*, ed. R. Lupacchini and G. Corsi, 81–94. Berlin: Springer.
———. 2014. The Status of Mathematical Knowledge. In: *From a Heuristic Point of View. Essays in Honour of Carlo Cellucci*, ed. E. Ippoliti and C. Cozzo, 73–90. Newcastle Upon Tyne: Cambridge Scholars Publishing.
Raatikainen, P. 2005. On the Philosophical Relevance of Gödel's Incompleteness Theorems. *Revue internationale de philosophie* 4: 513–534.
———. 2018. Gödel's Incompleteness Theorems. In: The Stanford Encyclopedia of Philosophy, ed. E.N. Zalta. https://plato.stanford.edu/archives/fall2018/entries/goedel-incompleteness/.
Rathjen, M., and W. Sieg. 2018. Proof Theory. In: *The Stanford Encyclopedia of Philosophy.* ed. E.N. Zalta. https://plato.stanford.edu/archives/fall2018/entries/proof-theory/.
Rescorla, M. 2017. The Computational Theory of Mind. In: *The Stanford Encyclopedia of Philosophy*, ed. E.N. Zalta. https://plato.stanford.edu/archives/spr2017/entries/computational-mind/.
Rodin, A. 2014. *Axiomatic Method and Category Theory*. Berlin: Springer.
Schickore, J. 2014. Scientific Discovery. In: *The Stanford Encyclopedia of Philosophy*, ed. E.N. Zalta. http://plato.stanford.edu/archives/spr2014/entries/scientific-discovery/.
Shapiro, S. 2016. Idealization, Mechanism, and Knowability. In: *Gödel's Disjunction*, ed. L. Horsten and P. Welch, 189–207. Oxford: Oxford University Press.
Sørensen, M.H., and P. Urzyczyn. 2006. *Lectures on the Curry-Howard Isomorphism*. Amsterdam: Elsevier.

Sparkes, A., et al. 2010. Towards Robot Scientists for Autonomous Scientific Discovery. *Automated Experimentation* 2: 1. https://doi.org/10.1186/1759-4499-2-1.

Sterpetti, F. 2018. Mathematical Knowledge and Naturalism. *Philosophia*. https://doi.org/10.1007/s11406-018-9953-1.

Sterpetti, F., and M. Bertolaso. 2018. The Pursuit of Knowledge and the Problem of the Unconceived Alternatives. *Topoi. An International Review of Philosophy*. https://doi.org/10.1007/s11245-018-9551-7.

von Plato, J. 2018. The Development of Proof Theory. In: *The Stanford Encyclopedia of Philosophy*, ed. E.N. Zalta. https://plato.stanford.edu/archives/win2018/entries/proof-theory-development/.

Wadler, P. 2015. Propositions as Types. *Communications of the ACM* 58: 75–84.

Weyl, H. 1949. *Philosophy of Mathematics and Natural Science*. Princeton: Princeton University Press.

Wiggins, G.A. 2006. Searching for Computational Creativity. *New Generation Computing* 24: 209–222.

Williamson, T. 2016. Absolute Provability and Safe Knowledge of Axioms. In: *Gödel's Disjunction*, ed. L. Horsten and P. Welch, 243–253. Oxford: Oxford University Press.

Wos, L., F. Pereira, R. Hong, et al. 1985. An Overview of Automated Reasoning and Related Fields. *Journal of Automated Reasoning* 1: 5–48.

Zach, R. 2016. Hilbert's Program. In: *The Stanford Encyclopedia of Philosophy*, ed. E.N. Zalta. https://plato.stanford.edu/archives/spr2016/entries/hilbert-program/.

Part II
Automated Science and Computer Modelling

The Impact of Formal Reasoning in Computational Biology

Fridolin Gross

Introduction

Computational methods have become increasingly prevalent in molecular biology over the last decades. This development has been met with mixed expectations. While many people are optimistic and highlight the potential of these methods to shape a new and more adequate paradigm for the life sciences, others worry that computers will gain too much influence and erase the human element from research. Yet, everyone appears to agree that we are witnessing a deep transformation of the epistemic and methodological conditions of the life sciences.

Some authors argue that computational methods are philosophically interesting mainly because they bring in a new and qualitatively different element to the methodological repertoire of science, in addition to theory and experimentation. This difference is sometimes taken to be related to the transition from analytical to numerical methods enabled by the use of digital computers. In some disciplines, such as physics, meteorology, or geology, computational methods are applied to a familiar theoretical framework. In this context it is plausible to assume that computational methods mainly represent an *extension* of previous capabilities (cf. Humphreys 2004). Numerical techniques allow scientists to go beyond analytically solvable problems and to investigate larger and more complex models. The philosophical questions that arise as a consequence of this transition concern the epistemic status of computer simulations as models of "second order" (Küppers and Lenhard 2005).

Another area in which computational methods have had a large impact is data analysis. Modern computers enable scientists to handle and process large amounts of data and to analyze them automatically. Here, philosophers have mainly discussed

F. Gross (✉)
IFOM, Milan, Italy
e-mail: fridolin.gross@uni-kassel.de

M. Bertolaso, F. Sterpetti (eds.), *A Critical Reflection on Automated Science*,
Human Perspectives in Health Sciences and Technology 1,
https://doi.org/10.1007/978-3-030-25001-0_7

the idea of a shift in scientific methodology from a hypothesis-driven to a data-driven style of research (e.g. Leonelli 2012).

While these are clearly interesting perspectives, they risk to distract from a more fundamental way in which some scientific disciplines, such as molecular and cell biology, are being transformed by computational methods. Differently from physics, meteorology, or geology, molecular and cell biology are disciplines that have largely been dominated by experimentation and by what one might call an *informal* style of reasoning. Instead of merely leading to an extension of the analytical power of existing theoretical tools, computational methods have a deeper impact because the domain of study first needs to be adapted to the application of formal methods. This difference with respect to other disciplines also explains why it is more controversial whether the application of computational methods is necessarily beneficial in these contexts, which is reflected, for example, in the polarized debates among scientists around *big data* or *systems biology*. The question of the general impact of introducing formal methods deserves attention and should be investigated in its own right. Philosophical discussions of more specific applications of computational methods might benefit from such an analysis.

In this contribution I investigate the influence of computational methods in molecular biology by focusing on the very meaning of the concept of computation. At the most general level, computation can be understood as the transformation of a given sequence of symbols according to a set of formal rules. On this view, computational methods are not restricted to processes carried out on a (digital) computer, but may include other practices that involve the use of formal languages. The basic idea, though, is that computational methods are in some sense *mechanical*, i.e. procedures that can in principle be carried out by a machine. This broad characterization, which is elaborated in section "Formal and informal reasoning", allows me to pin down some differences between computational methods and the informal cognitive methods of experimental scientists. In section "Informal reasoning in molecular and cell biology", I discuss the prevalence of informal reasoning in experimental molecular biology, while section "Examples of computational methods" looks at different examples from computational biology to investigate the ways in which formal methods come into play and interact with informal kinds of reasoning.

Formal and Informal Reasoning

One might think that the problem that arises with the application of computational methods is simply one of translating the informal language of experimental biologists into a language that can be understood by computers. Moreover, since biology, like science in general, is assumed to be based on the principles of logic, formalization should be more or less straightforward. Looking at concrete examples, however, it turns out that the application of computational methods invariably leads to non-trivial problems of translation. These problems can be traced back to the general challenge to introduce formal methods.

It has been noted that even among philosophers and logicians the term "formal" is often used in ambiguous and imprecise ways. Dutilh Novaes (2011) observes that there are two main clusters of meaning: one associates the term with forms ("formal" as opposed to "material"), the other with rules ("formal" as opposed to "informal"). In the first sense, "formal" refers to the result of an abstraction from matter, content, or meaning. A logical argument can be formal in this sense if it includes variable placeholders (such as *A*, *B*, or *x*) instead of concrete terms with associated meanings. In the second sense, "formal" refers to something that is in accordance with certain explicitly specified rules. We speak of a "formal dinner party" in this sense, for example. An important instance of this second sense is the idea of the formal as *computable*, which concerns rules of inference that are specified in such a way that they could be followed by anyone without specific insight or ingenuity, even by a machine. This sense of "formal," which in its modern formulation goes back to the work of Turing and others, will be the one most relevant for the present context (Floridi 1999). It is not fully independent from the formal in the sense of abstracting from content, however, because in order to specify rules that can be followed mechanically, it is plausible that one may refer only to the syntactical structure of the sequences of symbols to be transformed.

Computational methods extend or replace certain parts of scientific activity by introducing formal procedures. In many cases using a computer for a task previously carried out by a human is unproblematic, at least epistemically. If the task is already formally specified, e.g. to perform a linear regression for a set of quantitative measurements, one can easily delegate it to a computer that, if properly instructed, will be both faster and more reliable. Philosophically more interesting cases arise when computers assist or replace humans in tasks that were previously carried out in informal ways. Before we can meaningfully approach these kinds of cases, we must clarify first what we mean by "informal" in this context.

One might be inclined to use this term to refer to a form of reasoning that does not strictly follow the rules of formal logic. Indeed, as work in cognitive psychology has shown, humans often do not reason in accordance with the principles of logic in certain cognitive tasks (Stanovich 2003). However, it would be wrong to consider informal reasoning simply as irrational or as defective with respect to some normative ideal since especially in science, as I will show, reasoning is informal in important respects. In fact, scientists often use informal reasoning in similar ways as humans do in everyday life. Even though in general scientific reasoning is more rigorous and standardized than reasoning in everyday situations, it has been argued that both scientific and non-scientific activities use the same cognitive building blocks, and the main difference lies in the goals and in the way the building blocks are put together (Dunbar 2004). To get a clearer idea of the ways in which informal reasoning works in scientific practice, one has to look at concrete examples, but as a working criterion I suggest that we understand "informal" simply in direct contrast with the idea of the "formal" as computational, that is, reasoning is informal if it deviates from a procedure that can be described as following some finite, explicitly given set of syntactical rules.

Note that by calling some forms of reasoning "informal" I do not want to suggest that it is impossible in principle to formalize it. Even though there may be parts of human reasoning for which it is very difficult or even impossible in principle (cf. Dreyfus 1972), I am not interested in drawing a fundamental metaphysical distinction. In other words, the question is not whether reasoning can be *reduced* to some kind of formal calculation, but why under certain circumstances scientists themselves *replace* informal reasoning by some kind of formal calculation (and why often they don't).

One reason in favor of formalization is that, apart from increases in speed and decreases in cost that are due to the possibility of delegating formalized tasks to computers, formalization can have important *epistemic* benefits:

> Because it allows us to counter the 'computational bias' of systematically bringing prior beliefs into the reasoning process, reasoning with formal languages and formalisms in fact increases the chances of obtaining surprising results, i.e., results that go beyond previously held beliefs. (Dutilh Novaes 2012, p. 7).

In a formal language there is no room for prior beliefs because all components of the reasoning process must be made explicit. The tendency to take into account prior beliefs is therefore a characteristic feature of informal reasoning. Depending on the circumstances, this tendency can make reasoning very powerful or be a source of bias.

To avoid confusion, it is important to distinguish between *formal reasoning* and a *formal model of reasoning*. The attempt to describe and justify in terms of a formalized system the kinds of non-deductive inferences that scientists draw has been one of the main themes in formal epistemology. A common strategy to approach this problem is to use probability calculus to model how scientists assess the degree of confirmation that new observations confer to a hypothesis given a set of prior beliefs. But this project, however useful in other respects, does not expose the underlying formal nature of scientific reasoning. On the contrary, the fact that this is a non-trivial philosophical problem together with the observation that scientists do not make use of such a formal model themselves, suggests that scientific reasoning is informal in relevant respects. In most contexts the cost of formalizing scientific inference would be intolerably high because it would require considerable work to explicitly generate and accurately quantify the set of relevant prior beliefs. Moreover, one would have a hard time justifying even broad confidence intervals for the assigned probabilities. But more importantly, even if this were possible, it is not obvious whether formalization would have any significant scientific benefit. If the calculations of the formal epistemological model were in agreement with scientists' informal results, one would not have gained much (from the scientists' perspective). And if not, it would not be obvious whether to put the blame on the formal model or on the scientists' informal reasoning. This consideration just serves to show that in order to assess the prospects of formalization of a part of the scientific process, one must take into account the potential benefits, but also the costs of formalization and evaluate them in light of a given scientific goal.

In line with Dutilh Novaes I consider formal reasoning as a cognitive *tool*, and as any tool it may not function properly if it is badly designed or applied in the wrong context. As will be illustrated in the following sections, both formal and informal reasoning can exhibit their respective strengths and weaknesses when applied in different scientific contexts. In what follows I want to take a brief look at experimental and computational molecular biology to investigate the ways in which formal and informal methods are applied and how they interact in practice.

Informal Reasoning in Molecular and Cell Biology

I want to argue that scientific reasoning in molecular and cell biology is informal in important respects. Part of what makes this task difficult is that we do not seem to be able to directly observe how scientists reason.

Here, I mainly address the informal aspects of reasoning by looking at the ways in which scientists present their findings in published work. One might object that to look at the finished results of scientific activity does not really provide any information about the underlying reasoning process. However, I am not really interested in capturing the thought processes as they unfold in the heads of individual scientists. Rather, I am trying to follow reasoning as it unfolds as a public and social activity. A finer grained analysis that looks at communication within a single lab or research group might add important detail to this story, but I think it is not entirely misleading to assume that an important part of the process actually occurs in the interactions of different ideas via the medium of scientific publications. Moreover, if informal elements are found at the level of published articles (which go through a rigorous process of peer review and revision), this is strong evidence that informal elements play a role in scientific reasoning in general.

The first thing to notice is that a typical article in a journal of experimental molecular or cell biology is mostly written in natural language and usually does not contain much mathematics, equations, or formulas. However, this is not sufficient to conclude that the expressed reasoning is informal because natural language and diagrams can be used to represent formal reasoning as well. In order to reveal informal reasoning, one has to identify relevant elements within the reasoning process that are not easily formalizable.

As mentioned before, an important aspect is that while the rules of logic do play a role in scientific reasoning, they are typically not applied mechanically, but taking into account collateral information. This has also been highlighted in the literature on the cognitive aspects of science. Koslowski and Thompson discuss how causal inference based on correlative data can be informal in this respect:

> [T]he strategy of relying on co-variation to identify causal factors is not applied mechanically; it is applied judiciously, in a way that takes account of collateral information. (Koslowski and Thompson 2004, p. 175).

For example, scientists consider a correlation between two variables more likely to be causal if there is a plausible candidate mechanism that connects them. Such recourse on collateral information is not restricted to reasoning about correlations, but pervasive in the kind of reasoning that one finds in experimental molecular and cell biology. In general, scientific hypotheses are not assessed in isolation, but in the context of a broad network of collateral information. The well-known claim that scientific confirmation is holistic (cf. Quine 1951) can in this context be re-interpreted as evidence of its informal character.

Apart from relying on collateral information, scientific reasoning is often probabilistic or otherwise *defeasible* (i.e. not deductively valid). Especially in experimental biology, scientists often use expressions such as "might," "this suggests that," or similar, to indicate that individual steps in a proposed mechanism are still hypothetical in character. The account of a mechanism is not presented as a formal model from which definite predictions can be derived. Specific predictions typically depend on which details of the mechanism are highlighted as well as the collateral information that is taken into account.

Consider as an example from cell biology the cartoons that are shown in Fig. 1, to which Tyson and Novák (2015) refer as "informal diagrams." They depict the basic molecular mechanism underlying the cell cycle in a frog embryo and in fission yeast, respectively. Arrows correspond to activities that connect molecular entities and that are more or less specified, such as *activation* or *synthesis*. Scientists can reason along these arrows and thereby gain understanding of the represented process. Obviously, there are certain interpretive conventions regarding such diagrams within the community of experimental cell biologists, but these are typically not made explicit. Importantly, no unique way of interpreting the individual parts of a

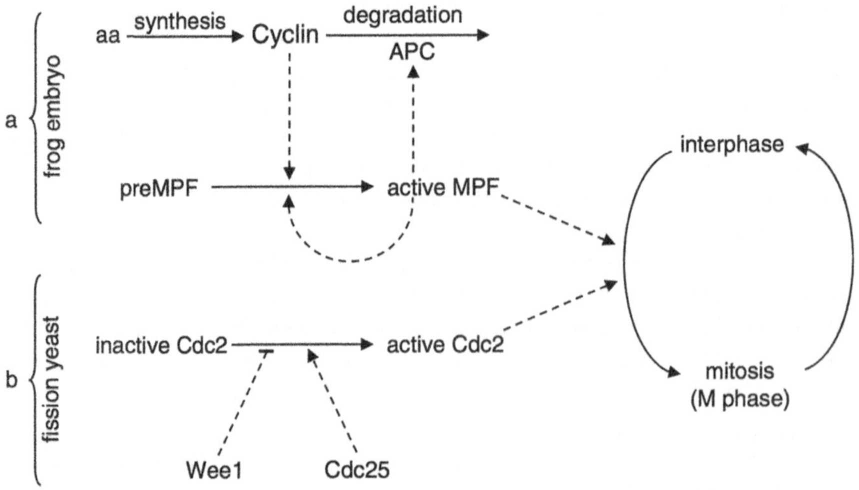

Fig. 1 Informal cartoon diagram of mechanisms of mitotic control. (Source: Tyson and Novák 2015)

diagram is prescribed. Therefore, scientists with different backgrounds or different available contextual information may derive different hypothetical causal consequences based on such a diagram.

Moreover, diagrams such as the ones presented in Fig. 1 are rarely considered complete. They are "mechanism sketches" in the language of Machamer et al. (2000). Experiments often do not conclusively establish whether the activity of a molecular entity is sufficient to achieve an observed effect, which is reflected in biologists' cautious formulations, such as the following:

> Although the destruction of cyclin is necessary to inactivate MPF, it is currently unclear whether it alone is sufficient for the inactivation of MPF. One additional protein that may be required for the inactivation of MPF is p13, the product of the fission yeast *suc1* gene [...]. (Murray and Kirschner 1989).

Reasoning informally with a diagram can take into account incompleteness. By contrast, using a diagram in a formal way requires unique rules of interpretation and effectively rests on the assumption that the diagram includes all relevant causal factors.

To be sure, some accounts of mechanisms in biology are considered essentially complete. At some point, often after decades of filling in the molecular details, the knowledge acquired of a phenomenon may reach a stage of maturity and consensus, and the community may agree on a more or less unique "textbook cartoon" of the mechanism. But even such a consensus cartoon (or its description in natural language) omits much of the actual detail of the underlying processes, even if now mainly for didactic purposes. Importantly, the diagram remains open in the sense that it is still embedded in a network of collateral information that has to be taken into account if one wants to understand more than just the basic aspects of the represented phenomenon. To be able to really understand and productively reason with an informal diagram in biology requires expert knowledge, and there are usually no syntactic rules to mechanically draw any conclusions from it.

I have used the example of a causal diagram to suggest that informal reasoning is pervasive in molecular and cell biology. Elsewhere, I have argued that the absence of formal reasoning should not be taken as a sign of immaturity of these disciplines. Instead, the particular informal strategies developed by experimental biologists are based on a particular idea of the underlying organization and complexity of living systems (Gross, 2018). This idea involves specific assumptions of modularity and a general perspective on molecular mechanisms as sequential processes of discrete information transfer. Given these assumptions, formalization is not expected to confer any significant benefits because biological organization is in a certain sense straightforward and can be described qualitatively. The scientifically difficult part, on this view, is to identify and describe the relevant macromolecular complexes and to demonstrate individual interactions between them. However, with some of the assumptions implicit in the traditional experimental approach being challenged by recent findings, formal methods will increasingly be found in the epistemic toolkit of biologists.

Examples of Computational Methods

In the following I briefly present examples from different areas where computational methods have been applied to problems within the domain of molecular and cell biology. Like in the last section these are not thought as full-blown scientific case studies. Instead, the goal is to show that an analysis through the lens of the distinction between formal and informal can be productive and reveals interesting parallels between the different examples. A particular focus lies on the way in which interactions between formal and informal reasoning play out in practice.

Computational Models in Cell Biology

An important part of computational biology consists in constructing and analyzing mathematical models of the candidate mechanisms that experimentalists propose as explanations for specific biological phenomena. Philosophers of science have discussed various modeling approaches in computational biology, emphasizing mostly the mathematical properties of specific kinds of models (e.g. Bechtel and Abrahamsen 2010, Brigandt 2013, Baetu 2015). Computational biologists, by contrast, often highlight the much more general role of models to provide a formal representation of biological processes:

> The resulting menagerie of ordinary differential equations, partial differential equations, delay differential equations, stochastic processes, finite-state automata, cellular automata, Petri nets, hybrid models, [...] each have their specific technical foibles and a vast associated technical literature. It is easy to get drowned by these technicalities, while losing sight of the bigger picture of what the model is telling us. Underneath all that technical variety, each model has the same logical structure. Any mathematical model, no matter how complicated, consists of a set of assumptions, from which are deduced a set of conclusions. (Gunawardena 2014, p.2).

In fact, features that are often considered hallmarks of computational modeling in biology, such as being dynamic or quantitative, are by no means present in all applications of modeling. Therefore, even if it seems tautological, it is worth pointing out that the only common denominator of all computational models in biology is their computational, i.e. formal, character.

In contrast with the type of informal reasoning described in section "Informal reasoning in molecular and cell biology", building a computational model requires that all assumptions and rules of transformation be laid out explicitly. The modeler must decide which biological entities to incorporate in a model and how to formally represent them (e.g. as numbers associated with molecular concentrations, or as discrete states of activity in a Boolean network). Causal interactions are represented as mathematical or logical relationships between those entities. The rules of transformation that a computer uses to derive the consequences from these assumptions arise from the specifications of the model together with the particular algorithm that is used to carry out the simulation of the model. These rules are not necessarily explicit in the sense that the modeler knows and understands them in detail. After all, a piece of software might simply be used as a black box. But they have to be explicitly specified at some level, such that the procedure can be carried out by a computer in a "mechanical" way.

Insofar as a model is taken as an accurate representation of reality, entities that are not included in the model are assumed not to make a relevant difference to the described phenomenon, i.e. to be causally negligible.[1] An important part of the use of computational models therefore consists in of turning implicit assumptions into explicit formal features:

> In general terms, building a model forces the modeler to lay out his or her assumptions clearly; analyzing and simulating the model determines the logical implications of these assumptions; and comparing the conclusions to experimental facts assesses the explanatory and predictive power of the mechanistic hypothesis. Where the conclusions of the model diverge from experimental facts may suggest problems with the 'working hypotheses' or missing components in the mechanism. (Tyson and Novák 2015).

Figure 2 illustrates computational modeling of the same cell cycle mechanism that was already discussed in section "Informal reasoning in molecular and cell biology". The diagram in Fig. 2 *a* is informal, but compared to Fig. 1, it involves some steps towards formalization. For example, the arrows shown in the diagram correspond in a unique way to the reactions included in the computational model, and some of the occurring interactions, such as the activation of Cdc2 are spelled out as transitions between a phosphorylated and a non-phosphorylated state. The plots in Fig. 2 *b* and *c* show the results of simulating experimental conditions. Notably, *b* involves an unexpected prediction of the model: the system suddenly switches from a low to a high state of MPF activity as the concentration of cyclin is increased, a phenomenon known as *bistability*. This illustrates the ability to generate surprising results by means of formalization. Of course, this prediction depends on specific mathematical features of the model and might be explained as a result of unwarranted idealizations. However, the result must be taken seriously if the assumptions underlying the formal model are accepted as plausible with respect to the target system. The element of surprise then springs from the fact that the consequences of these assumptions are derived with the reliability of a machine.

Working with computational models is not a purely formal affair, however. For example, the process of converting an informal cartoon into a formal model can itself not easily be formalized. Decisions have to be taken about which entities and causal interactions to include and how to represent them. These decisions depend in non-trivial ways on the particular goals that modelers are after. For example, in order to end up with useful predictions, modelers must strike a balance between model complexity and agreement with observations (Gross and MacLeod 2017). Building a computational model in this respect is no different from modeling in general:

[1] This has to be qualified: sometimes the goal of modeling is precisely to investigate the causal influence of a particular entity by not including it in a model.

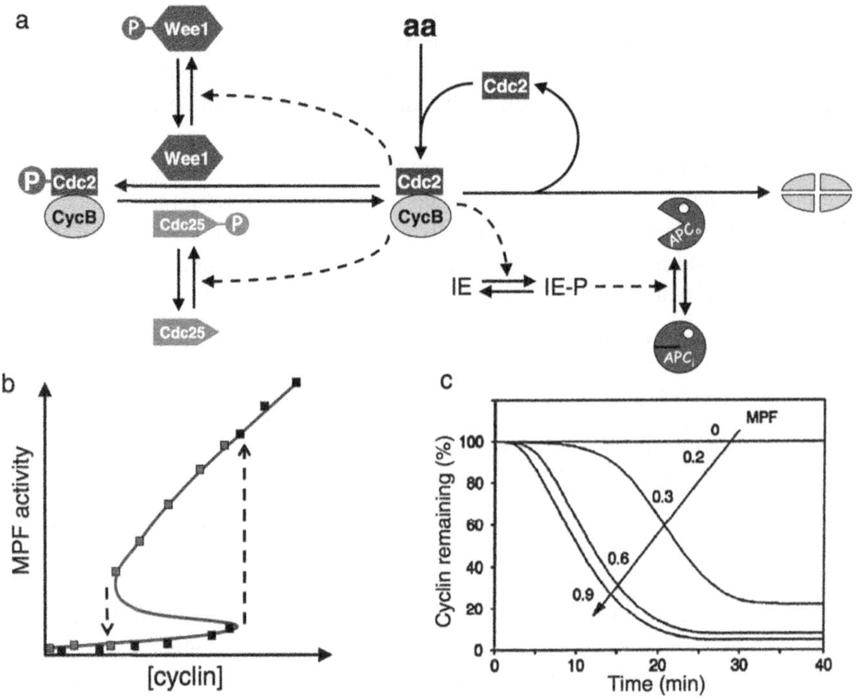

Fig. 2 A computational model of mitotic control. (Source: Tyson and Novák 2015)

> There appear to be no general rules for model construction in the way that we can find detailed guidance on principles of experimental design or on methods of measurement. Some might argue that it is because modelling is a tacit skill, and has to be learnt not taught. [...] This omission in scientific texts may also point to the creative element involved in model building, it is, some argue, not only a craft but also an art, and thus not susceptible to rules. (Morrison and Morgan 1999, p. 12).

Similarly, interpreting the results of a model brings scientists back into an informal setting in which findings are put back in and evaluated with respect to a context of collateral information. Computational modeling in biology is typically not regarded as a step towards transforming biological knowledge into a formal theory, instead models are often seen as heuristic tools (MacLeod 2016). More specifically, they act as tools that scientists use to "discipline" their informal reasoning about mechanisms, thereby allowing them to derive potentially surprising consequences of their assumptions. This works well if the right kind of information is available to generate a productive model (i.e. sufficient experimental data and accepted underlying formal principles, such as the law of mass action). Another potential benefit is the degree of abstraction in formal models that allows scientists to detect similarities between very different systems and to more easily transfer concepts from other "formalized" disciplines (e.g. network theory or engineering) to biology.

Image Analysis

Biology has a strong tradition of visualization. Scientific articles from molecular or cell biology are typically centered around a set of graphical figures that visualize key data in the form of photographs of western blots or of microscopic images of cellular structures in which features of interest are made visible by stains or fluorescent markers. In general, visual representation has been argued to play a significant role in scientific reasoning (Perini 2006). In science, as essentially everywhere else, digital methods have become indispensable in the production, processing and analysis of images. Digitalization has allowed scientists to delegate important parts of these activities to computers, but human visual inspection of images is still of great importance. Working with images is thus a paradigmatic area where informal elements of scientific reasoning interact with formal techniques.

Visual inspection of images presents a particular interesting case also because the human visual system has capabilities that cannot (yet) be matched by machines, for example in structure recognition. Such capabilities can partly be explained in terms of experience and context information and thus are analogous to the distinctive features of informal reasoning mentioned in section "Informal reasoning in molecular and cell biology". On the other hand, technical systems are superior, for example, when it comes to optical resolution and capturing absolute intensity values, not to mention processing speed and precision. It is thus obvious that there are trade-offs involved when using computers for tasks of image analysis that would otherwise be performed with the human eye. A recent textbook on autmated visual inspection in the technical and engineering sectors mentions various disadvantages for visual inspection carried out by humans. The authors note that:

> Manual visual inspection is:
>
> - monotonous,
> - laborious,
> - fatiguing,
> - subjective,
> - lacking in good reproducibility
> - costly to document in detail
> - too slow in many cases
> - expensive. (Beyerer et al. 2016, p. 4).

Thus, there are both pragmatic and epistemic reasons in favor of formalization. Although experiments in biology usually do not match the scale of industrial applications, there are many contexts in which pragmatic factors such as speed or fatigue become relevant. Time-lapse microscopy in cell biology, for example, typically gives rise to sequences of tens or hundreds of frames, each consisting of several images corresponding to different fluorescent channels or acquisition settings. But epistemic factors, such as objectivity and reproducibility, come to play an important role as well, as illustrated by the following example.

Charvin et al. (2010) investigated the decision process of yeast cells to commit to a new round of cell division using single-cell time-lapse microscopy. This transition is argued to be irreversible due to a positive feedback loop that involves the inactivation of a protein called Whi5. The upper panels in Fig. 3 show microscopic images of living cells at successive time points after exposure to a pulsed signal that can trigger the cell cycle transition. Inactivation causes Whi5 to exit the cell nucleus where it is otherwise localized in its active form. Whi5 was monitored with a fluorescent marker that gives rise to a signal that in the figure is represented in green and superimposed on phase contrast images showing the cell contours.

The difference between localized and delocalized signal can easily be detected by eye: some cells have a bright green dot in the center, while in others there is diffuse signal throughout the whole cell area. Nevertheless, Charvin et al. decided to develop an automatic algorithm to detect the nuclear fluorescence of Whi5. In particular, this involved the problem of detecting the nuclear shape when the signal is localized, which the human visual system appears to solve without any problem. The algorithm is based on the assumptions that the shape of the nucleus in the images is circular and that it has a typical size of around 100 pixels. The contour of the nucleus is then determined by minimizing a measure that takes into account the intensity of the pixels within the circle as well as the deviation from the typical nuclear size. In the lower panel of Fig. 3 the nuclear intensity determined according to this measure is plotted against time in cells that either fully inactivate Whi5 (blue), only partially inactivate it (green), or do not inactivate it at all (red). Only full inactivation is followed by the formation of a bud, suggesting that cells have to cross a certain threshold of inactivation in order to enter a new cell cycle.

Qualitatively, the plot in the lower panel provides the same information as the inspection of the images by eye. However, formalization comes with several advantages. First, by reducing the state of a cell to a single number, the results can be presented in a compressed form showing a larger number of cells and many time points.

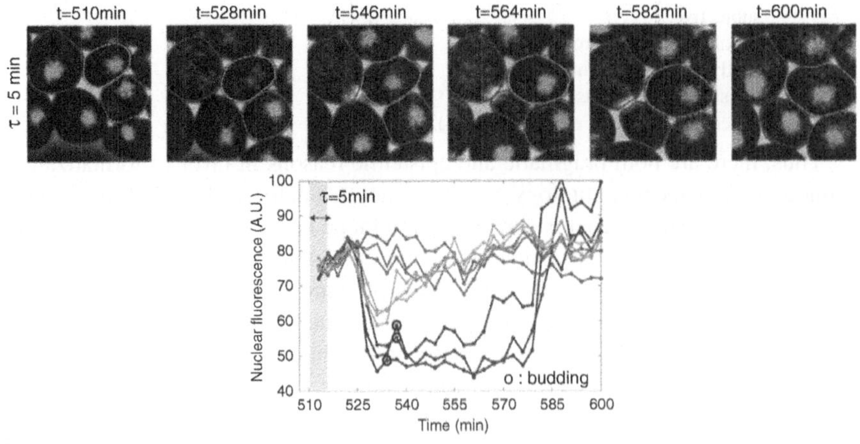

Fig. 3 Example of computational image analysis. (Source: Charvin et al. 2010)

Thereby, one can get an idea of the behavior of individual cells, but also of typical behavior and variability within a population. Second, the plot also provides quantitative information. Digital imaging technology directly converts light intensity into a number and thus naturally lends itself to a quantitative analysis. In this way differences in nuclear signal can be detected that are difficult to distinguish by eye. Third, formalization can be argued to be more objective in this context. The assumptions behind the algorithm are made explicit, and it can be executed mechanically, that is, without human intervention. In this way it counteracts possible biases of perception.

In spite of these advantages, formalization of image analysis has its drawbacks. The procedure will fail if the assumptions on which the algorithm is based on are not adequate, for example if there are cells with nuclei that are untypically large or that deviate from the circular shape. Moreover, the algorithm by itself would not be able to meaningfully compare images that were produced using different kinds of cells, a different fluorescent marker, or a different microscope because the intensity scale might be completely different. In all these cases human observers may still have no problems in analyzing the images by eye because they tacitly take into account information about the context. Furthermore, it is telling that, like in Fig. 3, the results of formal image analysis are typically shown side by side with selected images to be inspected by eye. The latter serve to justify the choice of formalization. Thus, human visual inspection in many scientific contexts retains the status of a gold standard.

It is an interesting question in this context whether increasingly sophisticated methods, such as machine learning procedures, will be able to eventually outcompete the human visual system altogether. It is likely, however, that formal methods as they become more sophisticated and epistemically opaque lose some of the advantages of simple formalizations such as the one discussed. Sometimes we value methods and consider them objective precisely because they are unsophisticated and inflexible.

Bioinformatics

One of the most dramatic changes in molecular biology in the last years has arguably been due to the large and genome-wide data sets that were made available by large sequencing projects and high-throughput technologies. In this area the pragmatic advantages of formalization, like speed, reliability, and cost, are undisputable. Bioinformatic technologies are typically based on a formalized kind of genetic information or on counting certain molecular entities of interest, such as reads of mRNA transcripts. Thus, like in the case of image analysis discussed in section "Image analysis", the technology is based on a digital representation of experimental measurements, and the application of formal methods is rather straightforward.

Genome- or system-wide experiments allow scientists not only to generate and store exhaustive data sets, but also to ask new kinds of questions. Of course, by itself there is nothing new about the application of formal methods in the context of experimental technology in biology. From measuring the size of specimens or counting cells to modern molecular techniques of qPCR, quantitative experiments have always been performed and evaluated statistically in the attempt to address biological research questions. What is different in modern bioinformatics, however, is that formal methods are applied not only to establish the validity of individual measurements, but to investigate large sets of such measurements and to look for biologically significant properties or patterns based on statistical methods. The epistemic role that formal methods play is thereby significantly altered. For purposes of illustration I will briefly discuss a concrete example.

An important problem in molecular genetics consists in understanding how certain proteins, so-called transcription factors, bind to specific sequences of DNA and thereby influence gene expression (Wasserman and Sandelin 2004, Mathelier et al. 2015). In order to investigate the role of a given transcription factor, bioinformaticians look for sequence properties that allow them to predict where it will bind. One method to achieve this is based on a set of already validated binding sites (often generated by means of ChIP-Seq technology) and consists in first aligning the corresponding sequences and then determining the frequencies of each DNA nucleotide (A, T, G, C) at each position. In this way one ends up with essentially a table of probabilities for observing each nucleotide at each position, called a *position frequency matrix* (PFM), as shown in Fig. 4. Based on this matrix, one can assign a quantitative score to each potential sequence and identify candidate binding sites. With the speed of modern computers it is possible to scan the sequence of an entire genome in a reasonable amount of time.

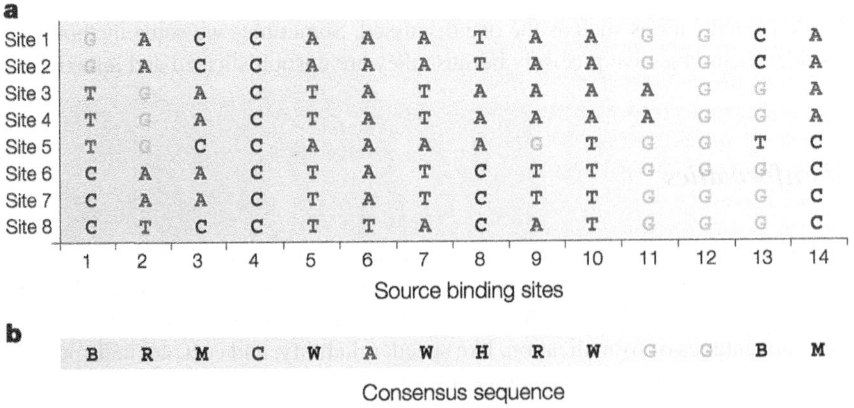

c Position frequency matrix (PFM)

	1	2	3	4	5	6	7	8	9	10	11	12	13	14
A	0	4	4	0	3	7	4	3	5	4	2	0	0	4
C	3	0	4	8	0	0	0	3	0	0	0	0	2	4
G	2	3	0	0	0	0	0	0	1	0	6	8	5	0
T	3	1	0	0	5	1	4	2	2	4	0	0	1	0

Fig. 4 Simple method for the prediction of transcription factor binding sites. (Source: Wasserman and Sandelin 2004)

The method just described exhibits at the same time advantages and disadvantages of formal methods. On the one hand, it automatizes the problem of finding promising pieces within sets of data that would be too large to be inspected by human researchers. On the other hand, it is based on a number of assumptions that make its results virtually useless, unless they are interpreted with great care. To look for binding sites based on sequence alone will result in a large number of false positives, that is, of sites that have a good score but no functional role *in vivo*. Wasserman and Sandelin call this the "futility theorem," according to which "essentially all predicted transcription-factor (TF) binding sites that are generated with models for the binding of individual TFs will have no functional role" (Wasserman and Sandelin 2004, p. 280), and they estimate that only around 0.1% of the predicted sites can be expected to be functional. The problem here lies in a decontextualization of the data that comes along with formalization: It is tempting to assume that a data set of sufficient size would allow one to build a powerful formal model, but such a data set is often meaningless if separated from the biological context. To overcome the problem, bioinformaticians must integrate sequence data with other kinds of information that is indicative of functionality, such as epigenetic marks, transcriptional activity, or chromatin interactions in 3D space.

Discussion

My aim in this chapter was to show that thinking about the differences between formal and informal methods can improve our understanding of the impact of computational methods in biology. Importantly, I have argued that informal reasoning should not be understood as deficient just because it cannot easily be delegated to a machine.

Experimental molecular biology is to a great extent based on informal reasoning. Reasoning about molecular mechanisms, for example, is informal in the sense that it is embedded in a large network of collateral information, which allows biologists to productively work with incomplete mechanism sketches. I suggested that the pervasiveness of informal reasoning should be explained in terms of a specific research context in which informal strategies appear promising, while formal methods are not expected to be productive.

In the last three decades this research context has been changing, which is probably due to a mixture of causes among which are the development of new experimental tools that can be well integrated with digital methods, the availability of large amounts of data, as well as the challenge of some of the assumptions that traditional molecular biology was based on. The three examples I discussed highlight different aspects of formalization in the area of computational biology and different ways in which formal and informal methods interact.

Computational modeling of molecular mechanisms is often used to "discipline" the informal reasoning of experimental biologists. The modeler is forced to make her underlying assumptions explicit and can then mechanically derive the consequences of these assumptions, which might deviate in surprising ways from what would be expected based on informal reasoning. However, model construction and interpretation are highly informal processes, and the success of modeling depends on various factors that influence the ease with which informal accounts of molecular processes can be translated into formal models.

Image analysis represents another case in which formal methods have been increasingly used. They exhibit clear strengths and weaknesses in comparison to the human visual system. While humans are good at pattern recognition and can easily switch between different contexts, computational methods are fast, reliable, and can easily be used for quantitative analysis. While the results of the latter are often considered more objective, human visual inspection is nevertheless considered as a gold standard in many contexts.

The example from bioinformatics brings out yet another aspect of formalization. In light of the large amounts of data to be taken into account in high-throughput experiments, it is practically impossible to approach their analysis without relying on formal tools. However, these tools have to be used with great care. The described method to predict binding sites would work well if the underlying processes themselves were based on simple rules. It fails to the extent that biological reality deviates from the idea that a transcription factor "reads" a sequence of DNA letters and binds wherever the sequence fits. By perhaps overstretching the meaning of the term, one might say in this case that it is the system itself that works in an informal way because the "meaning" of a binding site does not depend only on the "syntactic" features of the DNA sequence, but also crucially on features of the molecular environment.

In summary, the distinction between formal and informal methods can help to illuminate the complex ways in which scientists choose and combine different reasoning strategies depending on specific features of the research context.

References

Baetu, Tudor M. 2015. From Mechanisms to Mathematical Models and Back to Mechanisms: Quantitative Mechanistic Explanations. In *Explanation in Biology: An Enquiry into the Diversity of Explanatory Patterns in the Life Sciences*, ed. Pierre-Alain Braillard and Christophe Malaterre, 345–363. Dordrecht: Springer.

Bechtel, William, and Adele Abrahamsen. 2010. Dynamic Mechanistic Explanation: Computational Modeling of Circadian Rhythms as an Exemplar for Cognitive Science. *Studies in History and Philosophy of Science Part A* 41: 321–333.

Beyerer, Jürgen, Fernando Puente Léon, and Christian Frese. 2016. *Machine Vision. Automated Visual Inspection: Theory, Practice and Applications*. Berlin: Springer.

Brigandt, Ingo. 2013. Systems Biology and the Integration of Mechanistic Explanation and Mathematical Explanation. *Studies in History and Philosophy of Biological and Biomedical Sciences* 44: 477–492.

Charvin, Gilles, Catherine Oikonomou, Eric D. Siggia, and Frederick R. Cross. 2010. Origin of Irreversibility of Cell Cycle Start in Budding Yeast. *PLoS Biology* 8: e1000284.

Dreyfus, Hubert. 1972. *What Computers Can't Do: A Critique of Artificial Reason*. New York: Harper & Row.

Dunbar, Kevin N. 2004. Understanding the Role of Cognition in Science: The Science as Category Framework. In *The Cognitive Basis of Science*, ed. Peter Carruthers, Stephen Stich, and Michael Siegal, 154–170. Cambridge: Cambridge University Press.

Dutilh Novaes, Catarina. 2011. The Different Ways in which Logic is (said to be) Formal. *History and Philosophy of Logic* 32: 303–332.

———. 2012. *Formal Languages in Logic*. Cambridge: Cambridge University Press.

Floridi, Luciano. 1999. *Philosophy and Computing*. London: Routledge.

Gross, Fridolin. 2018. Updating Cowdry's Theorys: The Role of Models in Contemporary Experimental and Computational Cell Biology. In *Visions of Cell Biology: Reflections on Cowdry's General Cytology*, ed. Jane Maienschein, Manfred Laubichler, and Karl S. Matlin, 326–350. Chicago: Chicago University Press.

Gross, Fridolin, and Miles MacLeod. 2017. Prospects and Problems for Standardizing Model Validation in Systems Biology. *Progress in Biophysics and Molecular Biology* 129: 3–12.

Gunawardena, Jeremy. 2014. Models in Biology: 'Accurate Descriptions of Our Pathetic Thinking. *BMC Biology* 12: 1–11.

Humphreys, Paul. 2004. *Extending Ourselves: Computational Science, Empiricism, and Scientific Method*. Oxford: Oxford University Press.

Koslowski, Barbara, and Stephanie Thompson. 2004. Theorizing is Important, and Collateral Information Constrains How Well it Is Done. In *The Cognitive Basis of Science*, ed. Peter Carruthers, Stephen Stich, and Michael Siegal, 171–192. Cambridge: Cambridge University Press.

Küppers, Günter, and Johannes Lenhard. 2005. Computersimulationen: Modellierungen 2. Ordnung. *Journal for General Philosophy of Science* 36: 305–329.

Leonelli, Sabina. 2012. Introduction: Making Sense of Data-Driven Research in the Biological and Biomedical Sciences. *Studies in History and Philosophy of Biological and Biomedical Sciences* 43 (1): 1–3.

Machamer, Peter, Lindley Darden, and Carl F. Craver. 2000. Thinking About Mechanisms. *Philosophy of Science* 67: 1–25.

MacLeod, Miles. 2016. Heuristic Approaches to Models and Modeling in Systems Biology. *Biology and Philosophy* 31: 353–372.

Mathelier, Anthony, Wenqiang Shi, and Wyeth W. Wasserman. 2015. Identification of Altered CIS-Regulatory Elements in Human Disease. *Trends in Genetics* 31: 67–76.

Morrison, Margaret, and Mary S. Morgan. 1999. Models as Mediating Instruments. In *Models as Mediators*, ed. Mary S. Morgan and Margaret Morrison, 10–37. Cambridge: Cambridge University Press.

Murray, Andrew W., and Marc W. Kirschner. 1989. Dominoes and Clocks: The Union of Two Views of the Cell Cycle. *Science* 246: 614–621.

Perini, Laura. 2006. Visual Representation. In *The Philosophy of Science: An Encyclopedia*, ed. Sahotra Sarkar and Jessica Pfeifer, 863–870. New York: Routledge.

Quine, Willard v.O. 1951. Two Dogmas of Empiricism. *The Philosophical Review* 60: 20–43.

Stanovich, Keith E. 2003. The Fundamental Computational Biases of Human Cognition: Heuristics that (sometimes) Impair Decision Making and Problem Solving. In *The Psychology of Problem Solving*, ed. J. Davidson and R.J. Sternberg, 291–342. New York: Cambridge University Press.

Tyson, John J., and Bela Novák. 2015. Models in Biology: Lessons from Modeling Regulation of the Eukaryotic Cell Cycle. *BMC Biology* 13: 1–10.

Wasserman, Wyeth W., and Anthony Sandelin. 2004. Applied Bioinformatics for the Identification of Regulatory Elements. *Nature Reviews Genetics* 5: 276–287.

Phronesis and Automated Science: The Case of Machine Learning and Biology

Emanuele Ratti

Introduction

Recently, specialized and popular journals have emphasized the revolutionary role of big data methodologies in the natural and social sciences. In particular, there are two claims that fluctuate, in more or less explicit ways, in both expert and popularized pictures of science. First, big data generates new knowledge that traditional forms of the scientific method (whatever we mean by that) could not possibly do. Machine learning, it is said, is advancing and will advance science in unprecedented ways (Maxmen 2018; Zhou et al. 2018; Zhang et al. 2017; Schrider and Kern 2017; Angermueller et al. 2016; Sommer and Gerlich 2013). Second – and this is strictly related to the first claim – methodologies and tools associated to big data can automatized science in a way that humans will become dispensable. This is because humans' cognitive abilities will be rapidly surpassed by AI which will take care of every aspect of scientific discovery (Yarkoni et al. 2011; Schmidt and Lipson 2009; Alkhateeb 2017). However, it is not clear what automated science in this context means. In this chapter, I will show that the idea behind automated science is not substantial, *at least in biology* (and in particular molecular genetics and genomics, which are my main case studies). I will make one specific claim, namely that *ML is not independent of human beings and cannot form the basis of automated science, if by science we have in mind a specific human activity*.

The structure of the chapter is as follows. First, I will define the scope of ML (section "Machine learning and its scope"). Next, I will show a number of concrete cases where at each level of computational analysis human beings have to make a decision that computers cannot possibly make (section "Rules are not enough in machine learning"). These decisions have to be made by appealing to contextual

E. Ratti (✉)
University of Notre Dame, Notre Dame, IN, USA
e-mail: eratti@nd.edu

M. Bertolaso, F. Sterpetti (eds.), *A Critical Reflection on Automated Science*, Human Perspectives in Health Sciences and Technology 1,
https://doi.org/10.1007/978-3-030-25001-0_8

factors, background, tacit knowledge, and researcher's experience. I will frame this observation in the idea that computer scientists conceive their work as being a case of Aristotle's *poiesis* perfected by *techne*, which can be reduced to a number of straightforward rules and technical knowledge. However, my observations reveal that there is more to ML (and to science) than just *poiesis* and *techne*, and that the work of ML practitioners in biology needs also the cultivation of something analogous to *phronesis*, which cannot be automated (section "Experimental science and rules" and "Techne, phronesis and automated science") for the very simple of reason that its processes are completely opaque to us. Finally, I will frame this claim in psychological words by arguing (section "A possible objection and reply") that *phronesis* is exercised in terms of intuitions, and that how we develop intuition is not transparent. But even if knew how to frame the production of intuitions in terms of rules, still computers lack the biological constitution that is probably relevant to the way our intuitions are produced.

Machine Learning and Its Scope

Before delving into my analysis, let me briefly introduce ML and its scope. Let us start with a traditional definition of data science and then tailor it to ML. Dhar defines data science as "the study of the generalizable extraction of knowledge from data" (2013, p 64). This definition implies that there is something that we call 'knowledge' which is present in data sets. Knowledge here is not meant in any philosophical sense. Knowledge is understood as 'pattern', namely a discernible regularity in data. When Dhar says 'generalizable', he means that patterns detected will somehow occur in data sets which are similar to the one from which patterns have been extracted. The regularity is numerical and devoid of content. Similarly, ML is the study of generalizable extraction of patterns from data sets. However, this does not tell much about ML works, and it does not tell exactly what 'regularity' really means. Let me address these issues.

It is not that ML is presented with a data set, and then it extracts patterns. ML can extract generalizable patterns present in a data set starting from a *problem*. This is defined as a given set of inputs variables, a set of outputs which have to be calculated, a sample (previously input-output pairs already observed), and a set of real-world situation assumed to be similar to the one described in the sample – this is like the scope of the problem. What ML aims to calculate are the outputs, which stand in a quantitative relation of *predictive nature* with inputs – anytime that there is x (the input), then the probability of having y is such and such (the output). The sample will instruct the algorithm about this relation by providing previous observed pairs of "having x and y". Therefore, a first goal of ML is to generate *predictions*. Put it more precisely, ML generates *predictive models* where a prediction is defined as *the computation of the values of outputs for an input variable whose outputs are unknown*. The easiest example is e-commerce. ML is applied to find out quantitative relations between inputs (consisting of a set of customers' characteristics) and outputs (defined as buying specific products). The problem will be then composed of an input (i.e. the customers' characteristics), a set of input-output pairs (customers' characteristics and the products they bought), while the scope of the problem will be probably something like the class of products we are interested in. The solution to the problem will be to find out exactly the quantitative relation between inputs and outputs. Let's make another example, this time from biology. A common example of ML in biology is the class of algorithms aimed at identifying transcription start sites (TSS).[1] Let's say that we have an algorithm that we think is suitable to identify TSSs. The algorithm is provided with a large data set of TSS sequences. The algorithm processes such sequences and then it generates a model. At this point, the inputs for which we want to calculate the output (let's say, a whole-genome sequence) is provided and the model identify the sequences that are likely to be TSS.

Another thing worth to emphasize is that predictions are about a specific context, where the context is defined by the scope of the problem. In the case of TSSs, the scope of the problem is DNA sequences and not, for instance, amino acids sequences, even though these are somehow related. Moreover, an algorithm that is trained to recognize TSSs in human whole-genome may not be appropriate for other organisms, or it will not be functioning well for recognizing, let's say, promoters. A related goal of ML is *classification*, namely the use of classes for categorizing data points, where classes are sets of objects which share some features such that they can be relevant for some purpose. Usually, data points are assigned to a label, which indicates what the particular data point is about (e.g. a TSS). Related but different, *clustering* is the determination of which data points are similar to each other without having a labelled data set. There are other similar goals as well.

Given these aims, there are different modes of 'learning' for algorithms in ML. First, there is *supervised learning*, which is when ML is provided with labelled data sets, namely when data points are already classified in groups. Labelled data points may function as training data for ML algorithms, and from this structured data sets ML learns how to predict the way data points labelled are correlated in new (but similar from the standpoint of a problem) data sets. The example of predicting TSSs is supervised learning, since the data set of sequences provided to the algorithm contain labelled data points (i.e. sequences that are known to be TSSs). Classification is an example supervised learning. Differently, *unsupervised learning* analyzes unlabeled data sets. This means that ML in this case has to *cluster* data points in various groups or categories. Clustering is an example of unsupervised learning.

Automated Science

What do we mean by 'automated'? Which are the aspects of scientific practice that we are talking about? By drawing from Humphreys (2011), let's distinguish two scenarios. First, there is the present situation (called the *hybrid scenario*), where science is carried out both by humans and machines. The idea is that AI optimizes

[1] To simplify a bit, a TSS is where transcription of a gene starts, precisely at the 5'-end of a gene sequence.

certain aspects of scientific practice – mainly the quantitative aspects. This is uncontroversial. Much more controversial is the *automated scenario*, where computers replace all humans' activities and humans' epistemic authority in scientific practice. The idea is that in *science as we know it*, AI will replace humans and their expertise and their ways of gathering and producing knowledge.

In this section I will argue against the *automated scenario*. My claim is that *machine learning is not independent of human beings*. My argument on this issue is to argue that, as it happens in experimental biology, also in computer science the contextual situation in which a scientist acts underdetermines the technical rules she is equipped with, exactly those technical rules that a computer will follow to solve a scientific problem. Therefore, ML needs human judgment at every significant step of its discovery path. I will spell out the idea of 'judgment' by appealing to a virtue ethics/epistemology terminology and the difference between *techne* and *phronesis* in Aristotle.

However, this argument is not just another version of the strategy "computers cannot do *x*". My idea is to challenge the idea of the *automated scenario* itself. First, I will frame the idea of acting in the *phronesis* sense by appealing to the psychology of intuitions. Acting (well) by using intuitions seems to provide a contemporary psychological framework to the idea of *phronesis*. What emerged from the psychology of intuitions is that the way we educate our intuitions and the way we appeal to them to choose the right course of action is *opaque*. Therefore, because we do not know how our intuitions work, and scientists appeal to intuitions more than they are ready to admit, then we cannot automatize science. But – and here comes a possible objection – one may say that it may happen that one day we will make such a process transparent and hence elaborate a set of rules about it. However, the way we educate, accumulate and use intuitions is strictly connected to our biological constitution. Science is a human activity, and the way it is practiced and the shape of the results we obtain are dependent on the way humans are, not only cognitively but also biologically. But computers are not humans even if they could simulate human activities; computers are just different. This is of course applied to science only if science is seen as an intrinsically human activity; however, we do not know any other science that is not human.

Rules Are Not Enough in Machine Learning

Can scientific discovery be automatized so as to make human beings dispensable? The answer to this question is rather complex. I will show first the problems of automated science in the practice of ML and then I will conceptualize such problems in a bigger philosophical picture.

Any ML task is faced with a number of issues that require humans to take a decision based on background knowledge, experience and other factors hardly reducible to computational abilities *tout court*. In a sense, I will claim that humans have intuitions formed on an experiential ground which we do not know how are formed (Hogarth 2001). This means that ML cannot be completely automated because

Table 1 Aspects of Machine Learning practice that cannot be automated

Technical Issue in ML	Description of the Issue	What Cannot be Automated
Choice of the algorithm	Is there a strict set of rules for choosing between supervised and unsupervised learning?	Whether one judges the divergence of training and test sets as impacting significantly the reliability of results.
Encoding Prior Knowledge	Shall we frame the problem in terms of regression or classification?	Choice between regression and classification is a function of the task, background knowledge, and experience of the researcher.
Feature Selection	Selection of the features of a data set that are relevant to the problem one is facing	Choice of features is a function the task, background knowledge, and experience of the researcher.
Data Curation	How do we choose significant tags to label a data set?	Choice of labels depends on the foreseeable uses of the database, which depend in turn on the task, background knowledge, and experience of the curator.
Labels interpretation	Semantics must be manually assigned to labels	Interpretation of labels in a biological sense requires collaborative research.
Evaluation of the performance of an algorithm	Should we value precision or sensitivity?	Choice is a function the task, background knowledge, and experience of the researcher.

humans' judgment and intuitions[2] are required anytime uncertainty emerges, and an acceptable level of uncertainty will change according to the context and who will take the decision in the first place. Such issues are summarized in Table 1 and discussed below.

The first case where human's judgment is required is in the choice of the type of methodology. As Libbrecht and Noble reports (2015), sometimes you do not have much of a choice. For instance, if no labeled data set is available, then unsupervised learning is the only option. But, they note, having a labeled data set does not necessarily imply that supervised learning is the best option. The reason for this is that "every supervised learning method rests on the implicit assumption that the distribution responsible for generating the training data set is the same as the distribution responsible for generating the test data set" (Libbrecht and Noble 2015, p. 323). This means that sometimes data sets are generated differently, and they underlie different characteristics even if they are labeled with the same tags. As an example, Libbrecht and Noble report the case of an algorithm trained to identify genes in human genomes; this probably will not work equally well for mouse genomes. However, it may work *well enough* for the purpose at hand. In such situations the researcher has to identify, first of all, the fact that there might be a divergence between training and test sets. This divergence in ML applied to biology may also have a biological meaning, and it has to be scrutinized biologically and not

[2] I will equate 'intuition' and judgement' because I am thinking about 'judgements' as 'intuitive judgements'.

necessarily numerically. Next, one has to judge whether such divergence is such that a supervised learning might generate unreliable results. But unreliable results are a function also of the type of expectation and ability of interpretation that scientists are employing. Hence, there is not a clear-cut threshold between being reliable and unreliable. In other words, the researcher has to judge whether the level of uncertainty is acceptable or not.

Another issue worth to mention is the problem of *encoding prior knowledge*. Especially in the case of supervised learning, this might be understood as the choice between framing the problem in terms of either regression or classification. Regression is when the output of the problem is a real value, while classification is when the output is a category. Let's consider the analysis of nucleosome positioning from primary DNA sequences. Example of handling such problem in terms of regression is determining the coverage base of nucleosomes. However, we may want to frame the problem in terms of *classification*, namely predicting nucleosome-free region. It is important to note that, in light of researcher's judgment and interests, we may switch from classification to regression and *vice versa*, and this will also change the way prior knowledge is encoded. The issue is, again, how one will understand the problem to be solved and how framing it as either regression or classification will be the best option to obtain a valid result. Related to encoding of prior knowledge there are other close issues. First, there is *feature selection*, namely the set of decisions related to the selection of the features of a data set that are relevant to the problem one is facing. It is a case of implicit priori knowledge, as in any application of ML, "the researcher must decide what data to provide as input to the algorithm" (Libbrecht and Noble 2015, p. 328). But the problem is not just how to deal with data you have; rather, it can problematic also to judge data that you *do not* have, or data sets with poor quality. This is the problem of *missing data*, which can be of two types. First, values that are missing randomly or for reasons that have nothing to do with the problem one is trying to solve (e.g. poor data quality). Next, there are values that, when missing, are important for the task one is executing. Different strategies derive from the different types of missing data.

How to select data and how to deal with tags is not a problem that only users of databases have. There is compelling empirical evidence that computer scientists creating databases (i.e. database curators) have similar problems (Leonelli 2016, Chapter 4). Rules for standardizing heterogeneous data sets, label them, and make them available are underdetermined by the contexts and the specificities in which such procedures are actually used, because databases are curated on the basis of the foreseeable uses that other scientists will do of the database, the type of user that one imagines will use the database, etc. and all these factors will vary substantially, even within well cohesive scientific communities. Also, the very procedure of curating data and thinking about effective metadata has been described by Leonelli as a procedure to codify *embodied knowledge* of experiments and their environments and this "involves making a decision about which aspects of the embodied knowledge involved in generating data are the most relevant to situating them in a new research context" (Leonelli 2016, p. 97). In order to do this, being a computer scientist is not enough; you need also an expertise of the particular research context the database will be relevant to. Because of the importance of contextual judgments on the side of both curators and users, "human agents remain a key component of the sociotechnical system, or regimes of practices, that is data-centric biology" (Leonelli 2016, p. 112).

Furthermore, there is an additional problem if one is provided with a predictive model generated by unsupervised learning. Usually a solution from unsupervised learning provides a set of labels which are not interpreted – they are empty. A ML practitioner has to interpret such labels and connect them to the content of the science to which she is applying the algorithm – in our case, molecular biology. This may require collaborative research and integration of different epistemological cultures, which can be hardly automated. But even if one possesses both expertise, still there can be many ways of interpreting biologically results, especially in light of different epistemic aims. This is a case where pluralism should be cultivated or, as Chang says, "the great achievements of science come from *cultivating* underdetermination, not by getting rid of it" (Chang 2012, p. 151).

Another issue worth to be mentioned has to do with the evaluation of the performances of algorithms. According to ML practitioners, there is a tradeoff between *sensitivity* and *precision*. Let's say we want to predict the location of enhancers in a genome. We can either *value sensitivity* – the fraction of positive examples (enhancers) identified – or *value precision* – the percentage of predicted enhancers which are truly enhancers. These values stand in a tradeoff relation, and hence according to the task at hand we may end up valuing either one or the other. The upshot of all these issues is that working in ML requires making important decisions all the time. Algorithms are not independent, and sometimes in addition to ML knowledge of theoretical and practical constraints one needs to be proficient also in biology. Being an expert in both fields is very difficult. This is why bioinformatics is a collaborative enterprise; collaboration requires the ability to mediate and negotiate between very different areas of expertise; these are skills which are cultivated through experience, and they cannot be standardized in rules.

How do we frame philosophically all these problems? Which are the consequences that we should draw from such technical issues? What I will do is to interpret the view (and its denial) of automated science in light of some virtue ethics/epistemology basic distinctions.

Experimental Science and Rules

In order to frame the analysis of the problems mentioned above in light of some basic virtue-ethics/epistemology concepts, we need to take a step back to experimental biology.

Experimentation in molecular biology is characterized by an intense material manipulation of biological systems. The development of standardized techniques and the ingenious tinkering of biological entities are hallmarks of molecular biology (Rheinberger 1997; Ratti 2018). Experiments in molecular biology are rarely performed to test theories in the strict sense of *crucial experiments*. Usually the work of a molecular biologist is to develop a working model or a hypothesis about a biological system. Molecular biologists start from a general hypothesis about how a biological entity influences a set of processes. On the basis of a general hypothesis, they derive predictions and instantiate experiments to make sense of these predictions. Results provide biologists with other predictions that call for other experiments. Ensuing experiments refine the initial model by providing specific information about the entity under scrutiny. The experimental system is materially stimulated to reveal increasingly precise information:

> This process of shaping and refining hypotheses through experiments continues, virtually, *ad libitum*. This is a sort of progressive and ramified (but not-linear) deductive process, developed by poking and prodding experimental systems. (Boem and Ratti 2016, p. 150).

The aspect to emphasize is that, at each step, biologists have to stimulate the experimental system in unexpected ways, depending on the task at hand, the availability of resources and the nature of the experimental system itself. Even though there are plenty of standardized experimental techniques in molecular biology, still biologists have to modify their protocols to adapt them to the characteristics of the experimental system at hand. It seems that biologists have to 'feel' the experimental system or have to (metaphorically) 'dialogue' with it and interpret its responses accordingly (Keller 1983). Shannon Vallor makes sense of this complex dynamics by appealing to the cultivation of a specific scientific virtue called *perceptual responsiveness*, which is defined as

> *a tendency to direct one's scientific praxis in a manner that is motivated by the emergent contours of particular phenomena and the specific form(s) of practical and theoretical engagement they invite*. (Vallor 2014, p. 271).

In other words, perceptual responsiveness is a disposition to direct scientific praxis towards a proper appreciation of the affordances that a phenomenon under investigation yields. The virtuous experimental scientist is one who "properly reads or 'decodes' all of the salient invitations to measurement implied by the phenomenon (…) and creatively finds a way to take up just those invitations whose answer may shed the most light" (Vallor 2014, p. 276). However, an aspect that Vallor does not emphasize enough is the dependence of the scientific praxis on the experience of the researcher. I understand the phenomenological (Husserlian) concept of 'invitation' in terms of 'affordances'. The idea of 'affordance' comes from psychology, and it can be defined as a property of an object that suggests how to interact with or use that object (Ratti 2018). While 'affordances' exist independently of the observer, the fact that they can be 'exploited' depends on the actor interacting with the environment. The scientist must be sensitive to the ways affordances emerge from phenomena, and for this reason she must cultivate other important virtues such as perseverance, diligence, insight, adaptability, etc.; *perceptual responsiveness is an umbrella virtue because it implies many other virtues*.

The way a phenomenon is modelled is the result of the interaction of scientist's background, experiences, aims, the ability to properly see the affordances and know how to deal with them, and the phenomenon's characteristics. Scientific inquiry on this matter looks like value judgment rather than a mere clear-cut procedure by applying rules. Perceptual responsiveness is the disposition to see and respond appropriately to the affordances of biological phenomena. This is exactly the excellence that the experimental biologist strives to cultivate; by starting from a general hypothesis or an incomplete model, the biologist poses certain questions in forms of experiments to an experimental system, and by interpreting its answers in light of her experience and the components of the experimental system itself, she poses more accurate questions, until she elaborates a model of the phenomenon which is adequate – or, to use Rheinberger's terminology (1997), until she brings into light the opaque aspects of the *epistemic thing* (i.e. the phenomenon under investigation) and she turns it into a *technical object*.

An aspect that Vallor emphasizes is that perceptual responsiveness is a virtue because it is not rule-driven, but it is cultivated as a habit of seeing and acting in specific (both theoretical and experimental) contexts with the aim of achieving epistemic goals. Moreover, such habits are what make "a scientist exceptionally praiseworthy and a model of excellence" (Vallor 2014, p. 277).

My claim is that a similar virtue has to be cultivated by ML practitioners as well. This claim is motivated by the fact that even a quantitative enterprise such as ML requires a deliberation which is not just technical, but it depends on many other factors that will change from context to context, depending both on the nature of the data sets analyzed and the characteristics of the ML practitioner herself, and hence cannot be expressed fully in an algorithmic form. The issues emphasized in Table 1 are just a few of the contextual decisions that ML practitioners have to take, and that each of them will take in a different way, depending on their background and other contextual factors. In order to spell out this view properly, let me characterize the position of automated science in light of some basic distinctions.

Techne, Phronesis and Automated Science

In Aristotle, there is a distinction between theoretical knowledge and practical knowledge.[3] Theoretical/scientific knowledge is *episteme*, which can be understood as demonstrative knowledge of the necessary and the eternal (EN VI.3). Theoretical knowledge has no practical import (Dunne 1997, p. 238). On the other hand, practical knowledge includes *praxis* (usually translated as 'acting' or 'doing') and *poiesis* (usually translated as 'making'). The distinction is often put in the following terms; *poiesis* is related to the production/fabrication of things, and the success of an action is measured by the goodness of its product, while in the case of *praxis* the aim of an action is the good of the action itself (EN VI.4-5).[4] Examples of *poiesis* would be making a chair or building a bridge. Examples of *praxis* would be actions that are

[3] My reading of these aspects of Aristotle is mostly drawn from (Dunne 1997).

[4] "[T]he end of production is something other than production, while that of action is not something other than action, since doing well in action is itself action's end" (EN VI.5, 11140b).

constitutive to the good life, to human flourishing etc., which are usually understood in the context of morality or ethics. *Poiesis* is perfected by *techne*, which is understood as the knowledge of a set of rules and standards that are applied in order to make a well-constructed and well-formed external product. *Praxis* is perfected by *phronesis*, which is a very complex excellence, being an umbrella virtue as Vallor's perceptual responsiveness. *Phronesis* is practical knowledge, namely being able to recognize the salient features of a situation and act according to the choice of the right mean between extremes. Dunne (1997) puts the distinction in terms which are useful for the argument of this section:

> Techne provides the kind of knowledge possessed by an expert [...], a person who understands the principles underlying the production of an object or state of affairs [...]. Phronesis, on the other hand, [...] is acquired and deployed not in the making of any product separate from oneself, but rather in one's action with one's fellow. It is personal knowledge in that, in living of one's life, *it characterizes and expresses the kind of person that one is*. (p. 244, emphasis added).

Whether Aristotle implies that we can sharply isolate *poiesis* from the constraints of *phronesis* – in the sense that 'making' will be independent from 'ethics' and from all the tacit and implicit constraints that praxis is subjected to – is an open question, which I do not seek to answer. Rather, I want to focus on the consequences that insulating *poiesis* from *praxis* have for contemporary science.

When I talk about the constraints of *praxis* I have in mind a specific interpretation of this claim. Dunne (1997) tries to make sense of *praxis* with the idea of *incommensurability*. This emphasizes the plurality and heterogeneity of measures and factors to which we appeal to make sense of one's action. Incommensurability insists "on the disparateness of the factors that will weigh with *different people* faced with the same situation" (Dunne 1997, p. 47). Dunne connects this idea both with Aristotle's notion of *praxis* and deliberation and, interestingly enough, with Kuhn's remark that there cannot be algorithms for theory choice in science because every scientist will value different epistemic desiderata in different ways (Kuhn 1977). This happens because background and tacit knowledge, experience, goals and other contextual factors will make each scientist take a specific and unique turn towards theory.[5] Understood in this way, *praxis* has a parallel in the numerous factors through which perceptual responsiveness is exercised in the experimental sciences.

With the distinction between *techne* and *phronesis* framed in terms of the former being dictated by technical knowledge in the form of rules to produce something, and the latter as actions resulting from judgments which are influenced by a myriad of factors, let's turn to the natural sciences. How do we understand them? An interesting proposal is to see natural sciences as *poiesis*.[6] In fact, "[o]ur contemporary

[5] It is also important to point out that when it comes to the components that make human judgment, values of various sorts emerge, not only epistemic values. In philosophy of science this has been documented in several ways, both theoretically and empirically.

[6] A much more straightforward way to solve the issue would be to think about natural sciences as *theoria*, but if we pay attention to the way Aristotle meant this term and its cognate *episteme*, it is easy to see that natural sciences have not many things in common with *episteme* and *theoria*.

conception of scientific research [...] is deeply *poietic* [...]. As with poiesis more generally, the goal of scientific research is a product [...] external to the scientist herself, and that product can be detached from the scientist and evaluated independently" (Stapleford 2018, p. 4–5). It is not a coincidence that standards of scientific knowledge are usually expressed in form of adherence to protocols and technical skills which can be meaningfully understood in light of the typical *techne* language. Interestingly, Habermas explains the confusion that may emerge in blurring *techne* and *episteme* in terms of a deliberate operation instantiated by modern scientists, in the sense that "the modern scientific investigation of nature set about to pursue theory with the attitude of the technician" (Habermas 1974, p. 61). In our case, ML practitioners think about science as *poiesis*, and *techne* as the recital of technical skills needed to increase its products. The fact that computer scientists think about a fully automated science is because they reckon that 'making science' can be fully isolated from all the implicit and tacit components that are required to deliberate in a *praxis* environment. Therefore, at the very core of the idea of contemporary science lies the idea of science as a form of *poiesis*; modern science can be conceived as *poiesis* that can be fully isolated from *praxis*. Possessing *techne* would imply only "a logical progression through means to a predetermined end" (Stapleford 2018, p. 9). Automated science is an algorithm that possesses *techne*, and it will progress towards the achievement of certain products without the constraints typical of *praxis* that I mentioned above.

However, my claim – which is analogous to Stapleford's analysis (2018) – is that contemporary science is not just *techne*. Actually, science aims at being *poiesis* dictated only by *techne*, but in fact it is more dependent on something analogous to *praxis* than scientists would admit. In my list of situations where contextual decisions are required for ML practitioners, the dominance of *techne* might be undermined. Stapleford draws an interesting parallel to *praxis* when it comes to scientific practice, and in the case of ML the parallel can be made even more precise. The ML practitioner wants to act justly, where 'just' is in relation to the adequacy of her model. But we have seen that judging what counts as adequate requires something analogous to *phronesis*. It is not just about following a set of predetermined rules to construct a model (namely, the technical knowledge required to handle data), but also to identify the best mean given the particular situation the practitioner is dealing with and the level of uncertainty. The issue of missing values is a good case in point; depending on the nature of missing values (e.g. poor data quality) the researcher must make a choice as how to cope with such missing values. Feature selection is another example; depending on what our interests are, and the way the data sets are structured, we may prefer a regression or a classification. But even

---

Stapleford (2018) rightly emphasizes that episteme is described as knowledge of the eternal, it is achieved through demonstration, and it is about contemplation; "what admits to be known scientifically is by necessity" (EN VI.31139b). On the other hand, natural sciences are knowledge of the contingent, demonstration has not necessarily an epistemic priority over other discovery strategies (e.g. material manipulation, statistical analysis, etc) and theorizing – rather than contemplating – is an important aim.

more basic choices such as shifting from a supervised to an unsupervised algorithm require a sort of *phronesis*; judging whether the provenance of a data set and the way it has been labeled can provide a sufficient level of certainty to assure the reliability of the results we may possibly obtain will depend on several constraints, many of which will be tacit and strictly connected to the researcher's background. Moreover, judging a numerical result such as a ML predictive model in a biological sense goes beyond a simple automatized task; it requires experience in the biological field that a researcher in computer science may not possess. If the researcher lacks such an experience, she may ask for collaboration with a biologist, and collaborative research is the result of the integration of disparate fields of knowledge that no computer can possibly achieve. Computers suggest, humans decide.[7] On this point, it is interesting to notice that both *phronesis* and Aquinas' *prudentia* require the cultivation of virtues (such as caution, foresight, docility, etc) that could apply pretty well in the case of ML. A practitioner may exercise *caution* when getting certain values, and she has to be open to learn (i.e. *docility*) in interpreting results by listening to what the biologist has to say about her result. A data curator must cultivate *foresight* in order to imagine the different ways in which her database could possibly be consulted. But this requires *docility* again (Bezuidenhout et al. 2018), because a dialogue with actual users may change the way a data set is labeled. Docility is also fundamental when it comes to interpret the labels according to which a data set has been partitioned as a result of unsupervised learning; once the unsupervised algorithm has found the perfect way to partition a data set in terms of *n* labels, then a collaboration between biologists and computer scientists is in order if the aim is to *act justly* (i.e. to elaborate an adequate model).

Even though the predetermined end is the same (i.e. the construction of an adequate model), still the way this is achieved is the result of a complex network of constraints and decisions which are not 'end-directed automation'. What a highly sophisticated and technical field as ML needs is striking the right balance between *techne* and *phronesis*. A practitioner in ML needs both the technical knowledge in terms of protocols and rules provided by something like *techne*, and the experience, the capacity to 'sense' and 'feel' the context and a set of basic virtues that are provided by something like *phronesis*. It is no surprise that the situation is similar in an experimental field such as molecular biology. I am not claiming that what Vallor calls perceptual responsiveness is the *phronesis* required by computer scientists in bioinformatics. Actually, it does not matter how we call such a contextual *phronesis*. The important thing is that both biologists and computer scientists clearly need the technical knowledge to manipulate entities or data sets (e.g. protocols), but at the same time perceptual responsiveness and/or *phronesis* will dictate the way such technical knowledge is modified, adapted and used in specific contexts. *Techne* and *phronesis*/perceptual responsiveness here inform each other.

---

[7] This is analogous to Leonelli's remark that "[r]ather than replacing users' expertise and experience in the lab, metadata serve as *prompts* for users to use embodied knowledge to critically assess information about what others have done" (Leonelli 2016, p. 106).

A Possible Objection and Reply

My claim can be summarized by saying that the way a ML practitioner does research will depend not only on the technicalities of computer science, but also on the intuitions she has about her research object given her experience and research context – how she 'feels' and 'interprets' the specific context, and what her 'guts' say. Therefore, if the *automated scenario* requires that AI will perform science (as a human activity) by replacing humans in all their activities, then this is simply not happening and cannot possibly happen. The misunderstanding lies in conceiving science as *poiesis* driven exclusively by *techne*, while science is *informed* by both *phronesis* and *techne.*

However, one may buy this claim but not the consequences of this claim. One may say that yes; this is the way science is practiced today. But this does not mean that we cannot make the process of 'feeling' the experimental system less opaque, and somehow build a computer designed in a way that has 'intuitions' in the same way we do. And we should do this because our intuitions can be so good that we want to optimize them in a computer. In principle, this sounds right. But there are many things that in principle are legitimate but in practice are misleading. Let me be more precise of what is opaque and what is not, and then I will deal with the objection.

I have said that any scientist needs a sort of *phronesis*, namely a capacity to sense the situation and being able to recognize its salient features and act adequately to achieve a particular goal. Recognizing the salient features and act accordingly is having *an intuition* about the situation. Such intuitions are based on experience – we collect information about similar cases, somehow we internalize this set of information, and anytime we see an analogous case we almost automatically act in certain ways. However, the way we internalize such things and how these intuitions emerge is completely opaque – we have no clue about how we do it. Ironically, ML algorithms may be opaque, but human beings are opaque as well, and this is why we cannot turn into a set of rules the way a person possessing *phronesis* acts – or at least now.

What is exactly an intuition? Here I rely on some works in psychology, in particular (Hogarth 2001). Hogarth makes several examples of people taking decisions in their field of expertise without really deliberating about those things; "they have a feeling of knowing what decision to make [...] they depend on their 'intuition' or 'sixth sense', and this they cannot explain" (Hogarth 2001, p. 3). What is important to highlight is that the process of intuition is characterized by a lack of awareness of how it happens – unlike deliberative processes, intuitive processes are characterized by the impossibility of being made transparent. We can rationally reconstruct them, but rational reconstructions may be made in several ways. Interestingly, intuitions seem to depend on personal experiences, and hence each of us will have his/her own intuitions. The idea is that

> over time, we all build up a stock of ways of interpreting the world [...] [we] are typically not aware of how we acquired the knowledge that we use daily [...] the outcomes of intuitive processes are typically approximate and often experienced in the form of feelings. (Hogarth 2001, p. 9).

According to Hogarth, on pragmatic grounds we can think about two systems that control the processes by which we learn and we act accordingly to what we have learnt. First, there is the *deliberate* system, which deals with all processes that require attention and deliberation. Next, there is the *tacit* system, which controls all processes occurring with little or no use of conscious attention. In my understanding, intuitions are forms of expertise, in the sense that anytime an agent learns by associating things pertained to a certain domain or by acting in a certain way, then what has been learnt is somehow internalized in the tacit, and when similar situations emerge, intuitions about how to act or to understand the situation emerge as well. However, we do not know how these processes work. Interestingly enough, this is also what happens in ML; algorithms of ML – especially deep learning – evolve as they are trained with more and more data sets. Sometimes they are so complex that they are black-boxes – they generate results, but we do not know how they do it. In the context of human learning, we just know that, by having experience in a particular field, then most of the things we will do in that domain will be intuitive behavior. My idea is that *phronesis* or *perceptual responsiveness* is exactly the result of educating intuitions in specific fields.

However – and we return to the objection – one may say that we do not really know if these processes cannot be made transparent. Therefore, we can imagine a situation where psychologists identify clearly such processes, and a computer scientist tailors them into an algorithm to foster automated science. I do not think this is a very powerful objection, and for the following reason. As Hogarth recognizes, humans are composed of many information-processing systems, and hence our intuitions are the results of all these systems working together. We have no idea of how these complement each other, but what we can say is that they are deeply embodied *in our biological constitution*. Therefore, we should expect to have intuitions in a certain way and to act in a certain way also because we are biologically in such and such a way. We are not just *res cogitantes*. There is nothing unreasonable in thinking that also science is the result of the way we are. But computers are not like us, and hence we should not expect to have them the same intuitions and 'feelings' that we have; computers are just different. Therefore, automating science by looking at what humans and computers can and cannot do is like trying to teach to a dog to be a cat. Of course, this applied to 'anthropocentric' science, and it leaves open the possibility of a scientific practice AI-tailored, as some commentators suggested (Humphreys 2004),[8] even though I do not think I fully understand this statement. But even if I lack a clear understanding of these claims, still referring to science and outdistancing it from the human is quite strange, because we do not know any other science that is not human.[9]

To sum up, because (1) any scientific activity is dependent upon *phronesis*, (2) *phronesis* is the way it is as a consequence of the way humans are, and (3) comput-

[8] "[A]n exclusively anthropocentric epistemology is no longer appropriate because there now exist superior, non-human, epistemic authorities." (Humphreys 2004, p. 4–5).

[9] To use Humphreys' terminology, I do think that the *hybrid scenario* in which we cannot completely abstract from human cognitive abilities in science is here to stay even with ML.

ers are not like us, then attempts to automatize completely scientific practice are hard to understand.

Conclusion

In this article, I have argued that science as a human activity cannot be automated on the basis of ML because a look at the practice of science in this context shows that rules are not enough to lead scientific discovery. The philosophical lesson is that something similar to Aristotle's *phronesis* – the ability to sense and feel the specificities of the context and act (epistemically) justly – is required in dealing with complex quantitative models elaborated by ML algorithms.

References

Alkhateeb, A. 2017. Can scientific discovery be automated? *The Atlantic*.

Angermueller, C., T. Pärnamaa, L. Parts, and O. Stegle. 2016. Deep learning for computational biology. *Molecular Systems Biology* 12 (7): 878. https://doi.org/10.15252/msb.20156651.

Aristotle. 2014. *Nicomachean Ethics*. (C. D. Reeve, Ed.). Indianapolis: Hackett Publishing Company.

Bezuidenhout, L., Ratti, E., Warne, N., and Beeler, D. 2018. Docility as a Primary Virtue in Scientific Research. *Minerva* 57 (1): 67–84.

Boem, F., and E. Ratti. 2016. Towards a Notion of Intervention in Big-Data Biology and Molecular Medicine. In *Philosophy of Molecular Medicine: Foundational Issues in Research and Practice*, ed. G. Boniolo and M. Nathan, 147–164. London: Routledge.

Chang, H. 2012. *Is Water H2O? Evidence, Realism and Pluralism*. Springer.

Dhar, V. 2013. Data Science and Prediction. *Communications of the ACM* 56 (12): 64–73. https://doi.org/10.2139/ssrn.2086734.

Dunne, J. 1997. *Back to the Rough Ground - Practical Judgment and the Lure of Technique*. Notre Dame: University of Notre Dame Press.

Habermas, J. 1974. *Theory and Practice* (Theorie und Praxis, 1971). Trans. J. Viertel. London: Heinemann.

Hogarth, R. 2001. *Educating Intuition*. Chicago: The University of Chicago Press.

Humphreys, P. 2011. Computational Science and Its Effects. In *Science in the Context of Application (Boston Stu)*, ed. M. Carrier and A. Nordmann. Dordrecht: Springer.

Keller, E.F. 1983. *A Feeling for the Organism - The Life and Work of Barbara McClintock*. W.H. Freeman and Company.

Kuhn, T. 1977. Objectivity, Value Judgement and Theory Choice. In *The Essential Tension: Selected Studies in the Scientific Tradition and Change* (pp. 356–367). Chicago: University of Chicago Press.

Leonelli, S. 2016. *Data-centric Biology*. Chicago: University of Chicago Press.

Libbrecht, M.W., and W.S. Noble. 2015. Machine Learning Applications in Genetics and Genomics. *Nature Reviews Genetics* 16 (6): 321–332. https://doi.org/10.1038/nrg3920.

Maxmen, A. 2018. Deep Learning Sharpens Views of Cells and Genes. *Nature* 553: 9–10.

Ratti, E. 2018. "Models of" and "Models for": On the Relation Between Mechanistic Models and Experimental Strategies in Molecular Biology. *British Journal for the Philosophy of Science*.

Rheinberger, H.-J. 1997. *Toward a History of Epistemic Things: Synthetizing Proteins in the Test Tube*. Stanford: Stanford University Press.

Schmidt, M., and H. Lipson. 2009. Distilling Natural Laws. *Science* 324 (April): 81–85. https://doi.org/10.1126/science.1165893.

Schrider, D.R., and A.D. Kern. 2017. Machine Learning for Population Genetics: A New Paradigm. *Trends in Genetics*. https://doi.org/10.1101/206482.

Sommer, C., and D.W. Gerlich. 2013. Machine Learning in Cell Biology – Teaching Computers to Recognize Phenotypes. *Journal of Cell Science* 126 (24): 5529–5539. https://doi.org/10.1242/jcs.123604.

Stapleford, T. 2018. Making and the Virtues: The Ethics of Scientific Research. *Philosophy, Theology and the Sciences*. 5: 28.

Vallor, S. 2014. Experimental Virtues: Perceptual Responsiveness and the Praxis of Scientific Observation. In *Virtue Epistemology Naturalized: Bridges Between Virtue Epistemology and Philosophy of Science*, ed. A. Fairweather, 269–290. Springer.

Yarkoni, T., R.A. Poldrack, T.E. Nichols, D.C. Van Essen, and T.D. Wager. 2011. Large-Scale Automated Synthesis of Human Functional Neuroimaging Data. *Nature Methods* 8 (8): 665–670. https://doi.org/10.1038/nmeth.1635.

Zhang, L., J. Tan, D. Han, and H. Zhu. 2017. From Machine Learning to Deep Learning: Progress in Machine Intelligence for Rational Drug Discovery. *Drug Discovery Today* 22 (11): 1680–1685. https://doi.org/10.1016/j.drudis.2017.08.010.

Zhou, Z., J. Tu, and Z.J. Zhu. 2018. Advancing the Large-Scale CCS Database for Metabolomics and Lipidomics at the Machine-Learning Era. *Current Opinion in Chemical Biology* 42: 34–41. https://doi.org/10.1016/j.cbpa.2017.10.033.

A Protocol for Model Validation and Causal Inference from Computer Simulation

Barbara Osimani and Roland Poellinger

Introduction

Computational modelling in biology serves various epistemic goals by adopting different strategies. The main goal is to use models in order to simulate phenomena which cannot be observed *in vivo*: the integration of various pieces of knowledge serves the purpose of "observing", via simulation, the possible combined effect of various system components. Simulation is essential in that *emergent* behaviour is considered to result from the *iteration* of interactions among the individual components of the system, where the input conditions of iteration n depend on the interactions among the model components in iteration $n - 1$ (and in a recursive way on all previous calculation steps; see below our discussion of calculation and measurement in section "Measurement by simulation").

By modelling and simulating the population-level behaviour resulting from causal interactions at the unit level, this methodology essentially differentiates itself from other available methods for causal inference (such as causal search techniques like those performed via causal discovery algorithms – see (Pearl 2000; Spirtes et al. 2000), or standard statistical studies, such as randomised clinical trials, or observational cohort studies), which cannot model the micro-macro level interaction. We present here a case study in order to highlight such distinctive features with respect to more traditional methods of causal inference, and especially to advance a

This work is supported by the European Research Council (grant PhilPharm GA 639276) and the Munich Center for Mathematical Philosophy (MCMP).

B. Osimani (✉) · R. Poellinger
Ludwig-Maximilians-Universität München, Munich Center for Mathematical Philosophy, Munich, Germany
e-mail: b.osimani@lmu.de; r.poellinger@lmu.de

M. Bertolaso, F. Sterpetti (eds.), *A Critical Reflection on Automated Science*, Human Perspectives in Health Sciences and Technology 1,
https://doi.org/10.1007/978-3-030-25001-0_9

procedure for external validation of such techniques. In particular, we undertake a formal analysis of the epistemic warrant for such scientific procedures.

This account is motivated by several objections raised by philosophers regarding the scientific warrant of modelling and simulation for the purpose of scientific inference and discovery, especially in that this is at odds with the increasing use of such techniques as a complement and sometimes substitute to experiment, especially in the life sciences.[1]

This research is also urgent in view of the fact that much modeling and simulation research in biology has therapeutic purposes, e.g., the discovery and understanding of metastatic mechanisms in cancer. Also for this reason, we take as a case study, the discovery of the functional properties of a specific protein, E-cadherin, whose key role in metastatic processes has emerged from such simulation studies of cell growth and proliferation signalling.

In the second part of the paper we abstract from this case study and present a general theoretic scheme that relates the key concepts discussed in the philosophical literature on modelling and simulation in one and the same inferential framework (in particular, Parke 2014; Morrison 2015; Viceconti 2011). This is a diagrammatic map that represents the development process of the computer simulation:

1. The researchers start with (mathematically and/or logically) formalized assumptions about their object of interest (e.g., rules for an ABM, equations for an ODE system);
2. they translate these assumptions into a computer program and run simulations;
3. the simulation results are then compared to empirical observations and evaluated as to their quality in terms of reproduction of the observed behaviour of the target system;
4. the computer program is then repeatedly refined to generate results closer and closer to experimental findings (and biological background knowledge).

Nevertheless, in scientific practice, such a development process is rarely straightforward: first of all, revising the computer program might *increase* the discrepancy between (simulated) prediction and (physical) measurement. Yet, exactly by learning about such failures, scientists gain insights into the possible roots of such mismatches, guided by "how possibly" explanations of the repeated failures. If the revised model also fails to reproduce the target system, then the modeller has to find yet another explanation, while capitalising on the knowledge acquired through previous failures. In doing so, modelling failure helps refining the theory.

We present our protocol as a general, theoretical scheme which (i) visualizes the development and repeated refinement of a computer simulation, (ii) explicates the relation between different key concepts in modelling and simulation, and (iii)

[1]Although the focus of this paper lies on computational modeling in systems biology, we build on a rich discussion of modeling and simulation strategies in the philosophy of science; e.g., Parke 2014; Parker 2009; Casini and Manzo n.d.; Winsberg 2009; Weisberg 2013; Dardashti et al. 2014, 2015; Frigg and Reiss 2009; Beisbart and Norton 2012; Morrison 2015; Humphreys 2009; Viceconti 2011.

facilitates tracing the epistemological dynamics of model validation. In particular, we focus our analysis on the added epistemic value of modelling and simulation techniques with respect to causal inference of micro-macro-level interactions. Also, we emphasize that causal knowledge at the microlevel can be drawn upon in order to put constraints on the possible structure underpinning the macro-level behaviour of the system under investigation.

Our paper is organized as follows: section "Modelling and simulation in systems biology" provides an overview of the main modeling and simulation strategies in systems biology by focusing on the modal implications of the representation modes (discrete vs. continuous) and inference directions (top-down vs. bottom-up); section "Case study: cell proliferation modelling" is dedicated to the presentation of a case study: the computer simulation of cell proliferation by Southgate, Walker and colleagues through three stages. Section "Towards a protocol for causal inference from computer simulation" presents a formal account of the kind of inferential process illustrated by the case study, in view of providing a protocol for causal inference from computer simulation. In section "Computer simulation, causal discovery algorithms, and RCTS" we compare causal inference from computer simulation to other common techniques of causal inference available in the market, namely: causal discovery algorithms (see Pearl 2000; Spirtes et al. 2000) and randomised controlled trials (RCTs). In section "Causal inference from modeling and simulation" we embed the previous discussion in the debate around the epistemic nature of causal inference from computer simulation and present our concluding remarks.

Modelling and Simulation in Systems Biology

In systems biology, modelling is considered as a privileged instrument for knowledge integration and for the description/prediction/understanding of emergent behaviour. The main strategies available can be classified as a function of three fundamental categories: (1) mode of representation (continuum vs. discrete); (2) inference direction (bottom-up vs. top-down); (3) scale-adopted (single vs. multiscale).

Representation Mode

1. The "continuum" strategy: this is generally represented by a system of (partial) differential equations, which individually represent the different forces or quantities of elements at stake, and whose joint effect is mathematically studied by means of the system itself.
2. The "discrete" or "rule based" strategy: this is for instance implemented by applying Agent Based Modelling to the behaviour of cells or molecules.

In both cases iteration allows for the "observation" of emergent behaviour (e.g. oscillations, or particular patterns of behaviour), that may (or may not) reproduce the investigated system. What distinguishes the two approaches is that the former focuses on quantitative relationships between the entities involved, such as the rate of reaction in catalytic processes; in the latter instead, the focus is rather on rules-based behaviour, such as the definition of the conditions under which a certain reaction is triggered rather than another one.

Inference Direction

Model generation Mode and Modal character of the Model.

1. "Top-down" approach: starts with the observation of biological phenomena in the intact biological system (macro-level) and then constructs theories that would explain the observed behaviours, eventually uncovering the "underlying mechanisms" (meso-micro levels).
2. "Bottom-up" approach: begins with the study of systems components in isolation and then integrates the behaviour of each component in order to predict the behaviour of the entire system (aggregating knowledge at the micro-level to "predict" behaviour at the macrolevel). Bottom-up model are also considered to be "reductionist" to the extent that they rely on the assumption that higher-level phenomena can be analytically decomposed in modular, although holistically interacting, mechanisms.

The former proceeds via abduction (exploratory reasoning), whereas the latter by composition (integration of various pieces of consolidated knowledge). Both procedures are affected by uncertainty, in each case related to different kinds of underdetermination. In the former case, this refers to the various possible theoretical (and computational) models that can equally well reproduce the target phenomenon; in the latter, by the different possible ways in which the bits and pieces of local knowledge can be put together for global behaviour to be reproduced. One could say that in exploring the space of possibilities, both approaches draw on consolidated knowledge; however, whereas top-down approaches draw on tacit knowledge concerning the possible formal rules that can aid in reproducing the phenomenon phenomenologically, bottom-down approaches are instead constrained in their composition process by local specific knowledge related to the items of the target system that they intend to reproduce. Top-down models are "toy-model" in that they refer to the target system holistically; whereas bottom-up models intend to refer to it on a one-to-one basis.

Top-down approaches generally constitute early-stage explorations of the space of theoretical possibilities. However, in the case study below we will see that toy models may be constructed also in a later stage of investigation, that is, after a series of "how possibly" explanations have been excluded by unsuccessful models, and empirical knowledge has put additional constraints on the causal structure underpinning the model.

Scales

Models can also be combined in order to span several spatial and/or temporal scales: integration of various models which span across levels of the system and represent temporal dynamics in the medium/long run is known as multiscale modelling. This sort of modeling approach is also termed "middle-out" approach in that it starts with an intermediate scale (usually the biological cell, as the basic unit of life) that is gradually expanded to include both smaller and larger spatial scales (spanning micro, meso, and macro scales; see, e.g., Dada and Mendes 2011, p. 87).

Regarding the inferential mode, the middle-out approach does not differ substantially from the bottom-up approach, in that it proceeds via composition of pieces of consolidated knowledge and hence is affected by the same kind of uncertainty. Scales of different magnitudes may also be temporal, so as to allow the model to incorporate subsystems working at different time scales; as for instance is the case of the cell cycle process (which occurs in the order of magnitude of hours), vs. receptor binding (seconds to minutes), vs. intercellular signalling (less than seconds).[2]

Case Study: Cell Proliferation Modelling

We present a case study based on the research conducted by Jenny Southgate, Dawn Walker, and colleagues on epithelial cell homeostasis and tissue growth, as well as wound healing phenomena and malignant transformations (Walker and Southgate 2009; Walker et al. 2004a, b, 2008; 2010; Georgopoulos et al. 2010; Southgate et al. 1994). The investigated biological system is the set of mechanisms underpinning tissue growth and repair.

We analyse here a specific line of investigation in their research program: modelling and simulation of the dynamics (and mechanisms) underpinning cell proliferation. This line of investigation can be broken down into three main stages, relating to the three different computational models developed by the research team:

1. *ABM.* The first model was a single-scale, discrete, bottom-up agent-based model, developed for the simulation of the population dynamics of cell proliferation,[3] which integrated in a unified system various local rules derived from experimental data (Walker et al. 2004a). A comparison between the growth curves observed experimentally and those produced by the computer simulation revealed a significant discrepancy: two-way interaction in the experimental growth curve of the treatment setting, vs. a monotonic growth in the simulation curve. This discrepancy persuaded the scientists to revise the Agent Based Model by integrating it with a system of ordinary differential equations.
2. *Revised ABM + embedded system of ODE.* The second model was multiscale (cellular and subcellular level), multimode (discrete for the cellular and

[2] For a taxonomic review of computational modelling approaches in biology see Walker and Southgate 2009. See also Marco Viceconti's discussion of integrative models with respect to their falsifiability in Viceconti 2011.

[3] Cell proliferation is an ambiguous term in that it refers both to the *individual* cell reproducing itself through mitosis (see below), as well as to the growth of the cell *population*. Since the latter phenomenon is deterministically related to the former, this ambiguity comes with no serious epistemological consequence in this context. However, we will make a distinction between the two meanings clear whenever possible confusion may arise.

continuum for the subcellular level) model, which integrated the first model with a system of ordinary differential equations (ODE) representing a subcellular signalling pathway. This revised model investigated the hypothesis that the interaction effect observed experimentally could be explained by the interaction of the intracellular mechanism for cell proliferation (represented by the ODE), with the rules for proliferation dictated by cell:cell contacts (represented by the ABM) (Walker et al. 2008). However, in this case not only did the simulation not reflect the experimental curves, but not even the simulation "treatment" and "control" growth curve showed any significant difference.
3. *ABM, revised again, without ODE.* The third model eliminates the system of ODE introduced in the second one, which did not reproduce the expected behaviour, and adds a rule to the remaining ABM component. This "educated" toy model finally reproduces the growth curves observed experimentally.[4]

In this iterative process, empirical knowledge puts constraints on the kinds of rules that the model is allowed to have (that is, progressively reduced the set of possible worlds), and the simulation results put further constraints on the kinds of *explanatory hypotheses for the observed mismatch* between target and source system (at the macro-level), which further reduced the theoretical space for the possible mechanisms at work (at the micro-level). Eventually, a model was found which reproduced the target phenomenon with sufficient accuracy. Our philosophical focus here is not on the vexed question of the ontological status of models with respect to their target (see, e.g., Weisberg 2013; Frigg 2010a, b; Toon 2010; Teller 2009 or also Cartwright 1999; Downes 1992; Frigg 2006), nor on the kind of inferences that the similarity between target and source model allow (see, e.g., (Norton 1991; Reiss 2003; Sorensen 1992; Hartmann 1996; Humphreys 2004), but rather on the process of discovery carried out through iterative model revision by means of finding possible explanations for the mismatches found along the process of refinement. Our main objective here is to identify the distinctive epistemic role and inferential process of modelling and simulation with respect to discovering the laws underpinning system-level behaviour and population dynamics. In particular, we aim to identify the specific inferential bonus provided by modelling and simulation when dealing with explaining dynamic phenomena that other standard techniques of causal inference (such as causal discovery algorithms and controlled experiments) fall short of explaining (see our comparative treatment in section "Computer simulation, causal discovery algorithms, and RCTS").

[4]The agreement of the third model with the experimental data allows to learn, with reasonable confidence, that it is the temporal dimension of cell:cell contacts – rather than their spatial extension – that determines whether the involved cells will reproduce themselves or not. This constitutes the assumption on which a subsequent series of laboratory experiments is based and opens the way to more focused laboratory investigations, finally leading to the discovery of the complex context-sensitive regulatory role of E-cadherin in the homeostasis of cell proliferation (Georgopoulos et al. 2010).

First Model: Bottom-Up ABM Modelling of Epithelial Cell Growth

The model is intended to simulate cell proliferation, in particular, it aims to verify whether by putting together the behavioural rules of the individual system components, established experimentally[5], the overall system dynamics can be reproduced computationally. The reason for simulating such behaviour is to verify whether assembling such local rules (causal relationships) is sufficient to reproduce, *and therefore to explain on their basis*, the overall global system behaviour at the population level.

The left-side graph of Fig. 3 illustrates the causal structure being tested:

1. different environmental conditions (high/physiological vs. low calcium ions concentrations) determine different levels of the quality and functionality of the protein E-cadherin present in cell culture;
2. on its turn, presence vs. absence of E-cadherin, determines the type of cell:cell contact: stable vs. transient;
3. The number of contacts per cells at any given time determines proliferation ($<$ 4), or inhibition of proliferation ($\geq$ 4);
4. migration is inhibited by stable contacts.[6]

The computational model was matched to a lab experiment conducted on normal human urothelial cells (NHU), comparing cell growth in physiological (2.0 mM) vs. low (0.09 mM) calcium concentration medium. The comparison of the growth curves in the two settings showed similar patterns but also a significant discrepancy (compare the plots in Fig. 2 against the benchmark provided by Fig. 11 in Appendix C):

1. In agreement with what was "predicted" by the computational model, cell growth was *higher in the low calcium medium*, where the phenomenon of contact inhibition is significantly weaker (see Fig. 1).
2. In contrast to the curves representing cell growth in the computational simulation, where the lag phases in low and physiological calcium concentrations were similar (see Fig. 2 left), the results from the experiments showed that *the lag phase for the low calcium culture system was much longer than for the physiological calcium medium* (see Fig. 2 right). Or else, that *in low density populations, growth was higher in physiological than in low calcium concentration*

[5] Southgate and colleagues emphasize that their model distinguishes itself from previous models proposed for the same phenomenon whose "rules are selected to produce the desired outcome and are not linked to the underlying biological mechanisms" or whose scope covered "limited range of cell behaviours." The rules should represent "shortcuts" to *actual* controlling mechanisms. Hence, the model is not "phenomenological".

[6] Details about the model can be found in the Appendix B. The rules dictating cell behaviour can be found in Appendix A.

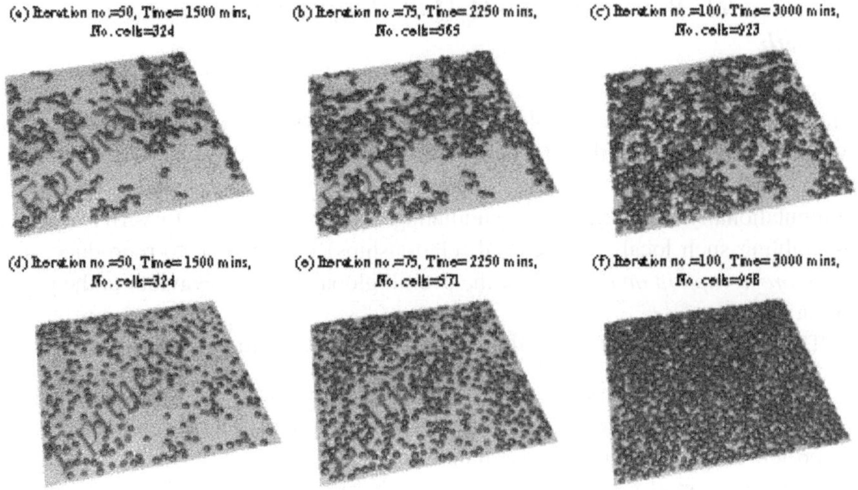

Fig. 1 Pattern of simulated culture growth in physiological calcium (**a**–**c**) and low calcium (**d**–**f**). Snapshots correspond to iteration 50 (**a**, **d**), iteration 75 (**b**, **e**) and iteration 100 (**c**, **f**). Red cells represent stem cells, blue cells are transit-amplifying (TA), pink cells are mitotic, and yellow cells are quiescent. The growth pattern in low and medium density populations (figures **a** and **b** in comparison with **d** and **e**) is relatively similar, because contact inhibition is less likely to have an influence in sparse populations. Instead as long as the population density increases the simulation shows diverging patterns. In the subconfluence phase (**c** and **f**) contact inhibition has a stronger influence in the physiological calcium environment (**c**), where contacts tend to be more stable, and this is reflected in decreased migration and the formation of cells clusters. Figure reprinted from Walker et al. 2004a with permission from Elsevier

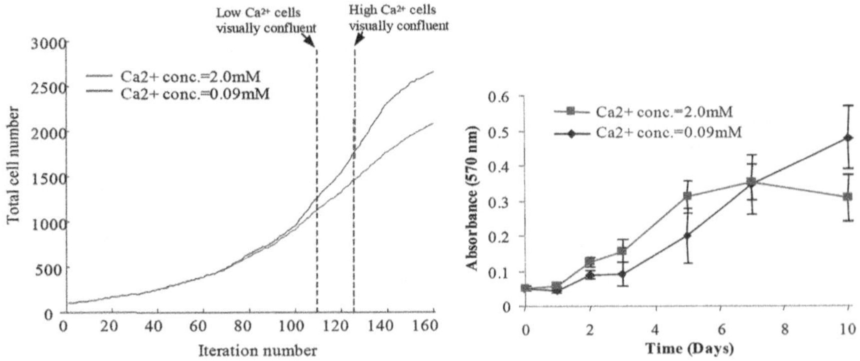

Fig. 2 Left: Growth curves produced by computational model (linear scale). Right: Results of MTT assay (error bars represent standard deviation of 6 samples). Figures reprinted from Walker et al. 2004a with permission from Elsevier

medium; in contrast to the hypothesis that, by inducing stable contacts, E-cadherin (present in high calcium concentrations) would inhibit proliferation.

Hence, Walker and colleagues looked for an explanation for such discrepancy, and searched for possible hints from the empirical literature: this suggested that autocrine signalling (a sort of cell self-signalling) induces growth transcription factors, such as Erk (Nelson and Chen 2002).

They hypothesised that, under the threshold for proliferation inhibition (≥ 4 cell:cell contacts), *stability of the contact itself might indeed induce proliferation through cell:cell signalling*. Thus, the effect of E-cadherine would be "bi-directional" and would induce:

1. both *inhibition of proliferation*; by promoting stable contacts and thereby increasing the probability that at any given time any given cell has ≥ 4 contacts – this was the main hypothesis underpinning the original model about cell growth rules.
2. and *promotion of proliferation*; through cell:cell signalling – this constituted the revised hypothesis.

The revised hypothesis does not replace the original one, but rather specifies it further, so as to possibly uncover the origin of the interaction between population density and low vs. physiological calcium conditions observed experimentally.

Second Model: Integration of the First Agent-Based Model and an ODE System into a Multiscale Model

In order to explore the "augmented" hypothesis, the scientists devised a second model where also intracellular signalling pathways could play a role in cell growth regulation. This was done by integrating the agent based model with a system of ordinary differential equations aimed at reproducing the (supposedly growth promoting) intracellular signalling pathway, and by iterating the ABM component and the ODE system in alternation.[7] Figure 3 compares the two hypothetical causal structures underpinning the first and the second model respectively.

---

[7]The set of equations of the ODE model representing Erk activation (i.e., Erk diphosphorylation: Erk-PP) through the EGFR signalling pathway, takes as initial value the number of receptors and ligands binded between two adjacent cells. The total number of activated receptors will depend on the number *n* of contacts and their areas, which is *fixed* in the low calcium setting, and *constantly grows* in the high calcium one). Each cell has a set of equations 1–5 associated with each cell:cell contact; these are then summed to provide the total number of activated receptors per cell, which in turn provides the relevant quantity for the subsequent equations in the system (Walker et al. 2008). The last equation of the ODE model (Eq. 29 in Walker et al. 2008) represents the level of Erk activation for that cell in that stage of the simulation. At the end of the ODE iteration, the updated profile of Erk-PP is returned to the memory of each agent of the ABM. Hence, the Erk activation type determines for each cell whether the cell divides and proliferates or not; this in turn will change the density (and thereby the topology) of the population of cells at the new start of

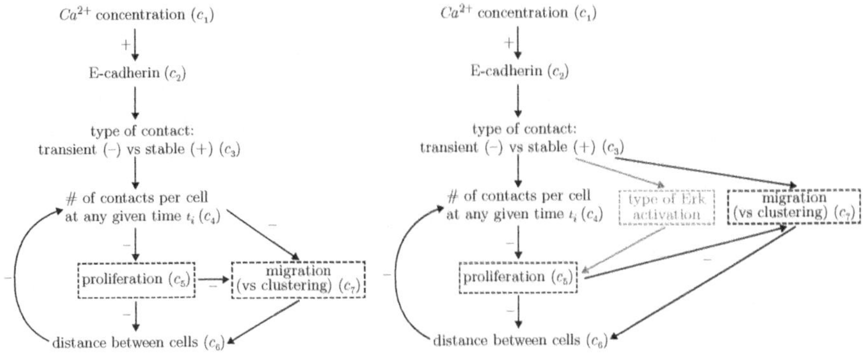

Fig. 3 Causal structure underpinning the first ABM of epidermal cell growth and proliferation (left) and extended causal structure where the type of ErK activation mediates type of contact and proliferation (right), all causal paths in the second model are represented by the same (or slightly modified) ABM rules of the first model, with exception of the causal path from "type of contact" through "type of Erk activation" to "proliferation", that is the intracellular signalling pathway, which is instead represented by a system of ODE standing for biological rates of reactions observed in laboratory experiments; see Walker et al. 2004a

The second model (Walker et al. 2008) was meant to explore the two counteracting effects of E-cadherin: (1) inhibition of proliferation through promotion of stable contacts, and (2) promotion of proliferation through sustained activation of Erk (exactly mediated by stable contacts) at low vs. high density population.[8] It was applied to two settings:

(a) *individual* cell:cell contact (for different rates of formation of the contact);
(b) a growing *population* of cells (in simulated low vs. physiological calcium conditions).

The first setting was meant to compare the effect of stable vs. transient cell:cell contact on Erk activation levels for one-to-one intercellular contacts. The different patterns are shown in Fig. 4 below.

In both cases the models are meant to reproduce the high/physiological calcium conditions vs. low calcium environments (with scarcity of E-cadherin mediated contacts). Whereas the individual cell:cell contact model (a) delivered results congruent with expectations (and with empirical data); the population level model (b) gave puzzling results, since not only did the model not reproduce the growth curve that it was expected to deliver according to the hypothesized causal structure, but

iteration, which, in turn, will determine which cells will have new contacts and whether these will determine proliferation (if these have been stable enough through iterations) or not (if they have been not stable, or if the number of contacts reached by the cells is ≥ 4 and therefore the "contact inhibition" rule trumps the signaling pathway effect). The constants for rates of reactions are all derived from experimental studies (Brightman and Fell 2000; Kholodenko et al. 1999; Adams et al. 1998; Waters et al. 1990; DeWitt et al. 2001; Lauffenburger et al. 1998; Oehrtman et al. 1998).

[8] In the new model however, the previous ABM component is not only integrated with an ODE system, but also modified – see Appendix D.

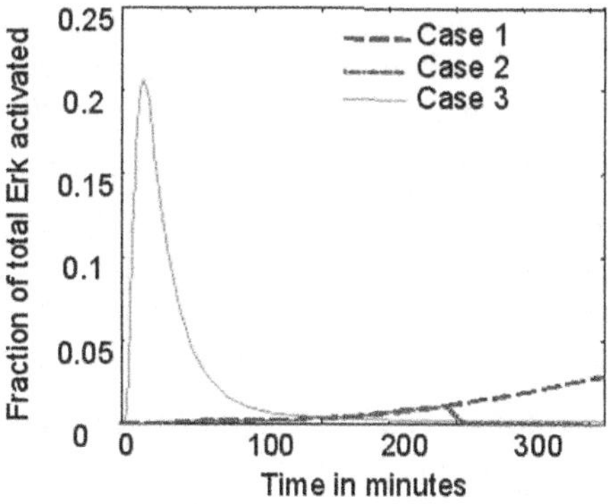

Fig. 4 Erk activation patterns (measured by normalized proportions of Erk-PP) for a single cell:cell contact. The dashed, monotonically increasing line (case 1) represents the case where two cells make an initial contact of 2 μm that increases linearly to 23 μm over 6 h. This type of behaviour has indeed been experimentally observed in (18). The continuous dark line (case 2) shows the same initial formation pattern as for case 1, but then has an abrupt break of contact after 4 h (when it has reached a contact length of 16 μm). The continuous light line (case 3) shows a strong initial signal which is "instantaneously" formed, but quickly decreases its strength already after 20 min., when it has barely reached 12 μm. Case 3 is typical of those contacts observed (by the authors) using time-lapse microscopy between migratory cells, maintained in subphysiological medium (Ca^{2+} concentration: 0.09 mM). Figure reprinted from Walker et al. 2008 under the terms of the Creative Commons Attribution License (http://creativecommons.org/licenses/by/2.0)

also the growth curves in the two conditions ("low" vs. "high calcium") did not differ from each other. This was puzzling at first, given that in both (a) and (b) the signalling pathways were modelled by the same set of ODE.

However, the initial conditions for model (a) and (b) where not the same.

In model (a), the initial conditions for three different settings were (see: 4)[9]:

1. moderate but sustained rate of contact formation;[10]
2. same initial formation pattern as for case 1, but with abrupt break of contact after 4 h;[11]
3. "instantaneously" formed contact, but quickly decreasing its strength already after 20 min.[12]

[9] Experimental evidence shows that the temporal characteristics of ERK activation (through the EGFR pathway) influence cell fate: moderate but sustained ERK activation lead to cell progression (and proliferation), whereas transient activation does not (Brightman and Fell 2000; Kholodenko et al. 1999, see Fig. 4 above and related explanation).

[10] Initial contact of 2 μm that increases linearly to 23 μm over 6 h.

[11] When it has reached a contact length of 16 μm.

[12] When it has barely reached 12 μm.

To these three different conditions correspond different patterns of Erk diphosphorilation (that is, of activation): only in the first case the cell reproduces itself.[13] Case 1 and 3 are supported by experimental evidence (Walker et al. 2008; Adams et al. 1998).

In model (b) the two initial conditions were:

1. linearly increase the area of cell:cell contact for each pair of cells entered into contact at each iteration until max length of 1,5 times the cell radius (and 5 μm height of the contact area);
2. do not increase contact area through iterations.

As can be noted, the difference in the duration of the contact is not encoded in this set of initial conditions. This revealed to be crucial to explain the mismatch between the individual level and the population level simulation, and also in devising the third model.

By observing the results of the two simulations, and considering the difference of initial conditions and related outputs, the scientists could infer that Erk activation depends on the temporal characteristics of contact formation only, *irrespective of its spatial extent*, and that it is through this dimension that the interaction effect in population growth dynamics is produced at the macro-level.

Hence the discrepancy is explained by uncovering a hidden, wrong, assumption, which is implicit in modelling the two conditions. Namely, the assumption that Erk activation would depend on the spatial extent of the contact; whereas, *as the one:one simulation showed*, this depended on the continuity of the contact in time only. The initial conditions for the individual level simulation fix the rate of growth in time exogenously, and therefore the ODE outputs three different curves depending on these initial conditions.[14] Instead, in the population level model, the two settings only differ for the growth of the *spatial* dimension of the contact in time, vs. no growth.

Since in the individual cell:cell contact simulation the different initial conditions give rise to different simulation outputs, whereas this does not happen in the population-level simulation, the explanatory hypothesis is that the ODE system is sensitive to temporal duration of the contacts only, and not to their spatial length (or only indirectly so). A stable (even if non-E-cadherin mediated) contact, may show exactly the pattern of Erk activation of type 1 in Fig. 4 if it lasts long enough.

The mismatch between the expected simulation patterns at individual vs. population level on one side, and the acquired knowledge that at the individual level the different patterns of Erk activation are determined by the temporal length of the intercellular contact only (consolidated by correspondence of three temporal patterns and Erk activation curves with independent experimental data, cf. Walker et al. 2008; Adams et al. 1998), finally led to a third (single-scale) ABM model.

[13] Large, instantly formed contacts induce a large peak in Erk activation at ca. 20 min. after the initial contact, but then a rapid decline to very low levels follows, even though the contact persists. Instead a contact which forms gradually but persists in time produces a monotonic, although moderate, increase in Erk activation. If the contact is lost (case 2), then the Erk-PP signal quickly goes down to 0.

[14] For an overview over how the modelling assumptions were tested, see Appendix E.

Third Model: Simplified Educated-Phenomenological Model

The third model appears in yet another study (Walker et al. 2010), which black-boxes the effect of the signalling pathway(s), and substitutes the mathematical modelling of the signalling cascade (the ODE system embedded in the ABM), with a simple agent-based rule, forcing the cell to increment its cycle step and reproduce itself whenever it has maintained a contact for $\geq$ 8 h. Hence this model consists of the ABM component of the second model only – updated with the additional rule that the cell should proceed in the cell cycle if it has maintained at least one contact, *whether "transient" in the sense of non-calcium mediated, hence with fixed dimension, or stable, i.e., calcium-mediated, hence growing in dimension*, for at least 8 h[15] (Walker et al. 2010). This means that the only parameter which is allowed to influence proliferation is the temporal length of the contact, whereas the spatial extension of the contact plays no role.

The model can be considered to be an "educated"-phenomenological model in that the additional rule aims to reproduce the expected cell population growth curve experimentally observed in the target system, with no assumption as to the biological mechanisms underpinning it. However, it serves the purpose of testing the hypothesis that the temporal length of contacts can account for 295 the growth curves observed experimentally.

In this case, the growth curves of experimental and simulated data finally matched each other. This means that whatever the signaling pathway at play, and whatever its intrinsic cascade, its effect on population growth may be (fully) explained by the temporal length of the contact. Is this conclusion warranted? If so, how and what are its implications?[16]

Towards a Protocol for Causal Inference from Computer Simulation

From Formal Model to Stable Code

When Margaret Morrison discusses the epistemic value of computer simulations (Morrison 2015), she emphasizes the twofold role of models in our attempt at grasping reality. She states that "[...] models are experimental tools capable of

[15] Please, note that in the second modelling attempt, cell proliferation (self-reproduction) was triggered in the individual cell:cell contact model (a), by a stable contact lasting more than 6 h.

[16] Keeping fixed the knowledge acquired through the various simulations – that is, the positive role of E-cadherin and of the temporal length of intercellular contacts in modulating cell growth – while questioning the role of the EGFR-Erk pathway in cell proliferation, the authors undertook then a series of experimental investigations on alternative E-cadherin related signalling pathways in normal human urothelial cells, finally coming to the conclusion that this protein has a complex, context-sensitive behaviour (Georgopoulos et al. 2010).

functioning as measuring instruments, but they are also objects of inquiry in cases where they, rather than physical systems, are being directly investigated and manipulated." This suggests an analogy between models and experiment, and "[...] it is this link that figures prominently in establishing the experimental character of simulations" (see Morrison 2015, pp. 228–229). Of course, when Morrison speaks of *models as objects of inquiry*, she does not mean to conflate the model and what is being modeled. She does not claim that manipulating the model is all we need to do to learn about the modeled object or system. Quite the opposite: Morrison identifies the methodological and epistemological characteristics of physical experimentation and applies this characterization to model manipulation, thereby providing an interpretation of models beyond the classical understanding as descriptive tools. In our conceptual framework we build on Morrison's vocabulary and aim to not only (i) tie real physical systems to their formal stand-ins, but also (ii) link such formal models to higher-level phenomena emerging from computer simulations. Our conceptual scheme, which spans all these epistemological/methodological levels, offers a unified way of tracing scientific inference – "upwards" from basic causal assumptions to complex causal claims, and "downwards" from emergent higher-level patterns to causal relations underlying the target system. Such higher-level phenomena are of special interest to us in this paper, since they facilitate causal inference in a special way: not only are emergent phenomena grounded in a simulation's basic rules, but they truly only arise in the dynamic evolution of the simulated system. Nevertheless, higher-level behavior evades both (i) Morrison's characterization of model manipulation as experiment and (ii) the concept of intervention as defining causal relations (as, e.g., formulated in interventionist causal theories). Still, although the relation between the simulation model and higher-level observations can be very opaque, emergent phenomena are nevertheless *generated* by aggregation/interaction of basic rules. So, when running a simulation leads to the detection of emergent patterns, the simulation acts as a *guide to causal understanding*. In particular, perceived macro-level phenomena will possibly suggest further manipulation tests at the basic micro-level. In this way, emergent phenomena help in tracing the interplay of the target system's mechanisms (and sub-mechanisms): i.e. the micro-macro-level interaction.

In the following, we abstract from the case study above and introduce a general theoretic scheme to visually relate the key concepts discussed in the philosophical literature on modelling and simulation. This scheme reflects the development process of the computer simulation: The researchers start with formal scientific assumptions about their object of interest (commonly mechanistic knowledge in mathematical form) and translate those assumptions into a computer program. Ideally, this computer program is then repeatedly refined to generate results closer and closer to experimental findings (and biological background knowledge).

Figure 5 traces the process from the initial formal model of the target system to its simulation in a repeatedly revised code base. The computer program can be understood as a set of structured equations or rules R, called in a predetermined order to update the model's parameters at each stage of the simulation. Since R is revised if simulation and empirical evidence deviate unacceptably, index k indicates

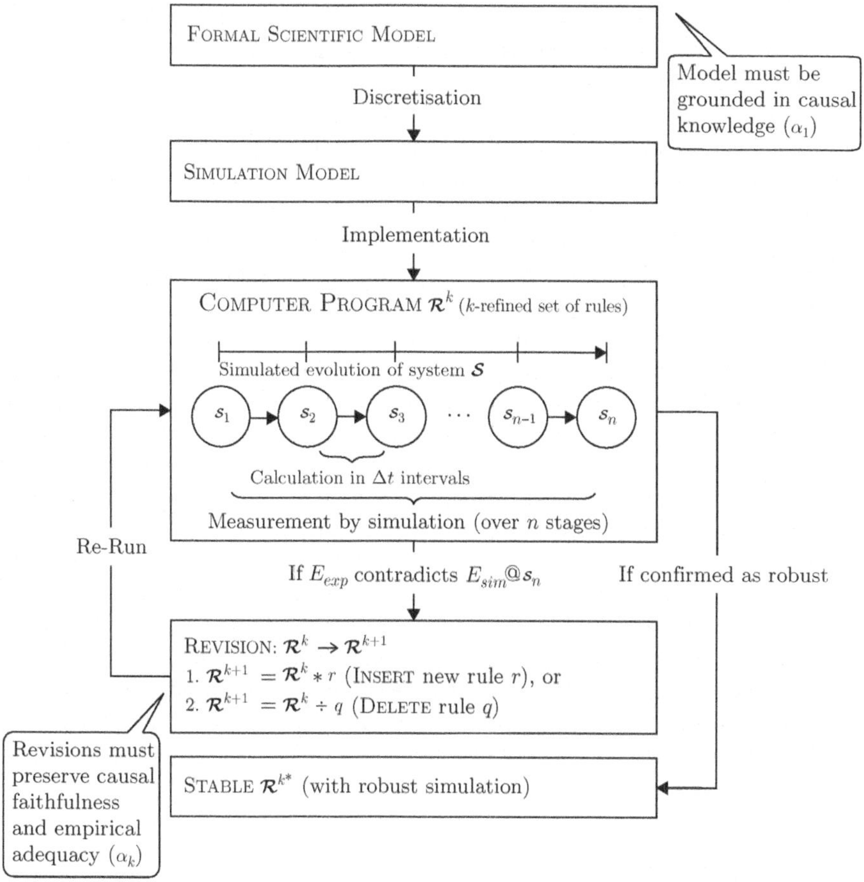

Fig. 5 Tracing the process from model to robust simulation in runs of the computer program R^{k*} (cf. Morrison 2015): A scientific model is transferred into a simulation model, implemented as a computer program, and revised until the simulation yields robust results

the "version number" of the program: The simulation model is implemented as R^1 and revised until the stable version k^* satisfies the researcher.

In scientific practice, such a development process is rarely straightforward: For example, revising the computer program might unexpectedly *increase* the discrepancy between (simulated) prediction and (physical) measurement and thus reveal gaps in the initial assumptions. Yet, learning about such theoretical gaps might then motivate the next revision of the computer program, and – in doing so – help in refining the theory. Our graphical scheme relates the computer program and the theoretical background by marking constraints for code revisions, grounded empirically and on abductive bases. As the case study presented above illustrates, revisions are made on the bases of "how possibly explanations" for the observed results constrained by empirical findings and consolidated background knowledge. On their turn, new simulations' results further constrain the space of possible explanations.

Our general theoretic scheme can not only be utilized (i) as a blueprint for tracking the development process of the computer simulation, but also (ii) as a basis for justifying causal claims drawn from such computer simulations. In sections "Measurement by simulation", "Verification, validation, revision", "Accuracy and robustness", and "Causal inference", we build on this scheme in order to outline the basis for a general protocol intended to support justified causal inference from computer simulations. We proceed as follows. Before we can start to compare quantities measured in simulation and in empirical experiment, we define the concept of *measurement by simulation* (section "Measurement by simulation"). In section "Verification, validation, revision", we present potential sources of discrepancy between experiment and simulation and discuss ways of revising the computer program while preserving consistency of the codebase. We then introduce the task of model validation as a question of accepting a model once it reaches the desired robustness, and we sketch this robustness condition in terms of multi-dimensional accuracy (section "Accuracy and robustness"). Our aim in section "Causal inference" is to go beyond reconstructing phenomenologically accurate prediction and facilitate understanding of the causal dynamics at work. To this end we formulate an anchoring condition a (sufficiently) robust simulation must satisfy (across code revisions), in order to warrant valid causal inference (i) from compound causal information, and (ii) from emergent phenomena. By explicating an epistemological concept of emergence, we finally have all the ingredients for a complete protocol for causal inference from computer simulation. We start by distinguishing calculation and measurement.

Measurement by Simulation

When the computer program is run, it simulates the evolution of the target system by iteratively updating the model's parameters in temporal increments of Δt intervals over n stages. To capture as much detail as possible, Δt should be as small as possible without unnecessarily increasing computing load or processing time. At each stage s_i $(1 < i < n)$ of the simulation, the parameter values are calculated from the results of the previous stage s_{i-1}. For instance, in our case study, each cell's properties such as size, location, cell cycle phase etc. are calculated given the parameter values calculated in the previous calculation step. This calculation cycle, generating interpretable data over n stages, can then be called measurement, even if the computer experiment lacks the material link to the target system.[17] For example, the ODE system in the second model of our case study was used to virtually *measure* Erk activation patterns for a single cell:cell contact (as illustrated in Fig. 4).

[17] This characterization follows Morrison's discussion of calculation and measurement in simulation and scientific practice in general, see Morrison 2015, chap. 6, pp. 199ff.

Verification, Validation, Revision

If empirical data is available, the physical, experimental evidence, abbreviated as E_{exp}, can be compared to the simulated quantity, abbreviated as E_{sim} (encoding the measurement result of n calculation steps). If the quantities deviate unacceptably (for a predefined minimum level of accuracy) the computer simulation is deemed invalidated, and R must be revised in order to improve the simulation.

The modeler is presented with the oftentimes difficult task of judging what such a discrepancy might be due to (see also our discussion of our case study's modelling assumptions in section "Testing the modelling assumptions of the second model"). Assuming that the experimental, empirical results are correct, one of the following situations might obtain:[18]

1. One of the idealizations (deliberately chosen or as constraint of the programming framework) applied to formulate the mathematical representation of the mechanistic knowledge has a severe effect.
2. The numerical solution of the mathematical form used to capture the mechanistic knowledge is inaccurate.

The above issues typically regard the verification phase of a computational model: Code verification ensures that the program runs without errors, and solution verification ensures that any deviations of expected outcome values are not due to how the program's functions are implemented. Discrepancies due to cases (1) or (2) can only be remedied by revisiting the simulation model or basic implementation decisions.

When we turn to validation issues, the discrepancy between simulated and experimental measurement might be due to one of the following cases:

3. The initial parameter values (at stage s_1) are inaccurate.
4. The mechanistic knowledge the model is based on is wrong or incomplete.
5. The mathematical form used to capture the mechanistic knowledge is inadequate.

Cases 3–5 refer to information encoded in elements of R, i.e., structural equations containing constants and variables. Discrepancies due to this second list of cases reflect problematic assumptions in the modeler's point of departure, i.e., her initial formal representation of the object of inquiry in the mathematical model. Emily Parke (Parke 2014) makes out discrepancies of different strength in her discussion of the supposed epistemic privilege of real experiments over computer simulations. She first distinguishes different degrees of surprise, as a researcher's reaction to unexpected observations: i) The researcher might find the observations unexpected, but in general agreement with background assumptions and theoretical

[18] We are grateful to Marco Viceconti for urging us to be explicit about the following range of sources of discrepancy.

considerations, or ii) she might be puzzled to such a degree that she is forced to reevaluate or completely overturn her theoretical assumptions.

Since this distinction is relative to the generality of the background assumptions, Parke shifts her focus to two relevant sources of empirical surprise *within the object under investigation itself*:

1. the investigated system evolves unexpectedly, or
2. the evolution of the investigated system reveals *hidden mechanisms* or *hidden causal factors*, thereby exposing implicit basic assumptions or detecting gaps in the simulation model.

In our case study, it is precisely such a gap in the simulation model that prompts further refinement, and "hidden" precisely refers to the role played by micro-level forces and entities in the output observed at the macro-level. Hence, the disagreement between simulation and empirical experiment which Walker et al. observe in their first model can be described as an instance of case 4 above. By looking for possible explanation of such a discrepancy and especially by looking at the form of growth curve of the cells population in the experimental setting, they *must* assume the presence of a factor interacting with the contact-inhibition rule. Then, by looking for possible candidates, they learn from cell biology literature about the possible positive effect on proliferation, caused by intracellular signaling pathways activated by cell:cell contact. Hence, they include a signalling pathway in the model in the form of a set of equations derived from empirical data, and reiterate the simulation. In doing so, the scientists add fine-grained detail to their assumptions about what is of relevance for a mechanistic understanding of cell proliferation.[19] In principle, revisions of R can be translated into revisions of the set of theoretical assumptions about the target system.

Our abstract scheme in Fig. 5 presents the revision step from one version of the program to the next, i.e., from R^k to R^{k+1}, in the syntax of the belief revision formalism (as, e.g., developed in Gärdenfors 1988):[20]

1. Insertion: $R^{k+1} = R^k * r$. The rule r is added to R while maintaining consistency, i.e., after any rule in conflict with r is removed from R. In belief revision terminology: R^k is *revised* by r.
2. Deletion: $R^{k+1} = R^k \div q$. R^k is *contracted* by rule q, which means that any rules implying q together with any rules implied by q are removed from R^k (i.e., all traces of q are eliminated). This operation might be realized in more than one way, since contraction does not have a unique solution in general. Causal

[19] We find it interesting to note how the modification of the ABM component in the third model remedies the observed discrepancy: The fact that it is a (phenomenologically motivated) local refinement of the (otherwise mechanistically motivated) ABM rules, which suffices to correct the higher-level measurement, seems to suggest that the ABM is overall adequate, yet underspecified (thus pointing to case 4 more than to case 5 among the sources of discrepancy listed above).

[20] This formalism provides ready-made operations for consistently updating sets of interdependent sentences (in our case: rules or lines of code). However, in principle, of course, any other formalism will do to formally capture the operations of rule insertion and deletion.

information, e.g. from empirical evidence, might be of help in disambiguating possible options (see also the remarks on condition α below in section "Causal inference").

Possible strategies for model revisions (through rule insertion and deletion as well as combinations thereof) are illustrated in the following sections. By revising the computer program, the modeler hopes to approach a stable code-base step-by-step. Is there a way for her to judge the quality of the simulation, though, and to validate the model conclusively?

Accuracy and Robustness

The goal of the modeler is to create a reliable and accurate simulation of the target system. Accuracy and reliability are verified in multiple dimensions:

1. *Point-wise comparison* of simulated outcome and empirical measurement might be deceptive, because agreement might merely be due to the simulation model being fine-tuned in calibration, i.e., due to the model crucially depending on its very input parameters (often referred to as "overfitting problem"); see our discussion of the comparison of the first model's predictions with experimental data (illustrated in Fig. 2).
2. If the modeler wants to confirm her understanding of the mechanistic structure though, she will have to check whether the simulation generates robust outcomes even under *variation of parameters* (within empirically testable ranges).[21]
3. Once the simulation reaches the desired level of accuracy even under such variation of parameters, the modeler may – in a further step – subject her model to *mechanistic intervention* and check whether the simulated evolution of the target system still corresponds to experimental evidence from the target system (again, with desired accuracy), for example, when the researcher blocks a mechanism in the experimental set-up and simultaneously disables the supposed corresponding rule in the model. In our case study, locally embedding the ODE system into the agent-based model can be understood as a starting point for such mechanistic tests.

The process of scrutinizing the model by varying the model's parameters in static comparison, and by intervening on the model's mechanisms, (revision), eventually leads to a stable code-base R^{k*} that *robustly* generates phenomena corresponding to empirical experiments with desired accuracy, thereby confirming the model – or, to be precise, the model's adequacy (and thus its applicability).

The link between simulation and target system can then be expressed in terms of *robustness*: The concept of robustness of a simulation captures a simulation's accuracy under parameter change *and* mechanistic intervention – robustness can thus be

[21] See our discussion of test 1 (sensitivity analysis) in the case study in Appendix E.

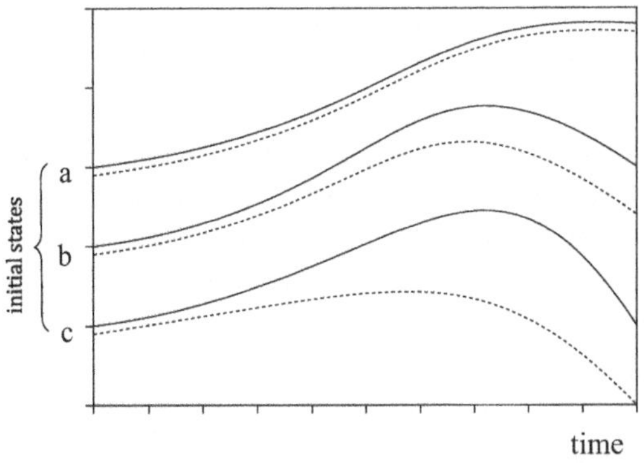

Fig. 6 Robustness by example: Solid lines indicate experimental evidence, dashed lines represent the simulated evolution of the target system. The simulation model is well-calibrated for initial states (**a**) and deviates with decreasing accuracy for initial states (**b**) and (**c**). The simulation thus reveals *non-robustness* under parameter changes

understood as "multi-dimensional accuracy." If the simulation returns results above a certain empirically motivated degree of accuracy, the researcher is justified in postulating an *analogous relation* between simulation model and target system. Optimizing this measure of robustness will thus *confirm* the simulation model as being *adequate* for making inferences about the target system.[22]

Example Figure 6 illustrates the concept of robustness by tracking the evolution of an unspecified example system and its simulation over time for three different initial states. Solid lines indicate experimental evidence, whereas dashed lines represent the simulated evolution of the target system for the same initial states. The simulation is obviously well-calibrated for initial state (a) (and for a short initial time segment across initial states), but parameter changes reveal its non-robustness: The simulation increasingly deviates over time from experimental results for initial states (b) and (c).

In our case study, the researchers want to achieve accuracy in two important respects: (i) not only for individual cell:cell contact (for which the second model seemed to be well-suited), but also at the population level (for which the second model failed); and (ii) not only in quantitative comparison between simulation and experiment (as by matching statistics), but also in qualitative comparison (as by comparison of structural macro-level dynamics, see also our discussion of inference from emergent phenomena below in section "Causal inference").

[22] In terms of Bayesian confirmation, increasing the robustness and thereby increasing the belief in *analogous evolution* of simulated (*SimS*) and target system (*TarS*), formalized as '*SimS* ≈ *TarS*', will boost the researcher's belief in the adequacy of the simulation model (*SM*), on which the simulation is based:

$$P(SM \text{ adequate} \mid \overbrace{SimS \approx TarS}^{\textbf{Robustness}}) > P(SM \text{ adequate}).$$

Now, where maximizing accuracy under parameter change may increase phenomenological agreement, maximizing accuracy under mechanistic intervention is the first step towards a mechanistic understanding of the underlying causal relations, since it leverages causal control for calibration. In the following we want to tie accurate prediction to basic causal knowledge by explicitly considering anchoring conditions for justified causal inference from the aggregate level.

Causal Inference

The philosophical skeptic might argue that simulations as virtual, artificial environments are detached from reality – they follow their source code and will behave like programmed, not like reality, which renders such computer simulations unusable for scientific – not to mention *causal* – inference. This claim is pointedly summarized by Diez Roux in her discussion of the distinction between observation-based and simulation-based causal inference in epidemiology:

> [...] there is a fundamental distinction between causal inference based on observations (as in traditional epidemiology) and causal inference based on simulation modeling. The traditional tools of epidemiology are used to extract (hopefully) reasonable conclusions from necessarily partial and incomplete (often messy) observations of the real world. [...] In contrast, when we use the tools of complex systems, we create a virtual world (based on prior knowledge or intuition) and then explore hypotheses about causes under the assumptions encoded in this virtual world. In the simulation model, we cannot directly determine whether X causes Y in the real world (because the world in which we are working is of our own creation); we can only explore the plausible implications of changing X on levels of Y under the conditions encoded in the model. In the real world, we have fact (what we observe) and we try to infer the counterfactual condition (what we would have observed if the treatment had been different). In the simulated world, everything is counterfactual in the sense that the world and all possible scenarios are artificially created by the scientist. (Diez Roux 2015, p. 101).

This topic has recently seen a surge of publications in the philosophy of science with some authors regarding computational modeling as simply one variant of modeling as such (e.g., Frigg and Reiss 2009), while others emphasize as a special feature of computer simulations the possibility to experiment with models as objects of inquiry (e.g., Parker 2009; Morrison 2015). In line with the idea of computational models as test objects we described the code-base of the simulation as constantly changing with rules inserted or deleted in attempts at increasing the accuracy of the program (see section "Towards a protocol for causal inference from computer simulation"). In the following we focus on the question how causal inference from iteratively revised simulated systems can be justified, if at all. We proceed in a bottom-up way: In order to guarantee the legitimacy of causal inference from computer simulations, the simulation model must be built on a solid knowledge base. The empirical adequacy of this foundation permits causal inference from the computer simulation with respect to two distinct dimensions of aggregation: (i) inference from compound causal information, and (ii) inference from emergent phenomena. Our argument has the following structure:

1. Causal inference from compound causal information: We first discuss how causal inference on the aggregate level must be anchored in empirical evidence, and how it can be guaranteed that the underlying causal assumptions, modularly encoded in the simulation model, are not distorted in the process of revising the codebase.
2. Causal inference from emergent phenomena:
   (a) We go on to explicate an epistemological concept of emergence we find particularly illuminating for our purpose, and we use it to distinguish micro-level and macro-level behaviour by drawing on the general scheme developed in Fig. 5.
   (b) We finally bridge the gap between the levels thus introduced again by linking emergent phenomena to the underlying generative rules, thereby once more emphasizing the value of simulation-generated higher-level observations for the understanding of mechanisms in systems biology.

1. *Causal inference from compound causal information.* If the goal of modeling a given system is to capture its mechanistic inner workings in a modular fashion, the structured set of R-rules will be as atomic as possible and as basic as needed for the given modeling depth.

Chaining atomic R-rules with causal content will yield "compound" causal information about the underlying causal web, i.e., information about the causal hypothesis or theory (encodable as chain or network like in Fig. 3). For this dimension of aggregation, causal inference will be valid if the model is grounded in valid causal knowledge and remains grounded in valid causal knowledge across revisions. We capture this *anchoring requirement* in the twofold condition α:

(α) Empirical knowledge supports both (i) the model's parameters (variables and their values) and (ii) the model's causal structure (i.e., the conditional causal independencies amongst the model's variables).

The first part of condition α captures the nature of bottom-up modeling, where a model is built on known behaviour of individual agents, i.e. causal knowledge about individual components of the system at the micro-level; the second part is also known as "causal faithfulness" and relates to the macro-level output of the whole system.[23]

[23] Example: Suppose the computer program implements the causal structure $X \rightarrow Y \leftarrow Z$ as a set of structural equations of the form $E = f_E(\vec{C})$ (i.e., the value of the effect variable E is functionally computed from the set of causes $\vec{C}$). The scientists are now interested in an additional parameter, mediating the influence of X onto Y (the collider node in the corresponding graph). This parameter is added to the underlying assumptions and implemented as U. If causal faithfulness is to be preserved, the modeler has to make sure that both of the following structural assumptions are implemented in R: (i) X still influences Y *indirectly*, and (ii) the variable U screens off Y from X's influence (in accordance with the *Causal Markov Condition*).

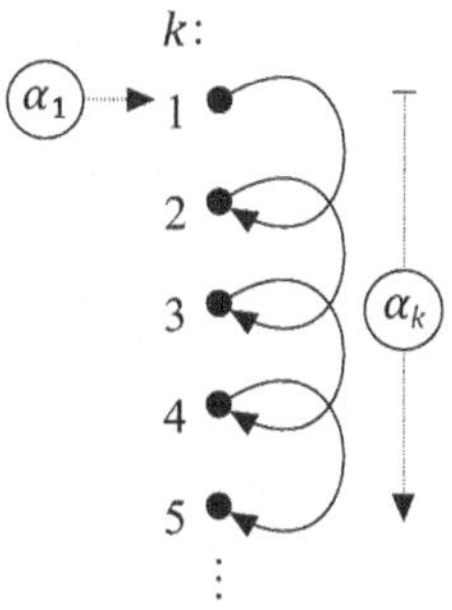

Fig. 7 The initial (α_1) and inductive (α_k) warrant for causal inference along the refinement of the computer program R^k for $k = 1, 2, \dots, k^*$

To track code revisions, we attach an index to α: α_1 ensures the empirical adequacy of the first simulation model (as a condition attached to the formal scientific model on which the first implementation is based). α_k (for $k = 1,2, \dots, k^*$) represents the inductive warrant for valid causal inference along the refinement of the computer program:

(Anchoring) R^k allows for valid causal inference iff it is *anchored* in empirical knowledge, i.e., iff α_k holds.

This translates directly into conditions for the revision operations: Causal inference on the aggregate level can be called valid across revisions if addition (revision) and deletion (contraction) preserve empirical adequacy both at the micro- and at the macro-level.

Figure 7 graphically relates the condition α to the process of revising code-base R. It is important to note that revising R can be done either (i) by systematically changing the codebase, triggered by mathematical intuition or by accuracy considerations during calibration, or also (ii) by introducing and exploiting new sources of empirical knowledge, triggered by abductive inference or by newly available empirical knowledge. Condition α is meant to track both the empirical adequacy of newly introduced empirical knowledge and the preservation of overall structural integrity (causal faithfulness) even under local modifications.

2. *Causal inference from emergent phenomena.* To distinguish the aggregation-based mode of causal inference above from emergence-based inference causal inference, we build on the following two facets of *epistemological emergence* which we find fruitful in the context of this paper:

(E1) *Unpredictability*: Macro-level features of a complex system are emergent if they could not have been predicted from the most thorough knowledge of the system's atoms.

(E2) *Irreducibility*: Macro-level features of a complex system are emergent if they cannot be reduced to fundamental physical theory.

(E1) and (E2) are formulated having in mind physical arrangement or biological (self-)organization and the like (e.g., the formation of crystalline structures). For the purpose of defining what it means to be an emergent feature of a computer simulation, we adjust and combine (E1) and (E2).

A phenomenon observed in a computer simulation is only due to predictable and repeatable calculations, i.e., to simple inference from the computational model (see also Beisbart and Norton 2012), who make this point precise even for Monte-Carlo simulations). Nevertheless, even if computer simulations cannot be understood as unpredictable in this strict sense, due to their narrowly confined sandbox nature, in many instances predicting a simulation's outcome is *practically* impossible due to the complexity of the modeled system (E2). For sufficiently complex systems, predicting the outcome is only possible by stepwise calculation of the system's evolution, which essentially amounts to running the simulation.

Predictability is often contrasted with *surprise*. As mentioned above, Parke (2014) distinguishes different degrees of surprise a researcher might show when encountering an unexpected result: In some cases, the researcher might even be prompted to revise her initial assumptions. Parke concludes therefore that, even if computer simulations can only be called *simple inference from the computational model*, there is no general epistemic privilege of real experiments over simulations concerning the capacity to generate surprise. What happens in the case study is precisely that the computer simulation points to a gap in the theoretical assumptions underlying the model itself and triggers further refinement. This is particularly evident in the second modelling attempt.

On the basis of these considerations we propose to merge (E1) and (E2), slightly adjusted, into a combined concept of emergence from computer simulation, in line with our case study:

(E$*$) Macro-level features of computer simulations are emergent i) if they are not describable in lower-level R-terms (i.e., if the higher-level description goes beyond R-vocabulary), ii) if they are not reducible to lower-level R-terms other than by presenting the full protocol of the simulation (or, respectively, an essential part thereof), and iii) if they are not predictable from the components statically (but only by running the simulation or tracing the protocol).[24]

Figure 8 illustrates the ingredients and their relations: Fig. 8a abstractly links the evolution of the simulated system (s_1, s_2, ...) to patterns on a macro-level that are not predictable from the static R-rules or from a snapshot of the R-parameters and not reducible to R-vocabulary, nevertheless generated by aggregation at the lower level.

[24] This concept of emergence is related to Bedau's concept of *weak emergence*, where a macrostate of a system is called "weakly emergent" if and only if it can be derived from the microdynamics of this system only by way of simulation (Bedau 1997). Our definition goes beyond Bedau's concept, though, in adding language (i.e., level-specific vocabulary) to mark the distinctive gap between what "macro" and "micro" refers to. Explicitly marking this gap allows for distinguishing causal inference as *combining atomic causal assumptions into compound causal claims* from causal inference as *recognizing behavioural patterns on the emergent level*. We believe this is an important distinction since it illuminates how inference from *genuinely* emergent phenomena in our sense may reach beyond language-bound causal learning algorithms.

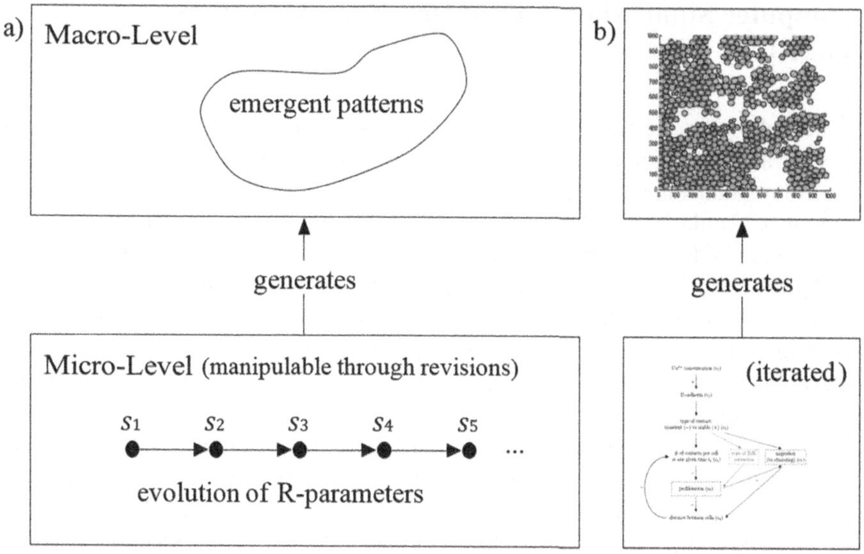

Fig. 8 Emergent patterns are induced through the evolution of the implemented system and only indirectly controllable through lower-level parameters. The figure includes material reprinted from Walker et al. 2008 under the terms of the Creative Commons Attribution License (http://creativecommons.org/licenses/by/2.0)

Fig. 8b interprets our case study along the abstract scheme: Iterative calculations of the systems parameters (encoding individual "social behaviour" of the cells) generate structural patterns on the macro-level which are compared at this higher level with empirical data (and found to be in "qualitative agreement" during validation of the third model).

Once condition α inductively warrants valid causal inference from compound causal information across revisions, emergent phenomena (as defined in E$*$) facilitate causal inference in a special way: not only are emergent phenomena grounded in the atomic rules and their interaction, but they truly only arise in the dynamic evolution of the implemented system.

Since the atomic level is not set up in macro-level terms, though, emergent phenomena cannot be directly subjected to intervention – *in other words, macro-level behavior can only be influenced indirectly by manipulating the basic micro-level.* Yet, although the relation between the codebase and higher-level observations can be very opaque, emergent phenomena are nevertheless *generated* by aggregation/interaction of atomic rules. So, when running a simulation leads to the detection of emergent events, objects, or patterns, the simulation acts as a *guide to causal understanding*, because – although systematic control over emergent cause-effect relations might not be feasible – perceived macro-level phenomena will possibly suggest discriminating tests (or revisions, respectively) at the atomic level. In this way, emergent phenomena help in tracing the interplay of the target system's mechanisms (and sub-mechanisms): micro-macro interaction.

Computer Simulation, Causal Discovery Algorithms, and RCTs

In the following, we want to look at the details of the epistemic role of computer models and simulations in causal inference by comparison with causal learning algorithms and randomized controlled trials. In this comparison, we understand computer simulations along the lines presented above, i.e., simulations based on scientific models of underlying mechanistic relations.

By "causal learning algorithms" we especially refer to automated inference techniques in the spirit of Pearl's IC∗ algorithm (cf. Pearl 2000) or the SGS, PC, and PC∗ algorithms developed by Spirtes, Glymour, and Scheines (cf. Spirtes et al. 2000). These procedures were created for the task of 'discovering' causal (Bayesian) networks from a joint probability distribution over a set of variables (joint frequencies, respectively). Their input can be a database, their output will be a network structure, or – to be precise – a class of (observationally equivalent) networks, that can only be narrowed down further if additional knowledge is available.

The third column in the table below looks at randomized controlled trials (RCTs) which have a certain methodological overlap with causal learning algorithms in drawing inferences from statistical data. To clarify this overlap and to make the distinctions between simulation, causal discovery, and RCTs transparent, we investigate the following seven dimensions: (i) the question asked by each method, (ii) the target (output) knowledge, (iii) the required input (and especially the causal information required), (iv) the investigative level(s) at which the method operates, (v) the type of dynamics exemplified by each method, (vi) the methodological characterization distinguishing each method, and (vii) the challenges a researcher is facing in the application of simulations, causal discovery algorithms, and RCTs. Table 1 summarizes the comparison.

Table 1 Features of computer simulation, causal learning algorithms, and randomized controlled trials in comparison

	Computer simulation	Causal learning algorithms	Randomized controlled trials
Question asked	What is the underlying mechanism, what are the generative causal dynamics? (micro-macro interaction)	What is the relational structure?	What is the effect (size)?
Target knowledge	"Thick": causation: possible mechanism and dynamics	"Thin" causation: (class of) network(s)	"Thin" causation: statistical value
Causal input	Fundamental mechanistic assumptions	Variable selection, distinctness assumption	Variable selection, no confounding
Inference level	Bottom-up (from unit to population)	Population level	Population level
Inference type	Dynamic	Static	Comparative
Methodology	Generative procedure	Global structure analysis	Local differential estimate
Challenges	Causal assumptions justified?	Sampling sufficient for analysis?	Sampling sufficient for extrapolation?

Question Asked The three methods can be compared and contrasted by looking at the type of question they answer (the teleological dimension): RCTs ask a very specific question about the relation between two targeted variables, namely about the impact of changing the value of one variable on the value of the second variable (the effect size in the study population, ideally transferable to the target population). Causal discovery algorithms take a broader view and look at the relational structure that can be read off a set of statistical data. And computer simulations really ask (i) about the underlying productive mechanism that supports an observable causal relationship between two (or more) variables, and (ii) the causal dynamics generating the observed phenomenon – in asking "how does X cause Y?", this third type of question requires and aims at knowledge of different quality, as we will discuss next.

Target Knowledge When Cartwright criticizes the formal approaches to causation put forward by Pearl or Spirtes, Glymour, and Scheines, she emphasizes the expressiveness of discipline-specific, connotationally rich terminology used to describe and analyze experiments close to practice. Her main point is that much of such "thick", useful detail knowledge is lost when formalizing causal relations in purely graph-theoretical, "thin" terms (cf. e.g., Cartwright 2004). Quite in Cartwright's sense, computational modeling aims at the generation of thick causal knowledge about *how* the parts of a mechanism underlying some phenomenon *possibly* interact to produce the observed behavior *in a dynamic way*. This is in stark contrast to causal discovery algorithms, which are concerned with *actual* data but merely order a set of variables w.r.t. their causal dependencies in thin fashion. The output of an RCT consists in mere difference-making information, i.e., in information about whether and to what degree one variable influences a second. Although this output is only a thin statistical value, the hope is that it might be closer to the ground truth than the information gleaned from discovery algorithms: The RCT was carefully designed to isolate the targeted relation, whereas we might not have much information about the quality of the database the discovery algorithm processes (see also Osimani et al. forthcoming for the related discussion about the distinction between real and virtual clinical trials).

Causal Input All three methods can be used for causal inference in one way or another, but interestingly, all three also require causal pre-knowledge as input in one way or another: Computer simulations are based on fundamental mechanistic assumptions and allow for causal insights at the macro-level, based on causal knowledge of the micro-level (see our anchoring conditions above). A similar finding is true for causal discovery algorithms: The input, a frequency or probability table, encodes basic causal assumptions in providing information about variables assumed to be causally sufficient for describing the targeted phenomenon and sufficiently distinct to permit representation in an interventionist Bayes net framework (with the Markov condition being a central modeling criterion). In contrast, causal pre-knowledge plays only a minor role in designing and evaluating RCTs: The goal of RCT methodology is precisely to avoid causal assumptions as much as possible and neutralize (or minimize), through randomization, all potential confounding factors. Nevertheless, implicit causal pre-knowledge enters in the choice of the variables and in their definition (e.g., the causal fact that all confounders are included in the variable set), and more obviously, in extrapolation.

Inference Level In this respect, computer simulation and RCTs stand in clear contrast: If the computational model encodes fundamental mechanistic entities and relations, the computer simulation takes the atom (or the unit) as a starting point and arrives at the population level in a bottom-up fashion, whereas RCTs necessarily operate at the population level, because they essentially require randomization and averaging. Causal discovery algorithms are agnostic as to at what level the input data are harvested, but in the context of systems biology or pharmacological research, the data typically capture population-level probabilities (i.e., frequencies) only.

Inference Type As we showed in our case study above, computer simulation techniques can truly be called *dynamic* inference methods: They simulate the dynamic behavior of a population by iteratively updating the virtual version of the targeted system (for an arbitrary amount of steps). Inference would not be possible without this dynamic property. They can also be called *dynamic* in an additional respect: For the purpose of validation, the computational model must be revised dynamically in order to increase the adequacy of the model and the predictive accuracy of the simulation. Causal discovery algorithms, on the other hand, can be called *static* in all respects: They take as input a static table and deductively compute a structural representation of this input. Our third causal inference method, RCTs, might be called *comparative*: The effect size as output is evidently calculated from statistical data, but much effort goes into defining the control and the treatment group (to be understood as an *intervention* in the population) in order to bring out the targeted contrast.

Methodology As we illustrated in the points above, computational simulation is indeed a *generative procedure*: It generates dynamic behavior from fundamental units and thereby generates a possible picture of the mechanism responsible for the observed phenomenon. In doing so, the simulation provides a *how-possibly* explanation whose strength must be tested in validation cycles. Causal discovery algorithms produce structures, but we would not call these algorithms "generative procedures", since they deductively calculate a network from a database. In this sense, they should rather be called "structure analysis", since the graphical structure they return as an output is contained in the probabilistic dependencies and independencies between the variables in the database. In contrast with RCTs, such causal discovery algorithms look at a set of variables in a 'global' way. RCTs focus on a specific causal link in a 'local' way – 'local' because (i) they are concerned with the (differential) influence of one variable on a second, and (ii) because they are conducted for a local set of population characteristics.

Challenges All three inference methods come with their own challenges that are closely related to the question of how their results can be validated. Computer simulations face the problem of justifying the causal assumptions on which their simulated evolutions rest. We showed above how the computational model as a theoretical basis can be validated in the cycle of simulation, experiment, and revision. In the case of causal discovery algorithms, the question arises whether the statistical data are sufficient for an analysis (in narrowing down the output class of networks as far as possible), but validation as such is not in the focus, because such discovery algorithms arrive at their results in a deductive way. One can of course ask additional questions about the data source, e.g., if the sampling is a valid representation of a larger population. This is precisely the kind of question, an RCT must answer before its results can be extrapolated: Once it is proven internally valid (i.e., the causal relation is indeed sufficiently isolated), the external validity of the trial must be checked by evaluating its representativeness (of a target population).

This comparative list makes the distinctive epistemic role of modeling and simulation transparent and conceptually demarcates its inferential process from other techniques aimed at learning about the causal features of our world. The added value of computational simulation consists in going beyond the phenomenological surface and providing insights into lower-level mechanistic processes. As a generative procedure, this technique certainly demands quite some scientific creativity of the researcher, but this precisely unlocks the capacity for scientific discovery from emergent phenomena. As a bottom-up constructive method, computational simulation is able to deal with complex phenomena which otherwise evade straightforward data-driven analysis. Nevertheless, this complexity and the intricate relation between unit and population-level behavior need not remain obscure, as we argued above by formulating our refined understanding of emergence.

Causal Inference from Modeling and Simulation

The philosophical literature on modelling is increasingly vast, however clear formal analyses of computational modelling in systems biology are still lacking. In many instances, the inferential power of computer simulations for causal assessment is called into question by comparing simulations to real experiments – in a misleading way, we think. As quoted above, Diez Roux summarizes her discussion of the epistemic value of computer simulations as follows:

> In the simulated world, everything is counterfactual in the sense that the world and all possible scenarios are artificially created by the scientist. (Diez Roux 2015, p. 101).

As a possible response to the charge of vacuousness, natural variability might be mimicked by introducing randomness in the computational model, as, e.g., in the case of Monte Carlo simulations. Yet again, are those simulations capable of delivering truly *novel* results? Beisbart and Norton (2012), for example, dismiss the idea of computer simulations as representing any kind of experiment. Their claim: Computer simulations represent inferences and not experiments, namely inferences "by ordinary argumentation of conclusions from assumptions built into the simulations" (cf. Beisbart and Norton 2012, p. 403). In their view, this extends to Monte Carlo simulations:

> It is true that Monte Carlo simulations appear, initially, to be quite different from other simulations because of the essential role of randomness. This appearance is deceptive, for randomness can have no role or only a minimal one in the final results if they are to be credible. Otherwise Monte Carlo simulations would be little better than forecasting by reading tea leaves or shuffling tarot cards, whose results are mostly expressions of random chance. The results of a good Monte Carlo simulation are not. They are repeatable and reliable. (Ibidem).

And they go on to comment on the idea of gathering novel evidence from computer simulations:

> In real experiments, the unpredictable and novel character of the data is the source of the final experimental result. In a Monte Carlo simulation, the unpredictable and novel character of the random numbers is minimized and plays almost no role in the final result. (Ibidem).

In other words, a phenomenon observed in the computer simulation is nothing but simple inference from the computational model. Nevertheless, to draw the conclusion from these observations that computational modeling cannot contribute to causal learning, would be false, as we argue with this paper: We emphasize the iterated interplay between scientific knowledge as input, modeling and simulation, and revision of the computational model in the dialog between virtual and experimental data-gathering, geared towards validation of the scientific assumptions. In our comparison of modeling and simulation with causal discovery algorithms and RCT methodology (Table 1), we carved 790 out the distinctive epistemic features and specific advantages of agent-based computer simulations: The bonus is twofold – not only offers modeling and simulation insights into the phenomenological black-box, it also allows tracing the causal dynamics at work.

As an illustration of these epistemic advantages, we took as a case study, the discovery of the functional properties of a specific protein, E-cadherin, which seems to have a key role in metastatic processes by way of influencing cell growth and proliferation signalling. The case study was presented in three stages, following the development of the computational model from its purely agent-based bottom-up version to a refined, multi-scale implementation and on to a final single-scale, agent-based formulation with the local insertion of a phenomenologically motivated rule (representing the exclusive role of temporal contact length in reproducing the growth patterns observed experimentally).

To capture the model's development, we followed Margaret Morrison in emphasizing the role of the computational model as object of inquiry, to be subjected to formal experimentation. The case study is particularly interesting in this respect, because reconstructing the process revealed both the chosen strategy towards validating the computational model and the value of the intrinsic link between the empirically grounded modular codebase and observations at the aggregate level.

A central part of our general reconstruction of a computer simulation's design process is the simulation–validation–revision cycle. In order to fruitfully integrate the phases of this cycle, we supplemented our scheme with formal notation as a potential basis of a protocol for valid scientific inference from computer simulations. By defining an anchoring condition, we explicated the formation of empirically grounded aggregate knowledge from atomic information. We sketched a robustness measure for evaluating a simulation's quality in terms of multi-dimensional accuracy and utilized belief revision terminology to specify how the codebase may be changed while preserving empirical adequacy (via anchoring).

We ultimately arrived at an analytic characterization of emergence that goes beyond outlining semantic constraints for scientific talk about emergent phenomena – our characterization demarcates different information-bearing levels and bridges this gap in showing how observations of emergent phenomena can fruitfully be utilized for directly testable causal inference despite their higher-level mediate nature. In lining up all ingredients for a validation–revision protocol we illuminated how well-designed computer simulations – though lacking the material link to their target systems as virtual, artificial constructs – are justified guides to a mechanistic understanding of biological systems.

Appendices

Appendix A: Rules Dictating Cell Behaviour

The set of relevant actions for cells in the model are: (1) bonding (to other cells and to substrate), (2) spreading (on the substrate), (3) migration (once bonded to the substrate), (4) apoptosis (if failure to bond to substrate takes place), (5) cell growth, and (6) division (mitosis). In addition to the internal parameters, each cell is assigned a set of parameters that defines the exact location of its centre in x, y, and z Cartesian coordinates. The agents (the cells) progress through the cell cycle in a sequence of phases: the interphase (I) – which comprises the growth phase 1 (G1), the synthesis phase (S), and the growth phase 2 (G2) – and the division (mitosis, M) and reproduction phase; the G0 phase represents a state of "quiescence" where the cell is neither dividing itself, nor preparing to divide, but merely performing its biological functions (see Fig. 9).

The ABM rules governing cell progression and proliferation are formalized as illustrated in Table 2 (cf. Walker et al. 2004a) and related to biological behaviour below (Fig. 10).

Bonding The rule for bonding is based on published data (Baumgartner et al. 2000), according to which the probability of two adjacent cells bonding to each

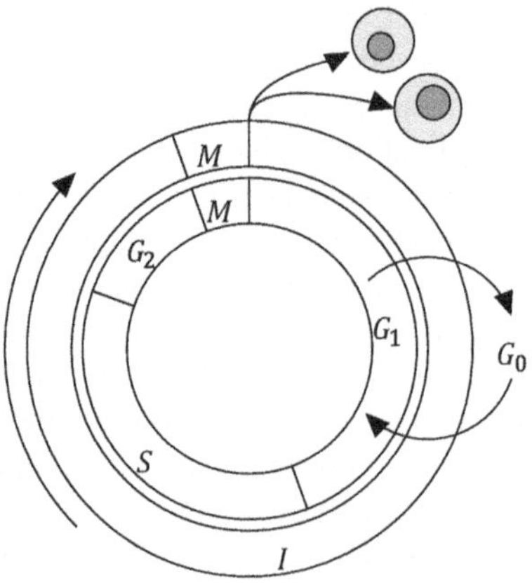

Fig. 9 Cell cycle: cell growth and division consist of two main phases: the interphase (I) during which the cell grows and synthesises a copy of its DNA in order to replicate, and the mitotic phase (M) during which the cell divides up into two daughter cells. The interphase is subdivided into the G1 phase, that is the first growth phase; the synthesis phase (S) during which its DNA is replicated in the nucleus, and the G2 phase during which it further grows before it enters the mitotic phase

Table 2 Summary of the agent based model rules where agents represent cells and their actions in the cell cycle phases

Current state		Action
Cell is in G1		Increase cell volume by set amount; Increment internal counter by +1
Cell is at G1/G0 check-point	if cell is bonded to 4 or more neighbours	Enter G0
	If cell is not sufficiently spread (radius <1.5x 'rounded' radius)	
	ELSE	Continue in G1
Cell is in G0	If cell is bonded to less than 4 neighbours	Re-enter G1
	If cell is sufficiently spread	
	ELSE	Remain in G0
Cell is in S/G2		Increment internal counter by +1
Cell is at G2/M boundary		Change to mitotic cell type
Cell is at end of M phase		Divide to create two identical daughter cells

other is inversely proportional to their distance, and directly proportional to the concentration of calcium ions in the culture (see Fig. 2).

Spreading Cells need to bond to the substrate in order to proceed in cycle progression, otherwise they enter into apoptosis; in order to bond to the substrate, they form numerous attachments thus assuming a flattened morphology. "This process is *emulated* in the model by increasing the surface area of attached cells by a small amount

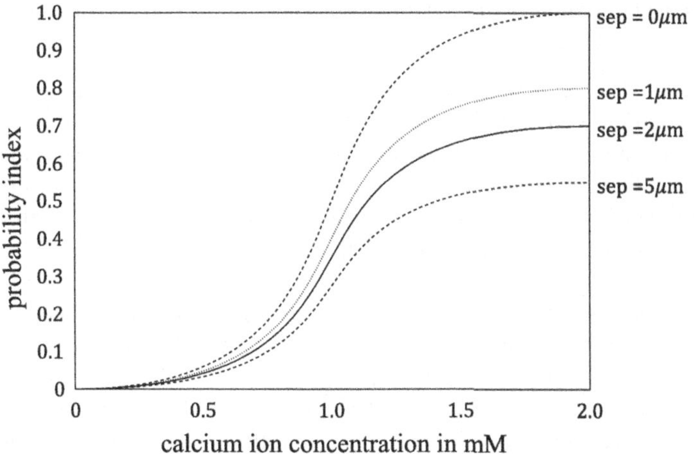

Fig. 10 Probability densities corresponding to four cell:cell distance ranges [0 μm, 1 μm, 2 μm, 5 μm] as a function of the Calcium concentration in mM (cf. Walker et al. 2004b)

at each iteration, whilst maintaining a constant volume. Cells continue to spread until their radius has increased by a factor of approximately 1.5" (Walker et al. 2004a), see also (Nelson and Chen 2002).

Migration Migration is also modelled on the basis of experimental observations: the allowed migration distance, 2.5 times the cell radius, was obtained from time-lapse recording of a wound closure assay. Direction of movement is chosen randomly within a 60° angle of the original trajectory; and if movement in the chosen direction is blocked by another cell, then the direction is changed up to 10 times, after which the cell will remain stationary.

Apoptosis Cells that fail to bond to the substrate before the inception of the first growth phase (G1) die (they enter apoptosis). Dead cells are known to be fragmented and phagocytosized by their neighbours. In models of cell growth, dead cells shrink and are removed from the model. However, for specific epithelial cells, such as urothelial cells, apoptosis rarely applies, and therefore this rule is not implemented in models for this kind of cells, as the one considered here.

Mitosis The final cell cycle phase results in cell division and replication. During cell division (mitosis), cells have been observed to round up and thus lose adhesion with their neighbours and with the underlying substrate to some extent; hence model cells undergoing the mitotic phase are assumed to alter their morphology from ellipsoidal to spherical and to break up all their bonds with neighbouring cells, eventually dividing up into two spheres. New cells are spherical and have no bonds with other cells. Furthermore, they do not form new bonds or migrate until mitosis is complete.

From the point of view of modeling cell proliferation, "quiescence" and "contact inhibition" are also essential.

Quiescence The cell cycle has a checkpoint, a sort of "bifurcation point" after which, based on its morphology at that given time, and on the number of contacts the cell has formed with neighbouring cells, it either proceeds through the remainder of the cell cycle and reproduces, or becomes quiescent, that is, stops proliferating. Quiescence may be caused by "contact inhibition"; this means that, after a certain amount of contacts with other cells, a cell stops proliferating.

Contact Inhibition In the model, the contact inhibition rule dictates that a cell becomes quiescent once it has ≥ 4 bonds with neighbouring cells (or, alternatively, if it has 3 bonds and is adjacent to the "edge" of the substrate). Quiescent cells remain in this state until enough of their intercellular bonds are broken so as to allow them to re-enter the cycle. This rule also follows observations showing that in the culture of epithelial cells, those situated at the centre of the colonies *in physiological or high calcium media* withdraw from the cell cycle and become quiescent. Since physiological calcium is necessary for calcium-dependent physical bonds such as those created by E-cadherin, it has been hypothesized that cells become "aware" of the presence of nearest neighbours through intercellular signalling, via E-cadherin bonds forming the intercellular matrix.

At each tick of the simulation, each cell is interrogated, and its response depends on the combination of states of all the above parameters.[25] The formation of extracellular E-cadherin bonds depends on the presence of calcium ions, therefore, the action of intercellular signalling on cellular growth is studied in experiments by putting cells in cultures with a physiological (around 2 mM) vs. a low concentration of calcium ions (around 0.09 mM), which is assumed to create a situation of higher vs. lower number of intercellular contacts (and, therefore, of growth inhibition vs. proliferation, respectively).

The simulation was run under the two conditions of high vs. low calcium environment (with related bonding probability) in order to verify whether these different settings would reproduce the kind of distinct behavioural patterns observed experimentally in high vs. low calcium environments.

[25] A numerical technique is applied to correct for any overlap between neighbouring cells that may have resulted from cell growth, spreading, or division during the previous agent based iteration: these may consist for instance in identifying any overlap between every pair of neighbouring cells and applying a force proportional to this overlap to separate them (Walker and Southgate 2009).

Appendix B: The Causal Structure Underpinning the Set of Rules

The assumed causal structure underlying the dynamics of cell growth through "contact inhibition" is presented in the left diagram of Fig. 3. The graph represents the hypothesized causal structure underpinning the phenomenon of cell proliferation as higher-level behaviour resulting from cell:cell interactions; this can be considered as the theory that needs confirmation through modelling, simulation and comparison with experimental data ("validation"). The main causal variables in the structure are:

- c_1) the characteristics of the environment in terms of Calcium ions (Ca^{2+}) concentration (physiological vs. low calcium condition);
- c_2) quality and functionality of the E-cadherin present in the cell culture as a consequence of the concentration of calcium ions;
- c_3) type of cell:cell contacts: stable (as promoted by the presence of E-cadherin) vs. transient;
- c_4) number of cell:cell contacts (for any given cell at any given time the cell goes into quiescence and ceases to proliferate whenever it is in contact with ≥ 4 other cells);
- c_5) cell proliferation as a result of the division and replication of each cell in the population;
- c_6) distance between cells as a result of the topology determined by cell proliferation; reduced distance increases the probability of contacts which in turn determine quiescence (i.e., inhibit proliferation) when they reach the number of four in each cell;
- c_7) migration is inhibited both by cell:cell contacts and is made more and more difficult as the population grows; in turn, migration also contributes to determining the distance between cells by changing the topology of the cell population at each iteration.

The culture medium (high vs. low concentrations of Ca^{2+} ions) determines a higher vs. low capability of E-cadherin to form contacts, which in turn determines the presence of intercellular junctions, contributing to stable vs. transient contacts respectively. The *number* of contacts that any given cell has at any time is obviously a function of how stable they have been up to then. In turn, a high number of contacts (≥ 4) signals the cell to stop proliferating.

Proliferation is part of a negative feedback in that the higher the proliferation, the higher the population density, and therefore the shorter the distance among cells on average, which leads to a higher probability of cell:cell contacts and therefore to inhibition of proliferation. Proliferation also negatively impacts on migration (high density of cells inhibits their movement), whereas the presence of contacts inhibits migration (and thereby promotes clustering).

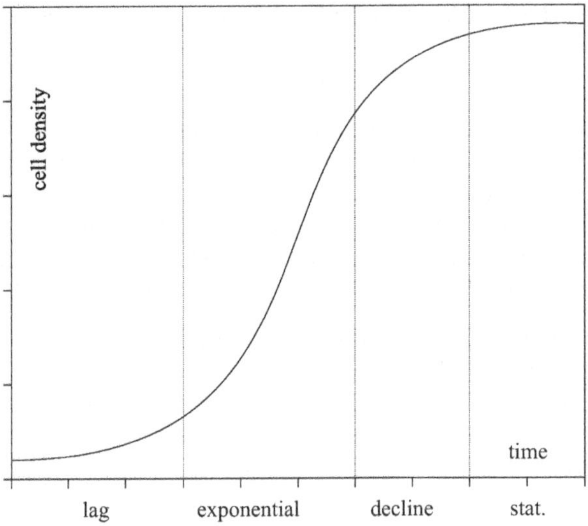

Fig. 11 Cell growth function with lag phase, exponential phase, decline, and stationary phase

Appendix C: Cell Growth Benchmark

Appendix D: Modifications of the ABM Component for the Second Model

In the second model, the previous ABM component is not only integrated with an ODE system, but it is also modified as follows.

E-Cadherin Mediated Contacts Whereas in the previous model, bonding probability was straightforwardly determined by the concentration of calcium ions in the medium and by the distance among cells, in the new version of the model the rule is expanded so as to qualitatively reflect more adequately observations in (Baumgartner et al. 2000; Chu et al. 2004). In addition to calcium concentration in the medium and distance between the cells, bonding probability is also determined by "adhesion probability", which is a function of the cell E-cadherin expression level: this varies between 1 (normal expression) and 0 (no expression, e.g., in malignant cells).

Transient cell:cell Contacts Whereas the first model only took into account the number of contacts as a basis for determining cell cycle progression, the new one also accounts for the contact *mode* (stable vs. transient) and allows for the formation of prolonged contacts also in low calcium concentration media (although significantly less in number). This makes it possible for cell:cell signalling to occur also through non-E-cadherin mediated contacts, although much less frequently so.[26] Moreover, E-cadherine mediated contacts may be also formed in low calcium concentrations environments, but, according to the bonding probability rule (see previous paragraph and Fig. 3), they too are much less frequent than in environments with physiological concentrations of calcium.

Migration Rules Whereas in the first model, cells where allowed to migrate only when they had 0 contacts with other cells, in the second one, migration results from the difference of strength between the "active migration force" possessed by a cell, and the intercellular "adhesion force" which binds it to other cells (this being a function of the stability/transiency of the contact, itself mediated by E-cadherin expression levels). When the net adhesion force is less than 20 nN, then the cell migrates.

Appendix E: Testing the Modelling Assumptions of the Second Model

When the behavior displayed by a simulation model contradicts the expected results, then various, *non-mutually exclusive* sources of discrepancy can be taken to account for this mismatch (see section "Verification, validation, revision" for a systematic discussion of such sources of discrepancy). Generally, a broad distinction between "internal validity" (generally referred to as "model verification") and external validity ("model validation") is drawn. The former regards the formal (mathematical/logical) consistency of the model itself; the latter its empirical adequacy. Scientists generally proceed by first testing the former, and, if no sources of error are found, then they go on to challenge the theoretical model underpinning the computational one, and change it more or less drastically, on the basis of explanatory hypotheses for the obtained results, and of new empirical knowledge possibly acquired in the meanwhile.

In this "Lakatosian" process, changes to the model follow an ordering which depends on the steadiness of the theoretical/empirical knowledge on which the model is based. The parts of the model which are more insecure are tested for their individual theoretical solidity and/or empirical adequacy; if these "pass the test", then other parts of the model are put to the test. However, given the holistic aspect

[26] Furthermore, whereas the old model only took into account the number of "non-dimensional" contacts, the new model also takes into account the physical length of the bond. The rationale for this rule would be that the longer the bond, the more receptors and ligands enter into contact and activate the intracellular signalling pathway. This aspect of cell:cell signalling is taken into account in Eq. 6 of the ODE model. However, as the individual cell:cell simulation will show, only the temporal length of the contact is relevant for patterns of Erk activation, whereas the spatial length is irrelevant. See section "Second model: integration of the first agent-based model and an ODE system into a multiscale model".

of computational modeling and simulation, it may happen that something goes wrong with the overall simulation, even if all its components "pass the test"; in this case the problem may lie in how such components are combined – this is especially true for multi-scale models, because of possible inconsistencies arising in the transformation of one submodel's outputs into the other submodel's inputs.

Instead, if the revised model delivers the expected results, then this is considered to provide evidence that the mechanisms underlying the target system it represents are analogous to the rules underpinning the simulation model. How exactly?

The final part of the case study will help us identify the essential epistemic tools used in model revision and confirmation. These will be formally explicated in the third section of the paper.

Test 1: Sensitivity Analysis of the Mathematical Model In our case, the scientists started by doing a sensitivity analysis of the mathematical model representing the signaling pathway:[27] each of the 50 parameters in the model was systematically varied by first dividing, then multiplying the value of each parameter by a factor of two.[28]

Although both amplitude and duration of the Erk-PP signals varied when the kinetic parameters (rate constants) were altered, none of the parameter variations affected the fundamental result that "transient" contacts produce a relatively large but transient response, compared to the smaller but sustained response associated with slowly growing stable contacts. Hence, the simulation result could not be ascribed to the particular contingent combination of parameter values that underpinned it in that specific setting. Interestingly, the *duration of the signals* was relatively insensitive to the model parameters: this means that the *temporal length of the signal, rather than its intensity*[29] was a robust predictor of the model outputs.

Test 2: Average Number and Spatial Length of E-Cadherin Mediated Contacts in ABM Another verification test regarded the information provided by the ABM component of the model to the mathematical model: number and length of the intercellular contacts in the two environmental conditions. This served the purpose of establishing whether through iterations, the ABM component would systematically deliver different initial conditions to the mathematical model depending on whether the simulation was run in low vs. high calcium conditions. The test showed indeed different patterns of growth in the number and in the perimeter of E-cadherin mediated contacts in the two settings (see Fig. 12).

[27] The signaling pathway model was solved both for a "transient" contact (with a fixed size of 15 μm) and for a "stable" contact (starting with 3 μm size and constantly growing).

[28] The rationale for testing the model for such a large perturbation was theoretically/empirically grounded on previous studies related to rate constants for binding, dissociation and internalization of epidermal growth factors (Waters et al. 1990).

[29] I.e., number of involved receptors and ligands on the cell surface at any point in time.

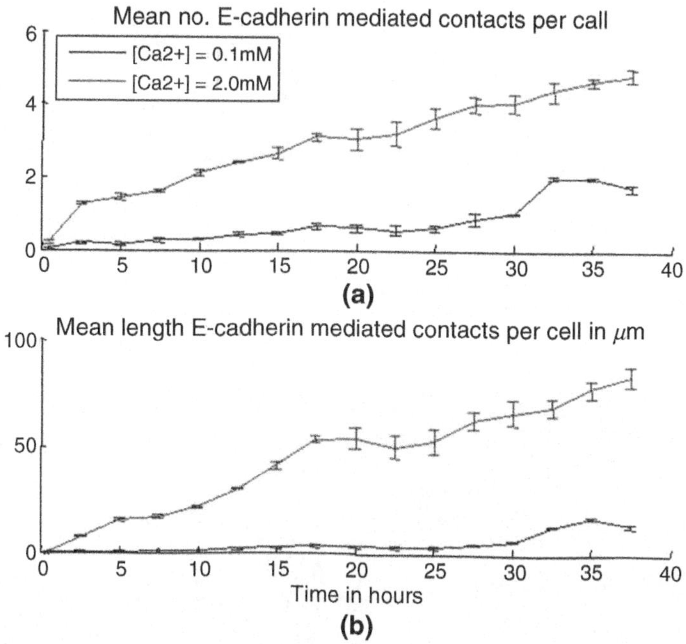

Fig. 12 Mean number (a) and mean total perimeter length (b) of E-cadherin mediated contacts per cell in high calcium (red) vs. low calcium (blue) environment. Figure reprinted from Walker et al. 2008 under the terms of the Creative Commons Attribution License (http://creativecommons.org/licenses/by/2.0)

Also tests for empirical adequacy were carried out. These served the purpose of verifying the empirical adequacy of at least some parts of the model.

Test 3: Partial Validity Test for Proportion of E-Cadherin Mediated Contacts in the Two Settings The revised version of the ABM component in the model prescribed "more realistic rules" for the formation of E-cadherin bonds, e.g., by allowing these to be formed also in low calcium conditions (although less frequently so). The incidence of E-cadherin mediated contacts in the two simulated settings was found to be in good qualitative agreement with the distribution of cells with low/moderate/high Erk activation in experimental settings. This is not a direct validity test, but one which draws on explanatory reasoning: The immunofluorescence of the distribution of Erk-PP in urothelial cells maintained in low vs. physiological Ca^{2+} conditions shows that in all cells there is some degree of Erk activation, but that there is also a clear heterogeneity between levels of Erk activation in the two cultures. Hence it is suggested that, in addition to the EGF stimulation explaining the activation of Erk tout court, the contact *mode* might be invoked as a possible explanation for such an heterogeneity of activation levels. And the most proximate cause for such qualitative difference is traced back to stable vs. transient contacts, as determined by the presence vs. absence of E-cadherin respectively.

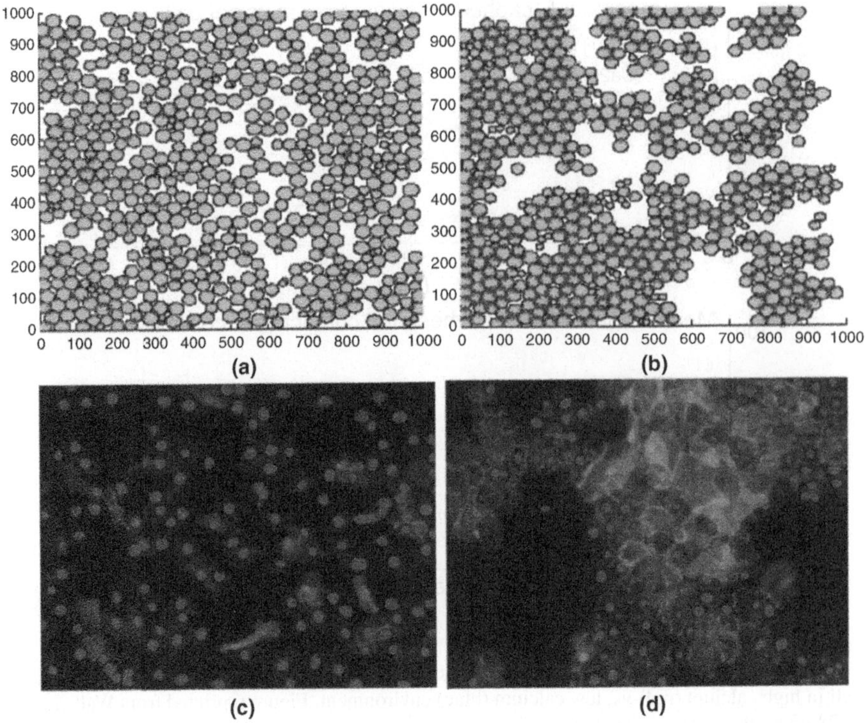

Fig. 13 Comparison of the distribution of E-cadherin in low vs. high calcium environment in computer simulation (a vs. b) and microscopy images (c vs. d). In the simulation, E-cadherin mediated contacts are in red; in the microscopy images, they are identified by immunofluorescence: cells are labelled with an antibody that specifically recognises E-cadherin (green areas of the picture). Figure reprinted from Walker et al. 2008 under the terms of the Creative Commons Attribution License (http://creativecommons.org/licenses/by/2.0)

Test 4: Validity Test for the Topological Distribution of E-Cadherin Mediated Contacts in the Two Settings Figure 13 show the qualitative agreement between the distribution of E-cadherin after 50 iterations in the simulated high vs. low calcium settings. These agree qualitatively with the microscopy images of normal urothelial cell cultures cultured in low vs. physiological calcium environments.

Test 1 establishes that the ODE system is working properly, and test 2 says that the ABM component is delivering the right inputs. So maybe the problem lies in the combination of the two. Tests 3 and 4 show that the role of low vs. high physiological calcium levels for the formation of E-cadherin and for Erk activation agree in the source and the target system. These constraints, together with the uncontested role of contact inhibition and with a deeper analysis of the inconsistency between the individual level and the population level simulation, provide the basis for the third model.

References

Adams, C.L., Y.-T. Chen, S.J. Smith, and W.J. Nelson. 1998. Mechanisms of Epithelial cell–cell Adhesion and Cell Compaction Revealed by High-Resolution Tracking of E-Cadherin–Green Fluorescent Protein. *The Journal of Cell Biology* 142 (4): 1105–1119.

Baumgartner, W., P. Hinterdorfer, W. Ness, A. Raab, D. Vestweber, H. Schindler, and D. Drenckhahn. 2000. Cadherin Interaction Probed by Atomic Force Microscopy. *Proceedings of the National Academy of Sciences* 97 (8): 4005–4010.

Bedau, M.A. 1997. Weak Emergence. *Philosophical Perspectives* 11 (s11): 375–399.

Beisbart, C., and J.D. Norton. 2012. Why Monte Carlo Simulations Are Inferences and not Experiments. *International Studies in the Philosophy of Science* 26 (4): 403–422.

Brightman, F.A., and D.A. Fell. 2000. Differential Feedback Regulation of the mapk Cascade Underlies the Quantitative Differences In egf and ngf Signalling in pc12 Cells. *FEBS Letters* 482 (3): 169–174.

Cartwright, N. 1999. *The Dappled World: A Study of the Boundaries of Science*. Cambridge: Cambridge University Press.

———. 2004. Causation: One Word, Many Things. *Philosophy of Science* 71 (5): 805–819.

Casini, L., and G. Manzo, n.d. Agent-Based Models and Causality – a Methodological Appraisal (unpublished manuscript).

Chu, Y.-S., W.A. Thomas, O. Eder, F. Pincet, E. Perez, J.P. Thiery, and S. Dufour. 2004. Force Measurements in E-Cadherin–Mediated Cell Doublets Reveal Rapid Adhesion Strengthened by Actin Cytoskeleton Remodeling Through rac and cdc42. *The Journal of Cell Biology* 167 (6): 1183–1194.

Dada, J.O., and P. Mendes. 2011. Multi-scale Modelling and Simulation in Systems Biology. *Integrative Biology* 3 (2): 86–96.

Dardashti, R., K.P.Y. Thébault, and E. Winsberg. 2014. Confirmation via Analogue Simulation: What Dumb Holes Could Tell us About Gravity. *British Journal for the Philosophy of Science*.

Dardashti, R., Hartmann, S., Thébault, K.P.Y. and E. Winsberg. 2015. *Confirmation via Analogue Simulation: A Bayesian Analysis* (unpublished manuscript).

DeWitt, A.E., J.Y. Dong, H.S. Wiley, and D.A. Lauffenburger. 2001. Quantitative Analysis of the egf Receptor Autocrine System Reveals Cryptic Regulation Of Cell Response by Ligand Capture. *Journal of Cell Science* 114 (12): 2301–2313.

Diez Roux, A.V. 2015. The Virtual Epidemiologist – Promise and Peril. *American Journal of Epidemiology* 181 (2): 100–102.

Downes, S.M. 1992. The Importance of Models in Theorizing: A Deflationary Semantic View. *PSA: Proceedings of the Biennial Meeting of the Philosophy of Science Association* 1992: 142–153.

Frigg, R. 2006. Scientific Representation and the Semantic View of Theories. *Theoria* 21 (1): 49–65.

———. 2010a. *Fiction in Science, in: Fictions and Models: New Essays*. Munich: Philosophia Verlag.

———. 2010b. Models and Fiction. *Synthese* 172 (2): 251–268.

Frigg, R., and J. Reiss. 2009. The Philosophy of Simulation: Hot New Issues or Same Old Stew? *Synthese* 169 (3): 593–613.

Gärdenfors, P. 1988. Knowledge in Flux. Cambridge (MA): The MIT Press.

Georgopoulos, N.T., L.A. Kirkwood, D.C. Walker, and J. Southgate. 2010. Differential Regulation of Growth-Promoting Signalling Pathways by E-Cadherin. *PLoS One* 5 (10): e13621.

Hartmann, S. 1996. The World as a Process: Simulations in the Natural and Social Sciences. In *Modelling and Simulation in the Social Sciences from the Philosophy of Science Point of View*, ed. R. Hegselmann. Dordrecht: Kluwer Academic Publishing.

Humphreys, P. 2004. *Extending Ourselves: Computational Science, Empiricism, and Scientific Method*. Oxford: Oxford University Press.

———. 2009. The Philosophical Novelty of Computer Simulation Methods. *Synthese* 169 (3): 615–626.

Kholodenko, B.N., O.V. Demin, G. Moehren, and J.B. Hoek. 1999. Quantification of Short Term Signaling by the Epidermal Growth Factor Receptor. *Journal of Biological Chemistry* 274 (42): 30169–30181.

Lauffenburger, D.A., G.T. Oehrtman, L. Walker, and H.S. Wiley. 1998. Real-Time Quantitative Measurement of Autocrine Ligand Binding Indicates that Autocrine Loops are Spatially Localized. *Proceedings of the National Academy of Sciences* 95 (26): 15368–15373.

Morrison, M. 2015. *Reconstructing Reality: Models, Mathematics, and Simulations*. Oxford: Oxford University Press.

Nelson, C.M., and C.S. Chen. 2002. Cell-cell Signaling by Direct Contact Increases Cell Proliferation via a 1100 pi3k-Dependent Signal. *FEBS Letters* 514 (2): 238–242.

Norton, J.D. 1991. Thought Experiments in Einstein's Work. In *Thought Experiments in Science and Philosophy*, ed. G.J.M.T. Horowitz. Savage: Rowman and Littlefield.

Oehrtman, G., H. Wiley, and D. Lauffenburger. 1998. Escape of Autocrine Ligands into Extracellular Medium: Experimental Test of Theoretical Model Predictions. *Biotechnology and Bioengineering* 57 (5): 571–582.

Osimani, B., Poellinger, R. and M. Bertolaso. Forthcoming. Real and Virtual Clinical Trials: A Formal Analysis, *Topoi*.

Parke, E.C. 2014. Experiments, Simulations, and Epistemic Privilege. *Philosophy of Science* 81 (4): 516–536.

Parker, W.S. 2009. Does Matter Really Matter? Computer Simulations, Experiments, and Materiality. *Synthese* 169 (3): 483–496.

Pearl, J. 2000. *Causality: Models, Reasoning, and Inference*. 1st ed. Cambridge: Cambridge University Press.

Reiss, J. 2003. Causal Inference in the Abstract or Seven Myths About Thought Experiments. In *Causality: Metaphysics and Methods Research Project*. LSE.

Sorensen, R. 1992. *Thought Experiments*. New York: Oxford University Press.

Southgate, J., K.A. Hutton, D.F. Thomas, and L.K. Trejdosiewicz. 1994. Normal Human Urothelial Cells In vitro: Proliferation and Induction of Stratification. *Laboratory Investigation* 71 (4): 583–594.

Spirtes, P., C. Glymour, and R. Scheines. 2000. *Causation, Prediction, and Search, Adaptive Computation and Machine Learning*. Cambridge, MA: MIT Press.

Teller, P. 2009. Fictions, Fictionalization, and Truth in Science. In *Fictions in Science. Philosophical Essays on Modelling and Idealisation*. London: Routledge.

Toon, A. 2010. The Ontology of Theoretical Modelling: Models as Make-Believe. *Synthese* 172 (2): 301–315.

Viceconti, M. 2011. A Tentative Taxonomy for Predictive Models in Relation to their Falsifiability. *Philosophical Transactions of the Royal Society of London A: Mathematical, Physical and Engineering Sciences* 369 (1954): 4149–4161.

Walker, D.C., and J. Southgate. 2009. The Virtual cella Candidate Co-ordinator for Middle-Outmodelling of Biological Systems. *Briefings in Bioinformatics*: bbp010.

Walker, D.C., J. Southgate, G. Hill, M. Holcombe, D.R. Hose, S.M. Wood, S. Mac Neil, and R.H. Smallwood. 2004a. The Epitheliome: Agent-Based Modelling of the Social Behaviour of Cells. *Biosystems* 76 (1): 89–100.

Walker, D.C., G. Hill, S.M. Wood, R.H. Smallwood, and J. Southgate. 2004b. Agent-Based Computational Modeling of Wounded Epithelial Cell Monolayers. *IEEE Transactions on NanoBioscience* 3 (3): 153–163.

Walker, D.C., N.T. Georgopoulos, and J. Southgate. 2008. From Pathway to Population – a Multiscale Model of Juxtacrine Egfr-Mapk Signalling. *BMC Systems Biology* 2 (1): 1–17.

———. 2010. Anti-social Cells: Predicting the Influence of E-Cadherin Loss on the Growth of Epithelial Cell Populations. *Journal of Theoretical Biology* 262 (3): 425–440.
Waters, C.M., K.C. Oberg, G. Carpenter, and K.A. Overholser. 1990. Rate Constants for Binding, Dissociation, and Internalization of egf: Effect of Receptor Occupancy and Ligand Concentration. *Biochemistry* 29 (14): 3563–3569.
Weisberg, M. 2013. *Simulation and Similarity: Using Models to Understand the World*. Oxford: Oxford University Press.
Winsberg, E. 2009. A Tale of Two Methods. *Synthese* 169 (3): 575–592.

[illegible]

Can Models Have Skill?

Eric Winsberg

Introduction

Chapter 8 of Working Group 1's report in the AR4, entitled "Climate Models and their Evaluation," contains "FAQ 8.1" which asks: "How Reliable Are the Models Used to Make Projections of Future Climate Change?" The answer begins:

> There is considerable confidence that climate models provide credible quantitative estimates of future climate change, particularly at continental scales and above. This confidence comes from the foundation of the models in accepted physical principles and from their ability to reproduce observed features of current climate and past climate changes. Confidence in model estimates is higher for some climate variables (e.g., temperature) than for others (e.g., precipitation). Over several decades of development, models have consistently provided a robust and unambiguous picture of significant climate warming in response to increasing greenhouse gases. (AR4, WG1, 8.2, FAQ 8.1).

The claim that "confidence comes from the foundation of the models in accepted physical principles and from their ability to reproduce observed features of current climate and past climate changes" is mean to characterize the methods that climate scientists use for evaluating what they often call the "skill" of their models. Climate scientists do not always mean the same thing when they speak of model skill, so when we talk about evaluating climate models, or evaluating a climate model's skill, we will be referring to the activity of evaluating a model's adequacy for particular purposes. In particular, we will be especially interested in philosophical issues that arise in thinking about three sources of evidence we can have for models' skill: their "foundation [...] in accepted physical principles;" their "ability to reproduce observed features of current climate and past climate changes."

E. Winsberg (✉)
Department of Philosophy, The University of South Florida, Tampa, FL, USA
e-mail: winsberg@usf.edu

M. Bertolaso, F. Sterpetti (eds.), *A Critical Reflection on Automated Science*, Human Perspectives in Health Sciences and Technology 1,
https://doi.org/10.1007/978-3-030-25001-0_10

Verification and Validation

Let's start with the first source of model confidence: their foundation in accepted physical principles. What, exactly, does that amount to? The core of all climate simulations is a set of five equations that govern the general circulation of the atmosphere which are, in turn, grounded in unassailable (in the relevant domains) physical principles like conservation of mass, Newton's laws, and the simple thermodynamics of heat transfer. How should we think about the role that these well-accepted principles play? One way of thinking about it is captured by what one might call the "linear model of computer simulation" (to emphasize how we move in a straight line from principles to results). This is closely related to what Nancy Cartwright has called "The vending machine view of theories." Since the linear model has implications for the epistemological significance of the fact that climate models are founded in accepted physical principles, we should scrutinize it here. The model is illustrated in Fig. 1.

The idea is rather simple. We can illustrate it with a simple example: the pendulum. We begin with a theory that governs the phenomenon of interest. For a pendulum, that would be Newtonian mechanics. A basic understanding of the system suggests to us a model of the pendulum. Such a model would consist of an abstract description of the physical system (a point on the end of a massless string), and some differential equations provided by the theory that describe the evolution of the values of the variables in the abstract model. The treatment consists of assigning values to basic parameters – in this case the length of the string, the mass of the bob, and the acceleration of gravity – and assigning initial values to the variables, which in this case would be an initial value for the angle of the pendulum and its angular velocity. Next, model and treatment are combined to create a solver – a step-by-step computable algorithm that is designed to be an approximate, discrete substitute for the continuous differential equations of the model – and the solver is run on a computer to produce results.

Despite the simplicity of the example, this schema captures a great deal about many computer simulations of physical systems. Associated with that schema is a

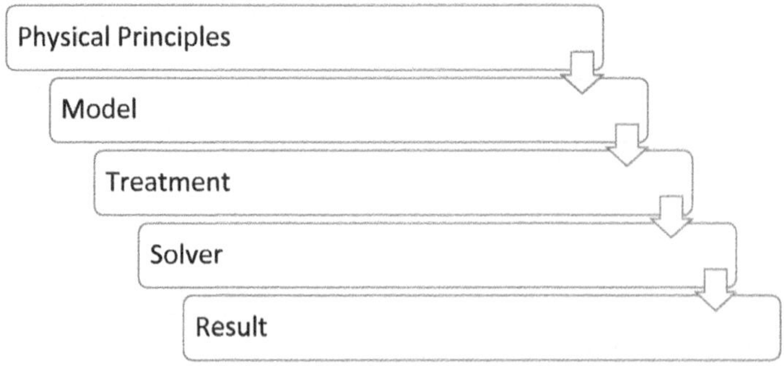

Fig. 1 The "linear model" of computer simulation

conception of the epistemology of simulation that is almost disarmingly simple and straightforward. The phrase associated with that idea is "Verification and Validation" or just "V&V" and it is built around the idea that there is a clean and simple distinction between the two "V"s that goes something like this:

- Verification: "the process of determining the extent to which the solutions generated by the computer simulation model approximate the real solutions to the original mathematical model equations."
- Validation: "the process of determining whether the original model equations are the correct or appropriate ones."

Alongside that distinction is a pair of claims about the relationship between the two distinguished processes:

- "Separability claim":
  - Validation and Verification are distinct activities that are carried out serially and give rise to distinct epistemological concerns that can be addressed on their own.
- "Disciplinarity claim":
  - Verification is a matter of *mathematics* and validation is a matter of *physics/chemistry/atmospheric science/etc.*

We can find these kinds of claims both in the scientific literature and the philosophical literature.

From the engineer Christopher Roy, for example, we get:

> Verification deals with mathematics and addresses the correctness of the numerical solution to a given model. Validation, on the other hand, deals with physics and addresses the appropriateness of the model in reproducing experimental data. Verification can be thought of as solving the chosen equations correctly, while validation is choosing the correct equations in the first place.

And from the philosophers Roman Frigg and Julian Reiss we get:

> We should distinguish two different notions of reliability here, answering two different questions. First, are the solutions that the computer provides close enough to the actual (but unavailable) solutions to be useful? [...] *this is a purely mathematical question* and falls within the class of problems we have just mentioned. So, there is nothing new here from a philosophical point of view and *the question is indeed one of number crunching*. Second, do the computational models that are the basis of the simulations represent the target system correctly? That is, are the simulation results externally valid? This is a serious question, but one *that is independent of the first problem*, and one that equally arises in connection with models that do not involve intractable mathematics [...]. (Frigg and Reiss 2009, p. 594)

In what follows, I will be arguing that some of these claims are misleading when they are applied to the epistemology of climate modeling. It is certainly true that climate scientists do engage in activities that can be labeled verification activities or validation activities and that, indeed:

1. Verification and Validation are conceptually distinguishable sorts of activities, and
2. In the actual practice of climate simulation, there are activities carried out that can or should be understood as either directed primarily at verification or at validation.[1]

But what I would like to make clear is that the general epistemological account of how simulation results are evaluated that goes by the expression "verification and validation" is very much tied to the separability claim and the disciplinarity claim. And neither of those claims are true of climate science and its credentialing activities.

The reason that climate scientists cannot genuinely verify and validate their models separately has to do with many of the features of climate simulation: the role of parameterizations, fuzzy modularity, and tuning. All of these features of climate simulation models render the "linear model" we discussed above inapplicable to climate modeling. Rather than undergoing "linear development", the development of simulation models in sciences that are as complex as climate science have more of a "life cycle."

The "life cycle" of a typical complex simulation model in climate science goes something like this.

1. Well-accepted physical principles suggest a model M1;
2. M1 is discretized and implemented in a computer simulation via scheme S1;
3. The output is compared to the real world. If the output is found to lack agreement with data, be unstable or otherwise unphysical, you proceed to:
4. Basic features of either the model or the discretization scheme are changed:
    - Either by parameterizing an effect that failed to be adequately captured, adding an effect that was not represented in the old model or by changing the way the underlying model is represented in the discrete algorithm;
    - The ways in which M1 and S1 can be modified are no longer suggested by well-accepted principles or theories, but come from other sources: physical intuition, phenomenology, local empirical findings, lore accumulated from parallel modeling successes, etc.;
5. M2 is discretized with S2;
6. Go back to 3, rinse, repeat;
7. Eventually, a model/discretization-scheme pair is found that can be sanctioned.

When a simulation model is developed in an iterative process like we see in Fig. 2 – one in which model development and model sanctioning are occurring hand in hand, there can be no justification of the final model that is independent of its

[1] Verification-directed activities include things like software quality assurance techniques, consistency and convergence tests (roughly speaking: making sure that as you make space and time grid cells smaller and smaller, the results you get start to change less and less) and doing comparisons of simulation results with known solutions. Validation-directed activities including comparing your model's output to the output of a model that has stronger theoretical underpinning, etc.

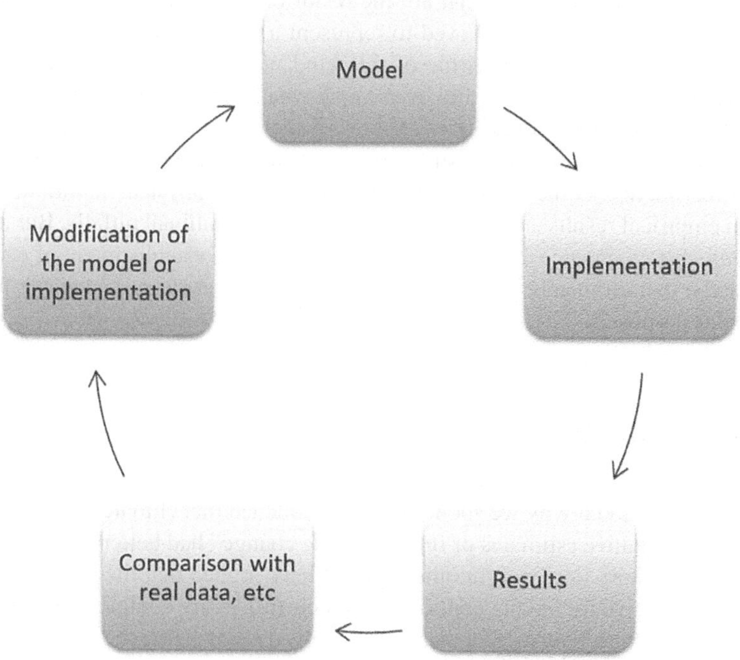

Fig. 2 The life cycle of a simulation

discretized implementation, and there can be no justification of the implementation that is independent of the model. In such a situation, the modelers have to settle on a model that they could never justify on a purely theoretically principled basis. There are, in the end, at best very weak theoretical arguments to be given for the ultimate choice of model. The ultimate justification of a set of model equations is that they work – that the final output of the simulation can be argued to be reliable (enough). There can be no independent "validation" of the model.

There are also, at best, weak mathematical arguments, weak in the sense that they only make the results seem plausible, rather than guaranteed, that can be given that the output of a simulation approximates the mathematical content of those models. This is because climate models are so complex, and rely so heavily on non-linear equations, that when these arguments are applied to the entirety of solution space, few guarantees can be offered. When models are sufficiently complex and non-linear, there can be no independent "verification" of the model.

When a climate model succeeds at passing whatever tests we subject it to, it might be because the underlying model is ideal and the algorithm in question finds solutions to that underlying model. Or it might be because of a "balance of approximations." This is likely the case when a model is deliberately tailored to counterbalance what are known to be limitations in the schemas used to transform the model into an algorithm. To the extent that climate modelers use parameter values to "tune" their models, particularly when they tune them to avoid radiation imbalances in the top of the atmosphere (TOA) that are the result of errors elsewhere in the model, these models are explicitly believed to represent a balance of approximation. And when success is achieved in virtue of this kind of back-and-forth, trial-and-error piecemeal adjustment, it is hard to even know what it means to say that a model is separately verified and validated.

In fact, the results of climate simulations are sanctioned all at once: users try to maximize fidelity to theory, to mathematical rigor, to physical intuition, and to known empirical results and to minimize known computational pitfalls. But it is the *simultaneous confluence* of these efforts, rather than the establishment of each one separately, that ultimately gives us confidence that the results are adequate for any particular purpose.

An Alternative Picture

The right story about why we should have "confidence that climate models provide credible quantitative estimates of future climate change" had better reflect this fact, and the V&V story does not. In other work, I have argued that the reasons simulation modelers give for having confidence in their results are complex and motley. A big part of the story I pursued in that work involved recognizing the "craft" of simulation. The move from basic accepted physical principles to the working model that is depicted in Fig. 2 is usually full of approximation, idealization, parameterization, and other techniques of transformation. Simulation models are most confidence-inspiring, I argued, when these techniques are deployed by practitioners that are well-acquainted with the techniques' strengths and weaknesses, and when they are knowledgeable not only about which physical principles are well-accepted, but also about which model-construction techniques are well-credentialed – and for what purposes and in what contexts. I argued that these kinds of techniques "have a life of their own" (Winsberg 2003, p. 121). This means that, "they carry with them their own history of prior successes and accomplishments, and, when properly used, they can bring to the table independent warrant for belief in the models they are used to build" (Ibidem, p. 122).

Not everyone will find a story like this entirely satisfying. William Goodwin, in particular, has expressed the worry that this kind of story has "no normative force." He worries that it:

> does nothing to establish the legitimacy of such techniques, but instead reports that their acceptance as legitimate is bound up with their detailed history of use. If this description is right, however, then it suggests that any convincing normative accounts of why particular simulation techniques are legitimate, or reliable, will be very local and detailed. Perhaps a simulation modeler could explain to his peers why it was legitimate and rational to use a certain approximation technique to solve a particular problem, but this explanation would – if Winsberg is right – appeal to very context specific reasons and particular features of the history of the technique. This does not suggest that such techniques are illegitimate or unjustified, but it does make it unlikely that the legitimacy of such techniques can be reconstructed, or rationalized, in terms of generally recognized, ahistorical evidential norms. As

> is often the case, it may be difficult or impossible to "step back" from the practice of global climate modeling and argue, independently, for its legitimacy; and this makes the absolute strategy for establishing the predictive reliability of global climate models unlikely to be successful. (Goodwin 2015, p. 342–343)

I have to confess that, unfortunately, I think Goodwin is right about the shortcomings of my account. It means that "a philosopher who hopes to address the epistemological concerns that have been raised about the reliability of climate models" isn't going to be able to tell a normatively grounded story that will secure the unassailable reliability of the results of climate modeling. But I still think the account is more-or-less right. I agree that it would be great if philosophers could do the above, but in fact the track record of philosophers in normatively grounding any kind of knowledge has been less than terrific. I think that when we look on the work of those who are in the business of modeling highly complex non-linear systems, the best we are ever going to be able to do is to arrive at a situation where "a simulation modeler could explain to his peers why it was legitimate and rational to use a certain approximation technique to solve a particular problem" by appealing to "very context specific reasons and particular features." The more time that I have spent talking to practitioners, the more my belief in this view has been reinforced. The responses that I hear to what prima facie sounded like plausible skeptical worries about particular conclusions often appeal to fairly esoteric details of the situation. My view is that philosophers do better to paint a picture in which we urge trust in the consensus of the scientific community, based on features of that community's social organization, then to try to provide a normative framework from which we can demonstrate the reliability (or its absence) of such-and-such modeling result.

Goodwin is also somewhat pessimistic about the potential for giving an "absolute" philosophical argument for the reliability of global climate models – that is, one that would ground their reliability in an ahistorical normative framework. Instead, he prefers a "relative" strategy, one where the philosopher shows that climate simulation models are on a par, epistemically, with scientific practice that is not typically called into question. Such a strategy "depends upon finding a set of predictions that are broadly recognized to be reliable and which can be plausibly argued to be relevantly similar to the sorts of quantitative predictions generated by GCMs" (Ibidem, p. 343).

Such a strategy is fine – it is certainly helpful to remind people that climate models are not in any deep and interesting way sui generis[2] – but tight analogies of the kind that Goodwin is looking for might be hard to come by. Given any other "relevantly similar" set of predictions one tries to use in order to "relatively" underwrite the reliability of climate models, there are likely to be relevant differences, alongside the relevant similarities. Certainly, if one looks to engineering contexts for instances of applied science, the ways in which models are sanctioned will display some salient differences with respect to climate science – particularly with respect to the extent to which each respective discipline seeks to apply techniques in ways

[2] Though of course this will be no help in responding to folks who think that all non-linear modeling is "embroiled in confusion."

that might turn out to be outside of the domain under which they have been directly tested. In some engineering contexts, moreover, the V&V framework is actually more or less applicable. Some engineers have made the point to me that in highly sensitive applications, such as simulations used to assure the safety of the nuclear stockpile, it is fundamental and important that the V&V framework apply (Oberkampf and Roy 2010).

We should also be careful about trying to secure more confidence in climate simulations than is possible *or* needed. We do not need to secure the reliability full-stop of climate models. Goodwin *is* correct to have no patience for philosophers and others who want to entirely undermine all confidence in the reliability of the simulation models and who insist that "basic climate science and the accumulated climate data are enough to establish the reality of climate change, diagnose its source(s), and differentiate between policy options (e.g. would halving our emissions make any difference?)" (Goodwin 2015, p. 341, fn. 5).

But for exactly the same reason that Goodwin thinks philosophers should avoid making their own determinations about how much evidential weight can be placed on basic science and data, I think philosophers should also be careful about making their own determinations of how much and exactly what forms of model reliability are needed to underwrite the well-accepted claims of climate science. One-line quotations from the IPCC, moreover, cannot be used to establish the truth about this. Scientists, in any case, are not epistemologists, and they are not necessarily experts on exactly what foundation their beliefs rest – even if they are the best experts regarding what *should be believed*. Those are not the same question. The epistemic role that climate simulations play in securing confidence in the IPCC's most important claims (for example) is more complicated than some of these one-liners make it out to be, as much as it is more complicated than the "basic science" proponents make it out to be.

What really needs to be secured is confidence in the claim that climate models make reliable projections or predictions, whatever exactly that would mean. What needs to be secured is confidence in the contribution that simulations make to our overall understanding of the climate system, which *in turn* underwrites our confidence in the various probabilistic claims that the IPCC is committed to. And that is no simple matter to tease apart. It is certainly not the same as having confidence in the predictive reliability of the simulations *tout court* (again: whatever exactly that would mean). Our most important climate hypotheses cannot be secured without any help from simulations, but they also aren't naively read off of the results of those simulations. *Attributing "skill" to models is not the same thing as taking human experts out of the picture. A science with skilful models is still human science.*

Simulations can, for example, gradually and in piecemeal fashion contribute to our process understanding of the climate system (Knuuttila, forthcoming). As such, they can contribute to our confidence in various hypotheses in a mediated way, rather than it being a case of "we have confidence in this or that hypothesis because it is a straightforward result in our simulations." An excellent example of this is the way in which simulations contribute to our knowledge of equilibrium climate sensitivity. Simulations can also contribute to our knowledge of transient climate sensitivity by providing inputs for values in back-of-the-envelope dynamical models which would otherwise not be useful. But each time such a simulation does contribute to this tapestry of evidence that supports well-accepted claims, it is done vis-à-vis the "very context specific reasons and particular features" that Goodwin, understandably, finds somewhat unsatisfying.

Tuning

Whatever turns out to be the true answer to the question of how reliable climate simulation models are, there is little doubt that whatever degree of reliability they have, they get at least some "from their ability to reproduce observed features of current climate and past climate changes." But given that these same simulation models are tuned against existing data concerning those same observed features, this raises an obvious question. How much reliability can we infer from the ability of a simulation to reproduce data against which it was tuned? A debate about this question, which is sometimes characterized as the question of whether or not "double counting" is permitted in model evaluation, has arisen in the philosophy of science. Katie Steele and Charlotte Werndl (2013) have argued that there is nothing at all wrong with using the same data to test a model as one used to tune it. They claim, in fact, that this practice is nothing other than the ordinary practice of using evidence to test a hypothesis. They argue that this conclusion is a simple consequence of the Bayesian approach to hypothesis testing. Mathias Frisch, on the other hand, has argued that Steele and Werndl overstate what the Bayesian framework commits us to (Frisch 2015).[3] Tuning is an important part of climate model construction. We should therefore say a few things about this debate.

Let's begin by canvassing some fairly typical remarks about this issue by climate scientists.

> Including quantities in model evaluation that were targeted by tuning is of little value. Evaluating models based on their ability to represent the TOA radiation balance usually reflects how closely the models were tuned to that particular target, rather than the models intrinsic qualities. (Mauritsen et al. 2012, quoted in Frisch 2015)

> Model tuning directly influences the evaluation of climate models, as the quantities that are tuned cannot be used in model evaluation. Quantities closely related to those tuned will provide only weak tests of model performance. Nonetheless, by focusing on those quantities not generally involved in model tuning while discounting metrics clearly related to it, it is possible to gain insight into model performance. Model quality is tested most rigorously through the concurrent use of many model quantities, evaluation techniques, and performance metrics that together cover a wide range of emergent (or un-tuned) model behaviour. (IPCC, AR5, WG1, Box 9.1, quoted in Frisch 2015)

---

[3] Much of the rest of this section follows Frisch (2015). The last part, in which I square off some of Frisch's claims against the stronger claims of climate scientists, is of course entirely my own.

> If the model has been tuned to give a good representation of a particular observed quantity, the agreement with that observation cannot be used to build confidence in that model. (IPCC, AR4, WG1, p. 596)

> It is incumbent upon those who develop models to know how they have (and have not) been tuned, in order to avoid inappropriate conclusions from successful tests, though the literature has historically been a little opaque on this topic. Though perhaps an obvious point, characteristics (or metrics) that are used to explicitly tune a model or its inputs should not also be used to evaluating the model – this would be a form of 'double counting'. A recent paper argued the opposite, that in fact some kinds of 'double counting' are both permissible and practiced. (Steele and Werndl 2013)

> On closer inspection though, both examples of 'double counting' addressed in that paper are simple versions of parameter tuning or model selection with no evaluation beyond the fitting procedure. The authors describe this as 'relative confirmation' (among models), but in our opinion that is irrelevant to assessments of model predictive skill which is the point that we are concerned with here. Specifically, Steel and Werndl ignore the fact that results that are predicted "out-of-sample" demonstrate more useful skill than results that are tuned for (or accommodated). (Schmidt and Sherwood 2015, p. 9)

Frisch points out that there are two readings of the concern. The IPCC quotation and the Schmidt and Sherwood quotation (which Frisch does not canvass) support Frisch's "stronger claim" that "data used in tuning simply cannot be used in model evaluation or in the confirmation of models at all" (Frisch 2015, p. 190), or as Schmidt and Sherwood put it, that data that have been used to tune a model are "irrelevant to assessments of model predictive skill" (Schmidt and Sherwood 2015, see above). Frisch's "weaker claim" is that "data not used in tuning can be more highly confirmatory of a model or provide a superior test in model evaluation than data used in tuning" (Frisch 2015, p. 190.) Frisch argues, pace Steel and Werndl, for the weaker claim, while Schmidt and Sherwood (obviously) argue for what Frisch calls the stronger claim. I will suggest below that the view of Schmidt and Sherwood is the correct view to hold, though I will also point out that their view is not *strictly* stronger than Frisch's "weaker claim."

Tuning is the process of adjusting a few key parameter values (associated with the sub-grid parameterizations discussed above) in order to get the behavior of the model *as a whole* to behave in an acceptable way. Here it is worth highlighting another feature of tuning that is especially relevant to assessing the permissibility of "double counting." First, "tuned" parameters are generally not adjusted to what we would ordinarily think of as their "correct" values. This is a tricky point because parameter values in climate models often do not correspond to a value of anything in the real world. But there are clearly cases in which parameter values are picked with the explicit goal of compensating for structural errors in other parts of the model in which those parameters live. Cloud parameters are used in order to correct for a structural error that persistently plagues global climate models: they "leak radiation" in a way that makes the top-of-the-atmosphere (TOA) energy balance incorrect. But because we do not know how to correct the structural error that leads to the TOA imbalance, we instead "fix it" by adjusting the values associated with another structural error: the representation of cloud dynamics in the model.

Another example canvassed by Frisch (2015, p. 175) is the tuning of the autoconversion threshold radius in the CMIP5 version of the GFDL CM3 model. This was done in order to achieve the best possible fit with twentieth-century temperature data. The best value that was found was determined to be one that was almost surely not realistic. As the authors of the study reporting that tuning note: "This might indicate the presence of compensating model errors" (Golaz, Horowitz and Levy 2013, p. 2249, quoted in Frisch 2015, p. 175).

It would seem obvious that, since most climate models are tuned *to get the TOA energy balance right*, the fact *that they get the TOA balance right* is not evidence of their skill at doing anything else. Given that the GFDL CM3 is tuned to achieve the best possible fit with twentieth-century data better than the CM2, the fact that CM3 does a better job of representing the twentieth-century data better than the CM2, is not, *by itself*, evidence that the CM3 has more skill than the CM2. Why, then, do Steele and Werndl dispute this?

Their argument is made in the context of Bayesian confirmation. Bayesian confirmation theory tells us how to calculate the "posterior probability" $P_f(S)$ of a hypothesis S from its "prior probability" $P(S)$ and from the "likelihood," the probability of E given S, or $P(E|S)$:

$$P_f(S) = \frac{P(E|S)P(S)}{P(E)}$$

This is what is known as "Bayes' rule". It is a simple consequence that the degree to which some new evidence E confirms some hypothesis S_1 relative to another hypothesis S_2:

$$\frac{P_f(S_1)/P(S_1)}{P_f(S_2)/P(S_2)}$$

is given by:

$$\frac{P(E|S_1)}{P(E|S_2)}$$

This mathematical reasoning, moreover, is independent of whether we consider the prior and posterior probabilities to be probabilities of the model themselves, of the hypothesis that the model is "valid" or "confirmed" or whatever, or of the hypothesis that the model has some degree of skill. It's a simple consequence of the assumption that Bayes' rule governs how we should update our beliefs in the light of new evidence – and few if any philosophers of science doubt this assumption. Importantly, moreover, it looks as if it shows that the relative degree to which two hypotheses are confirmed by some body of evidence depends only on the likelihoods, and the likelihoods do not have anything to do with the biographies of the hypotheses: they have nothing at all to do with whether the people who constructed the hypotheses (or the models that the hypotheses are about) used the evidence in question to construct them. So it look as if, by the light of Bayes' rule, there is no difference whatsoever between data used to tune a model, on the one hand, and other data, on the other, when it comes to the degree to which those data can evidentially support the model.

So what, if anything, went wrong?

One thing to note is that the version of Bayes' rule that we gave above is missing one element which is very frequently suppressed but is actually crucial to include in certain contexts: the background knowledge. After all, the probabilities in Bayesian inference are best thought of as something like degrees of belief. And it doesn't really make sense to talk about my prior degree of belief in a hypothesis, or of a likelihood, in a vacuum. These degrees of belief are always conditional on my background knowledge. Strictly speaking, the correct version of Bayes' rule should have all of the probabilities on both sides of the equality given as conditional on my background knowledge, which includes all of what I believe.

$$P_f(S|B) = \frac{P(E|S \,\&\, B)P(S|B)}{P(E|B)}$$

The problem is that on this (fully specified) version of Bayes' theorem, E is part of, and is thus a logical consequence of, our background knowledge B. But if E is logically entailed by B, then necessarily $P(E|B) = 1$. By similar reasoning, $P(E|S\&B) = 1$ as well. And it follows from this that the posterior probability of any such hypothesis is exactly equal to its prior probability. These results obtain for any E that represents evidence that was known at the time of model construction. And that means that no evidence that is known at the time of model construction shifts any of our degrees of belief about the model.

This is what is known in the literature on Bayesian inference as the "problem of old evidence." It is called the "problem" of old evidence because many have the intuition that this result is not right – that at least in some cases, evidence that was indeed known at the time of the construction of a hypothesis can in fact be used to confirm it. One example that is taken to be a canonical case of this is the relationship between the precession of the perihelion of Mercury and the theory of general relativity. Even though a detailed description of the precession was widely known prior to Einstein's formulation of the theory of general relativity, most people believe that this piece of evidence – the perihelion of Mercury data (PMD) – did allow physicists to appropriately raise their degrees of belief in GR.

And so of course the literature contains several possible solutions to the problem:

1. One possible solution is to divide our background knowledge into the part that is used in the construction of the model or hypothesis and the rest, and to call the

remainder of the background knowledge "use-novel." On this account, any part of B that is not used in model construction is allowed to be left out of the correct formulation of Bayes' rule. Let B' be the set of background beliefs in B that are not "use-novel" and set the new version of Bayes' to be:

$$P_f(S|B') = \frac{P(E|S \& B')P(S|B')}{P(E|B')}$$

In this proposed solution, because PMD is not part of B', the PMD can be used to confirm GR precisely because it is use-novel with respect to GR. On this view, however, since data used to tune a climate model are not use-novel, they cannot be used to confirm the model or any hypothesis regarding the model.

2. Another Bayesian solution to the problem of old evidence is to deny that Bayesian reasoners are logically omniscient and to suppose that it is a new thing that they can learn is that some piece of evidence is logically related to some hypothesis. One can argue, in other words, that one thing Einstein learned *after* he formulated GR is that PMD is a logical consequence of it. But note that on this view as well, data used to tune a climate model cannot be used to confirm the model or any hypothesis regarding the model, because climate scientists know what the relation between the data and model is as they do the tuning.
3. A third solution is simply to insist that all of the evidence that could possibly have been involved in model construction should be left out of the set B'. On this view (which is not, I would think for obvious reasons, a very popular one), it does turn out that data used to tune a climate model can be used to confirm the model or any hypothesis regarding the model.

Solutions 1 and 2 above correspond to what philosophers of science call "predictivism." Predictivism is the view that the amount of confirmation that some piece of evidence provides vis-à-vis some hypothesis can depend, in some cases, on whether the hypothesis predicted the evidence or accommodated it.

In *The Paradox of Predictivism*, Eric Barnes distinguishes several different versions of predictivism (Barnes 2008). According to "tempered weak predictivism," predictive success is not intrinsically superior to accommodation. But it *can* be a symptom of a property that the theory has that is hidden. That feature, in turn, can carry epistemic significance. According to strong predictivism, on the other hand, the history of the way a theory was put together can take central stage in the evaluation of the theory. The same evidence, according to strong predictivism, can provide different degrees of confirmation depending only on whether the theory's creator knew about that evidence at the time she constructed the theory. Strong predictivism is unappealing, according to Barnes, because it suggests that the degree of support afforded a theory depends on biographical facts about its creator. Weak predictivism, on the other hand, avoids this. It grants to the non-predictivist that what is ultimately responsible for a theory has to be a feature that is internal to the theory. If two theories are accorded different degrees of belief by us in the light of the same evidence, it can only be because one theory has some feature *F* and the other doesn't. It can't be merely because theory 1 predicted the data and theory 2 only accommodated it. But tempered weak predictivists explain how predictive success can still play an epistemic role. Predictive success by theory 1 can be evidence that theory one has some feature *F* that we were unaware of.

John Worrall's famous coin-flipping example helps to make this idea clear (Worrall 2014). Imagine a coin is flipped 100 times. Scientist A predicts all 100 coin flips. Scientist B only predicts the 100th flip but her model can accommodate all the previous flips. Should we trust Scientist A more than B to predict the next 20 tosses? The strong predictivist says "absolutely." The tempered weak predictivist says "hold on a minute! Are there any other significant differences between A's model and B's model? If not, then I am just going to conclude that A got lucky. But if there might be differences between A's model and B's model that I am not well positioned to evaluate myself, then I might use the fact that A made good predictions as a guide to what 'good-making' features A's model has."

In the coin-flipping case, for example, the tempered weak predictivist might conclude from the fact that A has such impressive predictive successes, that A's model probably has some good-making features: maybe a good model of the tossing mechanism and how it produces flip outcomes. In other words, to a tempered weak predictivist, predictive success is not intrinsically epistemically significant, but it can be epistemically significant as a symptom of some other epistemically significant feature – especially if the bystanding observer is not well positioned to evaluate all of A's model's intrinsic strengths and weakness.

There are two features of climate models and their application that make them especially good candidates for falling into the category of hypotheses that the tempered weak predictivist acknowledges get a boost for predicting, rather than accommodating, data sets. The first one is noted by Frisch, appealing to earlier work of mine and Johannes Lehnard (Lenhard and Winsberg 2010): climate models are, to some degree epistemically opaque to their users and developers. That is: the features that account for the successes and failures of climate models in simulating certain aspects of the climate system are not always fully understood and appreciated by modelers who use and make them. In Barnes' terms, this makes climate models especially strong candidates for being the kinds of entities that have hidden underlying features that are "good-making." If this is right (Frisch 2015, p. 190) argues in more detail than I do here that it is), then the concerns of even the tempered weak predictivist about when prediction vs. accommodation matters are relevant here, and there is an epistemic asymmetry between prediction and accommodation.

The second difference has to do with the difference between truth and model adequacy. What is typically being evaluated in the model evaluation process – the one that the climate scientists are discussing in the quotations above – is the model's adequacy for purpose, not its overall truth, or general fit with the world. Climate scientists usually already have a pretty good idea of which parts of their models are veridical and which parts are idealizations. What they are ignorant of when they do model evaluation, which they are keen to find out, is how much those idealizations impact the skill of the model for making predictions of various kinds.

As I pointed out earlier, all Frisch argues for regarding climate models is what he calls the "weaker claim" that "data not used in tuning can be more highly confirmatory of a model or provide a superior test in model evaluation than data used in tuning" (Frisch 2015). He does not endorse the "stronger claim" that "data used in tuning simply cannot be used in model evaluation or in the confirmation of models at all." But most climate modelers who have weighed in on this seem to endorse the stronger claim. What explains this? And who is right? I think the correct answer depends on paying closer attention to the fact that climate model evaluation is evaluation of *skill*, and not overall *fit*. Recall that we pointed out that the mathematical reasoning that Steele and Werndl employ is independent of whether we consider the prior and posterior probabilities to be probabilities of the model themselves, of the hypothesis that the model is "valid" or "confirmed" or whatever, or of the hypothesis that the model has some degree of skill.

When he is discussing Steele and Werndl's argument, Frisch is very careful to distinguish the question of the confirmation of a *model* from the question of the confirmation of *a hypothesis regarding model skill*. But I think he then goes on to fail to appreciate its significance. When we move from Steele and Werndl's maximally anti-predictivist framework to the framework of tempered weak predictivism, we might need to revisit the question of whether this distinction makes a difference. But Frisch does not do this.

Let's revisit the case of the coin-flip prediction model. Suppose that coin-flipping scientist Elizabeth is demonstrating the virtues of her model to Frank and Gertrude. Frank doesn't know any physics, but Gertrude knows all the fundamental physics that drive coin flips. She knows all the physics and modeling methodology that Elizabeth is making use of. (Let's suppose, just to keep things simple, that Elizabeth and Gertrude assign probability 1 to the fundamental physics they believe in.) The problem (for both Elizabeth and Gertrude) is that the physics is not entirely computationally tractable. But Elizabeth has come up with an idealized model that is partly based on the fundamental physics and partly based on a parameterization trick. Suppose further that her parameterization depends on details of the coin-flipping situation, and the only way she knows how to get her model to work with a particular coin-flipping situation is to tune it to the outcome of 100 trials. Frank and Gertrude both watch her flip the coin 100 times, tune the model, show that the tuned model can correctly retrodict the 100 past flips, and then successfully predict the next 100 flips.

Question: According to a tempered weak predictivist, did the fact that she could successfully retrodict the 100 flips confirm her model or are only the second 100 flips confirmatory?

Answer: It depends on if you are Frank or Gertrude. Depending on the details, if Frank is a tempered weak predictivist, he might very well say, "yes, the first 100 flips are confirmatory." Relative to his background knowledge, he might (again, depending on many details) think that the ability to retrodict is impressive. To him, the success might be a symptom of hidden good-making properties of the model (e.g. good grounding in physics) that he was unaware of. But Gertrude is not in this epistemic state. Let's suppose that she assigns probability one to the fundamental physics that guided the model, and probability zero to all the idealizations. She might smile knowingly at Frank and think to herself, "how quaint that to Frank, the fact that the model construction was guided by true physics is a hidden good-making feature. And how quaint that he uses Elizabeth's ability to retrodict as a guide to this." But she might have the relevant background knowledge to be in a position to be completely unimpressed by Elizabeth's "curve fitting." She thinks, "yes, of course Elizabeth can backfit a set of 100 flips. After all, her model has some veridical parts, and it has enough flexibility to fill in the wrong parts come what may." All she wants to know is: do the falsehoods that she knows are in Elizabeth's model offset in the right kind of way to give her model skill? Or do they only provide it with the flexibility to backfit (in a way that she never doubted)? If that is the epistemic state she is in, then she will adopt Frisch's "stronger claim" for herself, all the while conceding that for Frank, only Frisch's "weaker claim" is true.

When Schmidt and Sherwood say "characteristics (or metrics) that are used to explicitly tune a model or its inputs should not also be used to evaluat[e] the model" (2015, p. 159), I think we need to read them as speaking from the epistemic vantage of point of someone like Gertrude. Here is one way to make the point: the fact that a climate model can be tuned to some existing data set tells you some things about the goodness of the model. After all, a model with completely wrong physics in it, or one that was structurally defective in fundamental ways, would not be tunable. So, if you had poor background knowledge about a climate model, then the fact that the model could be tuned to an existing data set might boost your degree of belief in some hypotheses regarding some of the features of the model. As lay observers, this might very well be the state we are in. This is not, however, the epistemic state that climate modelers are in when they test their models for skill, after having already tuned them. At that point in the process, they are almost exclusively concerned with the degree to which features they are unaware of are helping or hindering model skill with respect to some variable or set of variables. (The degree, to use Barnes' language, to which the models have hidden good-making features with respect to model skill of some particular kind.) If so, then Frisch might be right that, from a particular epistemic vantage point (indeed the one that we lay observers have), the fact that a climate model can successfully be tuned boosts its credibility in various ways (albeit not as strongly as novel prediction). From that vantage point, only the "weaker claim" might be true. But from the epistemic point of view of a modeler who has built the model and is only interested in model skill relative to competitors, the stronger claim might apply. That is, such an epistemic agent might be right that no additional confirmation occurs as a result of being able to tune the model relative to where she was before.

Conclusion

In post-human science, there would obviously be a need for models with skill. But models with skill do not imply post-human science This is because evaluating model skill is not a matter of determining the "correctness" or "fit" of climate models, but rather their suitability for various prediction and projection tasks. While the "gold standard" for establishing model skill is the "verification and validation" paradigm, that paradigm is rarely helpful in climate science, because it does not apply well to modeling projects in which there is substantial back and forth between model evaluation and model construction (as illustrated in Fig. 2). Rather than following an articulable set of normative rules, establishing the credibility of model-based knowledge in climate science is local and detailed, even though this claim might not sit well with those who might hope for an epistemology that can be reconstructed or rationalized in terms of generally recognized, ahistorical evidential norms.

This does not mean, however, that nothing general, global or ahistorical can be said about the epistemology of climate models. We can, for example, ask questions like the one we discussed in section "Tuning": How much reliability can we infer from the ability of a simulation to reproduce data against which it was tuned? The answer we argued for was that the ability to predict novel data is always a stronger test of model skill than being able to re-accommodate data that were used in the process of tuning a model. But we also argued that the question of whether re-accommodation is *at all* indicative of model skill might depend on one's epistemic position.

References

Barnes, Eric Christian. 2008. *The Paradox of Predictivism*. 1st ed. Cambridge: Cambridge University Press.

Frigg, Roman, and Julian Reiss. 2009. The Philosophy of Simulation: Hot New Issues or Same Old Stew? *Synthese* 169 (3): 593–613.

Frisch, Mathias. 2015. Predictivism and Old Evidence: A Critical Look at Climate Model Tuning. *European Journal for Philosophy of Science* 5 (2): 171–190.

Golaz, Jean-Christophe, Larry W. Horowitz, and Hiram Levy. 2013. Cloud Tuning in a Coupled Climate Model: Impact on 20th Century Warming. *Geophysical Research Letters* 40 (10): 2246–2251.

Goodwin, William M. 2015. Global Climate Modeling as Applied Science. *Journal for General Philosophy of Science* 46 (2): 339–350.

Lenhard, Johannes, and Eric Winsberg. 2010. Holism, Entrenchment, and the Future of Climate Model Pluralism. *Studies in History and Philosophy of Science Part B: Studies in History and Philosophy of Modern Physics* 41 (3): 253–262.

Mauritsen, Thorsten, Bjorn Stevens, Erich Roeckner, Traute Crueger, Monika Esch, Marco Giorgetta, Helmuth Haak, et al. 2012. Tuning the Climate of a Global Model. *Journal of Advances in Modeling Earth Systems* 4 (3). http://onlinelibrary.wiley.com/doi/10.1029/2012MS000154/full.

Oberkampf, William L., and Christopher J. Roy. 2010. *Verification and Validation in Scientific Computing*. 1st ed. New York: Cambridge University Press.
Schmidt, Gavin A., and Steven Sherwood. 2015. A Practical Philosophy of Complex Climate Modelling. *European Journal for Philosophy of Science* 5 (2): 149–169.
Steele, Katie, and Charlotte Werndl. 2013. Climate Models, Calibration, and Confirmation. *The British Journal for the Philosophy of Science* 64 (3): 609–635.
Winsberg, Eric. 2003. Simulated Experiments: Methodology for a Virtual World. *Philosophy of Science* 70 (1): 105–125.
Worrall, John. 2014. Prediction and Accommodation Revisited. *Studies in History and Philosophy of Science Part A* 45 (March): 54–61. https://doi.org/10.1016/j.shpsa.2013.10.001.

Virtually Extending the Bodies with (Health) Technologies

Francesco Bianchini

Introduction

Self-perception of the body will increasingly change in the future because of new web technologies. The building of big databases concerning bodily features together with the creation of portals dedicated to the management of such data by individual users and by institutions or companies in charge of specific functions, such as electronic health record (EHR), or electronic medical record (EMR), will bring a new conception of the self as a body, and therefore as a cognitive embodied agent. I suggest a conceptual framework of the extended mind thesis is useful to analyze and understand the coming developments of such a trend. I deal with the extended mind thesis in section "The extension thesis" and build an analogy between the extended mind thesis and a sort of extended body thesis in section "The EMT bodily extension". Then, I speak about social and second bodies (section "Social and second bodies"), and changes in body schema and image, by supporting this study's claim with examples from health care technologies and introducing the virtual cyborg perspective (section "Extending the body and the health-extended bodies") to outline some potential further developments related to virtually extended bodies (section "Conclusion and what's next").

F. Bianchini (✉)
Department of Philosophy and Communication Studies, University of Bologna, Bologna, Italy
e-mail: francesco.bianchini5@unibo.it

M. Bertolaso, F. Sterpetti (eds.), *A Critical Reflection on Automated Science*, Human Perspectives in Health Sciences and Technology 1,
https://doi.org/10.1007/978-3-030-25001-0_11

The Extension Thesis

The extended mind thesis (EMT) refers to the traditional division between mind and what is out of the mind (especially, body and environment) and suggests a different position about the boundaries of mind. It states that mind is not restricted to what happens inside the head. Its main idea is that "if, as we confront some task, a part of the world functions as a process which, *were it done in the head*, we would have no hesitation in recognizing as part of the cognitive process, then that part of the world *is* (so we claim) part of the cognitive process. Cognitive Processes ain't (all) in the head!" (Clark and Chalmers 1998). This is the *parity principle* between something of outer world and what has the same cognitive function of a traditional cognitive process (for example, a notebook and the human memory).

Unsurprisingly, the EMT has been successful within the new trends of cognitive science (the embodied, embedded, and enactive cognition) because, even though EMT has a functionalist characterization, it goes beyond the traditional limit of the non-material mind, involving what seems to be relevant for cognitive science (also from a robotics point of view and a neuroscience one) of last two decades: not only the brain, but also the body, the body modifications, the artifacts, and the social and natural environment (Clark 2003, 2008).

Many arguments have been raised against EMT. The main ones are: (1) the coupling-constitutive fallacy (Adams and Aizawa 2008), according to which it is not granted that what is in a causal relation within (or is a causal part of) a process is constitutive of the process as well; and (2) the fact that there is no criterion to decide what is cognitive and what is not to establish a sort of mark of the cognitive. For example, Rupert (2009), held that against the externalist functionalism of EMT, there are specific cognitive phenomena that take place inside the organism thanks to "integrated cognitive architectures."[1]

Other criticisms specifically involve the role of body and the relation between body, brain, and environment. For example, the debate on the brain as a sort of prediction machine and the prediction error minimization framework gave rise to two positions: (1) on one hand, some external processes are part of the cognitive process and allow an interaction with environment without representing it or all its features (Clark 2013), which is consistent with EMT and an interpretation of it in the embodied cognition framework; and (2) on the other hand, prediction error minimization implies a boundary between cognitive systems and their body and environment (Hohwy 2016). The latter position involves the construction of models by the brain that generate predictions. Models have a quintessential representational nature both in standard prediction error minimization and in Bayesian inference. Upholders of the central role of the models claim that the body is just represented in the model and it rules out radical versions of embodied cognition as well as the hypothesis of

[1] For a discussion see Clark (2010).

an extended mind (Bruin and Michael 2016) and its embodied extensions, because every inference to the best explanation entails a boundary between brain and what is predicted/inferred.[2]

Even though all these arguments against EMT are relevant and raise crucial points, the criteria established for an external resource being cognitive, i.e. having a nontrivially involvement in a process, a *relevant* and *active* role in it; and are more fundamental than merely being part of the process. They concern the reliability and the portability of an external resource, which needs to be: (a) rapidly and constantly available, (b) automatically endorsed, and (c) easily accessible.[3] According to EMT, if these requirements are fulfilled by cognitive acts, a resource becomes as transparent as other intracranial entities involved in cognitive process. If we accept such criteria, *transparency* is what warrants that something non-mental and non-cerebral is constitutive of a cognitive process and extends mind.

The EMT Bodily Extension

With reference to this view, I will draw an analogy with the body, suggesting a conceptual framework with the aim to give rise to the idea of an *extended body* in step with EMT, which could be useful to discuss some possible developments of human-computer interaction.

Recent trends in cognitive science (the 4E: embodied, embedded, enacted, and extended approach) are mostly based on the belief that the Cartesian idea of a non-material mind is just a heritage of the past, an unfruitful concept to understand cognition. Also, with the support of phenomenological approach to cognitive science, the role of the body in the explanation of cognitive capabilities has changed. The notion of "extended body" has been used and developed to better explain intersubjectivity, mutual bodily influence (Fuchs and De Jaegher 2009), and social interaction (Froese and Fuchs 2012).

In the same period, we have witnessed the extremely quick web revolution, which has changed the world, making it more virtual and less material. Nowadays, humans live in a *virtual* reality of websites, web social communication, web services, and web information retrieval and storage. The immediate and everywhere widespread availability of all this virtual reality has undoubtedly extended the human mind as regards chances to do things. The *transparency* of interfaces and the new customs of web users (in fact, a larger and larger amount of people) make the

---

[2] A similar criticism involving the model of niche construction in which cognitive tools are used to scaffold the intelligent action is in Sterelny (2010).

[3] There is a fourth "arguable" criterion involving consciousness, but it can be considered more disputable. See Rupert (2004).

whole thing in line with EMT. Human minds are modified by the presence of web availability, and we could ask how much this development changes or will change the human cognitive capabilities, by becoming a crucial part of human cognitive processes (Sparrow et al. 2011). Web access is one of the best candidates to extend the human mind or to understand whether EMT is a plausible theory to explain intelligence (and its contextual nature). But, what about the body? Can we speak about it in the same way? Can we speak of an extended body theory (EBT)? And if so, how? I suggest using this notion in a way that is just partially overlapping with the notion of extended body that is typical of phenomenological approach. I propose and consider how it is possible that human perception of the body is extended and how this extension is able to affect human cognitive capabilities involving corporeity.

Social and Second Bodies

In the case of EMT, natural language has been the first and the main external scaffolding for epistemic actions within the framework of active externalism and human beings conceived as natural born cyborgs (Clark 2003). We might consider EBT view as an extension of the EMT not only in the sense of traditional artificial elements on, or inside, a biological body, or in the general perspective of posthuman (Braidotti 2013). EBT view can be seen as an extension connected with new web resources. In particular, we can make an analogy with the "re-materialization" of the mind on the web medium and information systems (the Internet and the world real time connection of online devices), which has taken place thanks to social networks and, even more, virtual worlds such as Second Life and the like. These two kinds of experiences have changed our concepts of social relation, by extending and transforming the cognitive range and characteristics of our identity and our concept of self. I am not speaking about the use of virtual worlds as a simulation to study cognition (Veksler 2012), but rather, I suggest that human bodies are *cognitively* extended by new web tools and technologies in line with recent proposals, such as De Kerckhove (2011).

In De Kerckhove's view of *augmented mind*, Second Life is seen as "a curious hybrid of objective and subjective", which "intertwines mind and screen". This is relevant because "our relationship to screens is a very complex one where we are actually delegating a great deal of our cognitive function" (De Kerckhove 2011). Although this view seems to be leading to a unified meta-mind, in which every single mind of individual users is absorbed, things are otherwise. In the augmented mind view, individual identities are not melded. There is no such a thing as a collective mind, but a connective one: augmented mind "augments one's mind without submerging it [...]. It connects without binding. The Internet is an extension of my mind but I am still in charge. I am not a mere node in a global mind." This view is still based on language to communicate and interact, but what is interesting is the fact that the language is considered as provided with the power to control the body, which "it normally enjoys in a uniquely oral context. The real issue is [...] whether there is still a body out there" (De Kerckhove 2011). So, such a different relationship between language and human being, through the new augmented mind, which we may consider a form of EMT according to the three criteria above mentioned: the constant availability (of other mind on the Web), the automatic endorsement (provided by the Web interactive tools and platforms), and the easily accessibility (of the Web itself and its social and informational contents), gives rise to new questions about the body and its place in the vision of human nature.

Extending the Body and the Health-Extended Bodies

The discussion on bodily extension is usually connected with body perception and body representation. One standard distinction of the phenomenological approach to embodied cognition and in line with these two aspects of bodily extension is between body schema and body image: "The body image consists of a complex set of intentional states and dispositions—perceptions, beliefs, and attitudes—in which the intentional object is one's own body. This involves a form of reflexive or self-referential intentionality. [...]. In contrast to the body image, a body schema is not a set of perceptions, beliefs, or attitudes. Rather it is a system of sensory-motor functions that operate below the level of self-referential intentionality" (Gallagher 2005: 25–26).[4] The former is mainly a conscious and reflexive representation of the human body, which collects information from every sensory channel as well as information stored in "offline" cognitive systems, such as long-term memory. The latter is characterized by its unconscious nature and by its involvement in subpersonal and automatic processes leading the action execution, having to do specially with proprioceptive and kinesthetic information.

The body schema is involved in the use of tools; in some cases, they end up by being incorporated in the body schema itself. Experimental neuroscientific evidence shows that the body schema is modified by the tool use, as the tool is considered as an extension of the body both in a perceptual and a physical sense (Iriki et al. 1996). Moreover, it is interesting that the extension is connected to an intentional use of the tool itself (Marivita and Iriki 2004), because, if the body schema controls the interaction between environment and body, "operations of the body schema may generate plastic changes in the body image" (Gallagher 2005: 37). Quite similar remarks could be done also for prosthetic devices such as artificial arms, hands, or limbs, which are used intentionally.

This kind of body extension is consistent with EMT and the idea that a tool or a device can become part of a cognitive process, at least of the sensory-motor processes. The modification of body image caused by body schema change suggests; however, that *also higher cognitive processes can be influenced by such incorporations* and that a tool or a device can become a part of the cognitive processes at that

[4] See Gallagher (2005) for a historical reconstruction of these two notions.

level as well. This is one sense in which we can speak about an extended body in analogy with EMT. Not only the mind changes by extending itself outside, but also the body changes by extending itself and therefore its schema and image. We may see it as a sort of *embodied EMT*, in which the body is a relevant part from a physical, perceptual, and representational standpoint.

Nevertheless, present-day tools are less and less materially based and more and more digitally based. In which sense are we allowed to speak about the extended body from this point of view? In other words, how do digital tools and devices extend human bodies? New web resources can be analyzed in the same way I analyzed tools extending body schema and image.

Digital tools for health services are good candidates for new kinds of extensions for the body. Let us consider an example of possible new tools, which extend human beings in some way, allowed by electronic health record (EHR) (or electronic medical records). Actually, this is a family of systems for collecting medical data about individual patients as well as specific populations. They are common in many countries, even though they are not used by all citizens, depending upon several social, practical, and cultural issues as well as the spread of web access and employment. Unsurprisingly, debates on ethical, legal, and privacy issues stem from the increasing use of EHR, together with the problems caused by the (unexpected) consequences on physicians and patients in relation to its use and usability (Ratwani et al. 2015).

EHRs usually consist of a portal in which every user has a homepage (the "My page") where subjective and objective data are interwoven. At present, subjective data are all health data of the patients/users: their medical history, the collection of examinations and treatments, etc. Such data are accessible online by *both* the users *and* medical staff and institutions. The objective data concern the medical institutions and medical knowledge, diseases, medications, treatments, tests, etc. among other related data. The connection between them yields self-monitoring data, instructions, or suggestions for a single user, provided by the institutions, and statistical data for different exploitations by institutions *and* users.

The question is: what will happen if new types of interfaces are achieved? What will happen if interfaces based on a sort of "active-iconic" representation of the body, reminding the body image and used in the same conscious self-referential way, are the virtual place of interaction between users and institutions on such a sensitive topic?

I call it "active-iconic" representation of the body because it is a sort of body image representation on the homepage. Suppose that the bodily representation (an icon of the body) changes in line with the online flow of data concerning the health issues of a specific population selected by age, gender, place, and common habits, or other parameters. This is what I mean by "active". The real-time changes in the active-iconic representation could concern the part of the body much more at disease risk, or under specific clinical tests, or much more affected by a specific illness. We may imagine that such active-iconic representations of the body, which can be regarded as *virtual bodies*, would become part of the bodily representation of the subject. The crucial question is: could the changes of the body image in the active-iconic representation generate changes in the body schema and therefore in human behavior at the sensory-motor and action-execution level? In other terms, is it possible to think a mutual influence in the plasticity of the body image and the body schema? And if so, what would the consequences of a massive use of digital and web tools on human body be?

Consider for example the fast development of virtual human technologies (Fleming et al. 2017). Many projects on this topic aim to produce 3D models of a human body able to move in the same way as real humans.[5] These indistinguishable dynamical representations of human bodies are used in virtual reality or in digital assistant technologies, but also in other fields such as psychiatric and psychological therapy, cognitive experimentation, and virtual social interaction, among others. These human avatars are self-representations that can exemplify the sort of active-iconic representations well, the ones I am thinking of when I speak about extending the body within the health context of EHR.

Now, let us consider an example to show in which sense the active-iconic representations of the body might turn out to be part of an extended body in the sense of an EBT mentioned above. In the example, the virtual body has to be imagined as the virtual human of the user, a technological development that appears not so hard and far to implement. The virtual body, participated by the institutions and statistically connected with other virtual bodies, would constitute, together with the real body, a *whole extended body*, subject to a different kind of manipulation and control. Such a whole extended body is allowed by the availability-endorsement-accessibility criteria for extension, thanks to the online connection through the usual channels, especially mobile ones, which make the virtual body and every dynamical and real-time modification of it, which is transparent to the user.

If the active-iconic representation became transparent to the user in the sense of the EMT criteria, the outcome would be probably a different way to perceive and treat the body, and the approach to reality by human users, because their role in cognition would be more and more interwoven with virtuality from a bodily standpoint. The dematerialization of human corporeity will then lead to a re-materialization of it in an active-iconic and virtual representation in interaction with (health) institutions and other virtual bodies. This generates a new representation concerning corporeity itself, that would affect human user bodies. The users would become a sort of *virtual cyborg* (Richardson and Locks 2014) provided with new information concerning parts of their body and represented on the interface and influences them. On one hand, such information would change user behavior; however, on the other hand, user behavior would affect and modify statistical data collected by the institutions and the whole meta-body of which the virtual and non-virtual user bodies would be just a part, without losing autonomy, as in the augmented mind perspective (see Fig. 1).

In this way, we could have an extended body and a functionalist extension of cognitive capabilities involving the elements and the powers of embodied cognition, much the same as the EMT is a functionalist extension of the traditional idea of

[5] See for example the MakeHuman Project at http://www.makehumancommunity.org (retrieved on June 19, 2018).

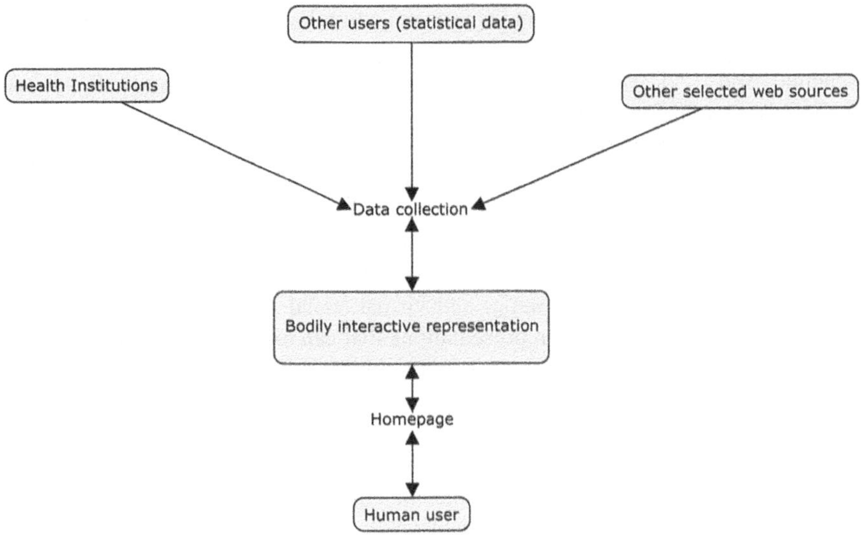

Fig. 1 The relation between users and institutions through a bodily active representation

mind in cognitive science. As an example, let us consider the impact of the availability of real time data on a specific disease in a specific place and of the risk percentage for single users on the basis of their peculiar physical characteristics. The real-time connection between these two kinds of (individual and statistical) data would cause a new and different way of thinking of and perceiving the body and its physical-ness.

One interesting question about interfaces in health digital tools concerns the nature of representation. We may ask whether different kinds of representations give rise to different effects on users. What are the better bodily representations to achieve the desired results? For example, using virtual human technologies that mimic the user may generate an uncanny valley effect, especially if the anthropomorphism is not connected to a human-like motor behavior (Damiano and Dumouchel 2018). It seems that interfaces like these could be also useful for collecting data and arranging experiment on present-day topics of human-robot and human-avatar interactions. In this sense, we may say that science will remain human, but just if we accept that the notion of "human" is a notion epistemologically evolving toward new meanings.

One more issue concerns the risk of dualism implied by the proposal of the EBT framework. Is it a new form of dualism? In light with such an objection, we say that it would be more correct to speak about an embodied EMT or EMT from the point of view of embodiment. Things change, however, if we consider the relationship between the body image and the body schema. It is unsurprising if the body schema will change along with the use of these technologies, as much as it changes with other tools, and these changes could change also the body image, as we said before in the case of material tools. Virtual and digital tools can be able to change the body image without the material manipulation and by means cognitive abilities. This is the idea at the basis of my proposal of EBT and some neuroscientific experiments that have addressed this issue since 1990 (Ramachandran et al. 1996).

The change in body image is interesting if it will be a virtual change. In other words, changes in virtual human images for the interfaces I spoke about above may generate a change in body schema; and, therefore, in the behavior of users. In this complex interaction, the interesting part is that the direction is the opposite of the usual one: not from the body schema to the body image, but from the (virtual) human body image to body schema. This could be the main result of extending the body through digital tools concerning the human corporeity.

Conclusion and What's Next

All such prospects are speculative, and the problems raised by them are important (privacy and ethical issues, data management, influence on individual practices and habits, etc.). Nonetheless, we are not so far from the possibility of having these kinds of interfaces. The impact on human bodily representation and human cognitive capabilities seem to be impressive in consideration of the analogy with the social network revolution for social relations and even for the perception of spatial geography. The idea of an extended body goes beyond the limits of health care and together with EMT, which is a good candidate to become a new theoretical resource for studying cognition from the point of view of applied artificial intelligence, human-computer interaction, and neuroscience.

Two examples of possible research topics are: research on the role played by sensorimotor representation of the body on virtual corporeity, represented through a virtual human avatar or an active-iconic image or scheme from the point of view of the personal state of health; and research on connections and interactions between ownership of the body, proprioceptive representation and interoception (Ferri et al. 2013). I did not address the issue of body sensors for the control of bodily activities. The technology of small biosensors implanted inside a human body is still partially undeveloped, but it appears to be useful and revolutionary for future individuals and social evolutions (Yuce 2010). However, my thesis is not in contrast with such technological developments, which, on the contrary, may be an integration and a powerful incentive on the way of body extension.

Nevertheless, experiments on the effects that interactive bodily representations in health digital technologies such as the EHR and the like are hard to put into practice. Ethical and legal issues are a big obstacle. We may collect data ex post, by detecting and analyzing the unexpected consequences of these kinds of interfaces if they will be developed and used.[6] Most likely, human bodies of patients will not change, but behavior, habits, and lifestyle could be influenced by these kinds of technologies. Understanding the role of an extended digital body, however, can also

[6] See Ash et al. (2007) for an example of this sort of issues in computerized medicine.

be useful to steer the patient behavior in the prevention of diseases and in avoiding the factors leading to a probable condition of illness. Such potential developments are worthy of further investigation.

In summary, in accordance with the EMT analogy, I claim that in the view of extended body just outlined, the three a-b-c criteria for establishing whether an external resource is cognitive (availability, endorsement, and accessibility) are satisfied, and the transparency conditions were as well. In other words, a bodily online interactive representation, such as a virtual human, can become something with the mark of the cognitive, or within the range of the cognitive system that every human being is.

Challenging questions remaining are: what changes should we expect from such developments? Would they be able to influence human body responses to external stimuli, both environmental and social stimuli? Would they be able, eventually, to modify also some bodily features as social web interaction has modified our conception of personal identity, by giving rise to new real, even though virtual, identities of single human beings, all of them coexisting and building the connective meta-identity of the single human beings themselves? Most likely, biomedical scientific research will benefit from the rising of such frameworks, which are in turn allowed by parallel research in cognitive science, artificial intelligence, and philosophy of science, where human and non-human elements closely coexist.

References

Adams, Frederick, and Kenneth Aizawa. 2008. *The Bounds of Cognition*. Oxford: Blackwell.

Ash, J.S., D.F. Sittig, E.G. Poon, K. Guappone, E. Campbell, and R.H. Dykstra. 2007. The Extent and Importance of Unintended Consequences Related to Computerized Provider Order Entry. *Journal of the American Medical Informatics Association* 14: 415–423. https://doi.org/10.1197/jamia.M2373.

Braidotti, Rosi. 2013. *The Posthuman*. Cambridge: Polity Press.

Bruin, L., and J. Michael. 2016. Prediction Error Minimization: Implications for Embodied Cognition and the Extended Mind Hypothesis. *Brain and Cognition*. https://doi.org/10.1016/j.bandc.2016.01.009.

Clark, Andy. 2003. *Natural-Born Cyborgs*. Oxford: Oxford university Press.

———. 2008. *Supersizing the mind*. Oxford and New York: Oxford University Press.

———. 2010. Much Ado About Cognition. *Mind* 119: 1047–1066.

———. 2013. Whatever Next? Predictive Brains, Situated Agents, and the Future of Cognitive Science. *Behavioral and Brain Sciences* 36: 181–204.

Clark, Andy, and David Chalmers. 1998. The Extended Mind. *Analysis* 58: 7–19.

Damiano, L., and P. Dumouchel. 2018. Anthropomorphism in Human-Robot Co-evolution. *Frontiers in Psychology* 9. https://doi.org/10.3389/fpsyg.2018.00468.

De Kerckhove, Derrick. 2011. *The Augmented Mind*, 40k Books, online publication.

Ferri, F., M. Ardizzi, M. Ambrosecchia, and V. Gallese. 2013. Closing the Gap between *the Inside* and *the Outside*: Interoceptive Sensitivity and Social Distances. *PLoS One* 8 (10). https://doi.org/10.1371/journal.pone.0075758.

Fleming, Reuben, Betty J. Mohler, Javier Romero, Micheal J. Black, and Martin Breidt. 2017. Appealing Avatars from 3D Body Scans: Perceptual Effects of Stylization. In *Computer Vision, Imaging and Computer Graphics Theory and Applications: 11th International Joint Conference, VISIGRAPP 2016, Rome, Italy, February 27–29, 2016*, Revised Selected Papers, 175–196. Springer International Publishing.

Froese, Tom, and Thomas Fuchs. 2012. The Extended Body: A Case Study in the Neurophenomenology of Social Interaction. *Phenomenology and the Cognitive Sciences* 11: 205–235.

Fuchs, Thomas, and Hanne De Jaegher. 2009. Enactive Intersubjectivity: Participatory Sense-Making and Mutual Incorporation. *Phenomenology and the Cognitive Sciences* 8: 465–486.

Gallagher, Shaun. 2005. *How the Body Shapes the Mind*. Oxford: Oxford University Press.

Hohwy, Jacob. 2016. The Self-Evidencing Brain. *Nous* 50: 259–285. https://doi.org/10.1111/nous.12062

Iriki, Atsushi, Michio Tanaka, and Yoshiaki Iwamura. 1996. Coding of Modified Body Schema During Tool Use by Macaque Postcentral Neurones. *Neuroreport* 7: 2325–2330.

Maravita, Angelo, and Atsushi Iriki. 2004. Tools for the Body (schema). *Trends in Cognitive Sciences* 8: 79–86.

Ramachandran, V.S., L. Levi, L. Stone, D. Rogers-Ramachandran, R. McKinney, M. Stalcup, G. Arcilla, R. Zweifler, A. Schatz, and A. Filippin. 1996. Illusions of Body Image: What They Reveal About Human Nature. In *The Mind-Brain Continuum: SENSORY PROCESSES*, ed. Rodolfo Riascos Linas and Patricia Smith Churchland, 29–60. Cambridge, MA: MIT Press.

Ratwani, R.M., R.J. Fairbanks, A.Z. Hettinger, and N.C. Benda. 2015. Electronic Health Record Usability: Analysis of the User-Centered Design Processes of Eleven Electronic Health Record Vendors. *Journal of the American Medical Informatics Association* 22: 1179–1182. https://doi.org/10.1093/jamia/ocv050.

Richardson, Niall, and Adam Locks. 2014. *Body Studies: The Basics*. London/New York: Routledge.

Rupert, Robert D. 2004. Challenges to the Hypothesis of Extended Cognition. *Journal of Philosophy* 101: 389–428.

———. 2009. *Cognitive Systems and the Extended Mind*. Oxford: Oxford University Press.

Sparrow, Betsy, Jenny Liu, and Daniel M. Wegner. 2011. Google Effects on Memory: Cognitive Consequences of Having Information at Our Fingertips. *Science* 333: 776–778.

Sterelny, Kim. 2010. Minds: Extended or Scaffolded? *Phenomenology and the Cognitive Sciences* 9: 465–481.

Veksler, Vladislav Dan. 2012. Second Life as a Simulation Platform. *AISB Quarterly* 135: 8–12.

Yuce, Mehmet Rasit. 2010. Implementation of Wireless Body Area Networks for Healthcare Systems. *Sensors and Actuators A: Physical* 162: 116–129.

Part III
Automated Science and Human Values

Behold the Man: Figuring the Human in the Development of Biotechnology

J. Benjamin Hurlbut

Introduction

One of the core problematics of the biotechnology revolution has been defining its relationship to human life: its value as a scientific and technological enterprise, its potential to enhance and advance human well-being, and its relationship to the figure of the human—of human nature, integrity and dignity—whom it promises to work for without re-working. The many technological transformations of life, including human life, that the era of biotechnology has ushered in have been accompanied by the need to give accounts of the forms of human life that benefit from, and thus demand, these transformations. Thus, for the last half-century, the question of how to govern biotechnology has raised the corollary question of to what figure of the human should governance be calibrated. Those who have lead developments in biotechnology have tended to characterize them as in the service of rather than as a danger to human life. Figuring projects in biotechnology as in the service of human life has generally entailed a corollary figuring of human life in ways that authorize those projects.

The question of the human has figured most explicitly in discussions about whether there are places that biotechnology ought not go lest it perturb or undermine essential features of human nature, for instance in controlling human reproduction, selecting or modifying human genetic traits, or otherwise modifying human biology in ways that pose risks to the integrity and dignity of human life. Yet the question is at stake in at least two other dimensions as well. First is the human in whose name projects in biotechnology are marked as progressive—that is, the figure of the human that new biotechnologies are meant to serve, and which is invoked to characterize advances in technological control of life as human progress. Second,

J. B. Hurlbut (✉)
School of Life Sciences, Arizona State University, Tempe, AZ, USA
e-mail: bhurlbut@asu.edu

M. Bertolaso, F. Sterpetti (eds.), *A Critical Reflection on Automated Science*, Human Perspectives in Health Sciences and Technology 1,
https://doi.org/10.1007/978-3-030-25001-0_12

is the way the human is at stake in the very processes of constituting human relations and modes of moral sense-making in order to contend with such questions of governance. Insofar as the figure of the human is a focus of the politics of authorization, judgment and deference, those forms of politics are also modes of figuring the human—that is, of figuring those forms of reasoning, action and self-governance that are appropriate to human political communities.

This paper explores the question of whether science can remain human by interrogating some of the ways that the figure of the human has been rearticulated and transformed in the context of the biosciences. Accounts of what the human is in debates about biotechnology have mostly focused on the human as an object of technological intervention and control, for instance in asking what biological forms constitute morally significant human life, or what technological transformations might undermine some essential dimension of humanness. Other chapters in this volume engage this question in a slightly different mode, asking whether science as a human undertaking loses something essential insofar as the tasks of experimental research and interpretation shift out of human hands and are delegated to nonhuman (artificial) intelligences. At stake in both of these questions is a falling away from a preexistent humanness that is potentially perturbed by shifting terrains in which the sovereign territory of the human is colonized simply because it has become technically possible to do so. In this chapter, I want to take the work of "figuring the human" itself as my object of analysis. I see "the human" as at stake both in the sense of quintessence and as a discursive figure that is, in some sense, made and remade in giving articulation to human quintessence and in calibrating action accordingly. Indeed, these two dimensions, call them the ontological and the discursive, cannot be separated because the modes of political engagement, the formation of institutions, and the authorization of practices that are undertaken in the name of protecting the integrity of the human are essential enactments of human life. Put differently, human being is not somehow separated from the ways of being human together, particularly in the political modes of self-reflection and self-governance that such being together demands. The human is no less at stake in the modes of moral sensemaking about technological transformations of the human than in those very acts of transformation themselves. Thus, in what follows, I explore the question "will science remain human?" by taking the political conditions of possibility for asking that question as an object of analysis, attending to the ways those political conditions have been transformed in conjunction with the development of the late twentieth and early twenty-first century biological sciences.

I take certain imagined technological futures that have been key sites of political transformation as case studies, showing how in these moments the biosciences have refigured the human in the name of figuring biotechnology as aligned with and not in violation of the human. In particular, I examine problem framings that seem distant from "the human" as an object of technological intervention and control. I take these as sites precisely because they are moments in which the figure of the human was pushed out of the picture in order to simplify and solve problems of governance. Yet, as I argue below, the conceptual orientations and practical consequences of framings of risk and benefit in biological research have been crucial sites for figuring "the human" in whose name projects in biotechnology are undertaken.

The chapter explores three registers in which these modes have played out: (1) in processes of characterizing and governing risk, for instance in notions of biological containment, where the figure of the human appears merely as a biological being potentially at risk from physical or public health hazards; (2) in a project of conceptual reorientation of the purpose of biological knowledge from ontological description to forms of control that invite black-boxing and automation of workflows, for instance in the synthetic biology's self-description as a reorientation from science to engineering; and (3) in the debates over what conceptual and discursive starting-point is appropriate to evaluating the potential of biologically transformative techniques for serving versus violating human integrity, for instance in current debates over human genome editing.

These registers developed more or less sequentially in the history of biotechnology, building upon (and thereby reinforcing) each other. They have been profoundly consequential for the development of biotechnology, both for the social, political and moral meanings associated with it, and in shaping the scientific and technological trajectories through which it has taken shape.

I argue that these registers are significant not only because they have informed the practical and normative character of the biosciences, but also because they are loci in which socially shared imaginations of human integrity are at stake. Also at stake, therefore, are the ways in which political capacity (or incapacity) has developed for asking whether and in what ways science will remain human.

Perfecting What?

For more than half a century, biotechnology has been one of the most important domains around which notions of human integrity have been elaborated and contested. Initially, this took the form of full-throated debate about the purpose and perfectibility of human life. In a 1962 conference on "Man and his future," the great evolutionary biologist Julian Huxley told a gathering of mid-twentieth century scientific luminaries that the question of how to employ technologies for reshaping human reproduction was, most fundamentally, a question of "what are people for?" (Huxley 1963). In 1969, Robert Sinsheimer, an accomplished and respected molecular biologist, wrote that technologies for the "defined genetic improvement of man" are a means "to carry on and consciously perfect" the human species (Sinsheimer 1969). One of the earliest and most insightful commentaries on the technological project of artificial control over human reproduction was offered by Hannah Arendt, who saw in it a "rebellion against human existence as it has been given," and the risk of enslavement to forms of know-how that would be de-humanizing. They would be de-humanizing, she suggested, not primarily because of the material transformations they would bring about, but because of the thoughtlessness such know-how would impose: the meanings of scientific postulates and technological proofs in which the human would be re-figured would "no longer lend themselves to normal expression in speech and thought" (Arendt 1958).

Although these ways of talking—in which the development of sciences and technologies of the human raises the question of what figure of the human those sciences and technologies serve—do persist, they have undergone profound transformation, particularly in elite spaces of scientific agenda-setting, bioethical deliberation and governance of biotechnology (Evans 2002). Some of these transformations trace back to the inception of biotechnology in the 1970s and have been largely conserved in the succeeding four decades. They are worth paying attention to because they have shaped the registers in which many political communities interrogate the meaning of developments in science and technology for human life, and at the same time figure and re-figure the human in whose name those technologies are made meaningful.

One of the central features of these transformations is that explicit questions of human integrity, purpose and perfectibility are often marked as vague and unwieldy and expelled from discourse, for instance in scientific and mainstream bioethical discussions of risk and benefit (Hurlbut 2018a, b). But those discussions remain central and consequential precisely by virtue of the fact that their assiduous efforts to draw down ethical debate to apparently more immediate and resolvable questions of risk and benefit entail a corollary, if subtle and often tacit, refiguring of the human.

From Understanding to Know-How

Hannah Arendt's reflections on the political meaning of science and technology's "rebellion against human existence as it has been given" offer a valuable conceptual entry-point for the project of this paper. Sixty years ago, Arendt opened her meditation on the human condition by reflecting upon the significance of the technological achievements and aspirations of her moment (Arendt 1958). Sputnik had been launched into orbit just a few months before, an event that was "second in importance to no other." Military implications notwithstanding, it was received with joy not because it represented an unprecedented human achievement that elicited "pride or awe at the tremendousness of human power and mastery," but because it elicited relief: "relief about the first 'step towards escape from man's imprisonment to the earth.'" This foray into a space long contemplated but never before colonized through human artifice marked the promise of radical human freedom: an unboundedness from the earth and a corollary emancipation from the place, character and form of human life that is, she argued, essential to the human condition.

Arendt oriented her project of elaborating "the human condition" as a response to this impulse to emancipation from the natural boundedness of human life. What worried Arendt was not primarily the advent of technologies like Sputnik, but the meanings of and for the human that the scientific aspirations behind them entailed—in effect, the visions of what human life could and should be that were insipient in the scientific and technological aspirations.

Sixty years later, the futures she imagined and the projects she pointed to as warranting such a meditation seem prescient. She pointed in particular to two projects that today are at the forefront of visions of technological promise and peril: the project of rendering life itself, and in particular human life, a product of artifice; and the project of displacing labor, essential for and constitutive of the most basic rhythms of human life, with technologies of automation. These projects were, in her view, more radical than seeking to escape the bounds of the earth because they aspired instead to escape the bounds of still more basic elements of the human condition: the givenness of life as an irrevocable bond with the rest of living nature, and the necessity of labor as the basis for transcending that necessity and working towards freedom.

Although much has changed in six decades, these technological futures remain imagined futures, even if perhaps less distant than they were in Arendt's day. In this sense, technological advances notwithstanding, these projects remain oriented by the same aspirations Arendt worried over.

Her central concern is worth recalling because it speaks to the question of this volume. The perils Arendt saw in these technological possibilities lay less with the material transformations of human life and labor that they could bring about than in the reimagining of the human that was insipient in the aspiration to these futures. Put differently, the aspiration to emancipation through artifice reflected an imagination of the human condition that departed from and thereby denigrated what Arendt saw as essentially human. Insofar as scientific knowledge and technique became the idiom in which the human beings would understand themselves and their condition, the capacity to speak of—and thereby live out—the human condition would be attenuated. "We who are earth bound creatures, and have begun to act as dwellers of the universe, will forever be unable to understand, that is, to think and speak about the things which nevertheless we are able to do" (Arendt 1958). Thus from this perspective, the question of whether science will remain human is less about science than about politics: it is less a question about the intrinsic character of knowledge and technique than about how knowledge and technique and the aspirations that animate them figure in (or displace) the basic repertoires of meaning and discourse in which humanity speaks of itself and its forms of life (Arendt 1958).

Arendt offers an important provocation that is worth engaging: the most consequential transformation of the human lies not first in its material remaking—the rendering artificial of the stuff of human life—but the conceptual reorientation in which the capacities "to think and speak" of what we are doing, and to calibrate action accordingly, are rendered subsidiary to the presumption that knowledge and technique are more real and foundational and therefore can and should be more determinative of the human condition than the modes through which we speak of them. Because speaking is also acting—for "speech is what makes man a political being" (Arendt 1958)—a fundamental transformation in the conditions for thinking and speaking is a transformation of the human itself.

Arendt responded to her own provocation by seeking to elaborate what is essential to the human condition, and thus what is threatened by the promise of technological emancipation from it. Yet if there is a limitation to Arendt's account, it is that she seems too prepared to assume that technical know-how entails a kind of technocracy; that is, that know-how necessarily displaces knowing-what we are doing (or, rather, the capacity to understand and orient ourselves by talking together about what we are doing), and thus engagement with the question of whether we ought to be doing it. This sort of displacement of democracy by technocracy (Horkheimer et al. 2007), of substantive rationality by instrumental reason (Habermas 1986), of lifeworld by technium (Kelly 2010) is not inevitable, and neither are the clean boundaries that such displacement presumes. For the meanings and modes of meaning-making that come into being alongside technological transformations are coproduced with them (Jasanoff 2004). Sliding too quickly into an *a priori* demarcation between the technical and the political, know-how and knowing-what, obscures the dynamics whereby valorizing know-how has entailed corollary transformation of practices of public moral-sense making and the figure of the human that at once underwrites and is elaborated through those practices.

In the remainder of the paper, I want to build upon Arendt's provocation that the sphere of shared meanings—and the political possibilities of collective sense-making that they are tied up with—must figure centrally in any answer to the question of whether science will remain human. The project of my analysis is to explore what emerges if we take Arendt's provocation as an historical rather than purely philosophical thesis. It asks: in the six decades since Arendt wrote, what has become of our capacities to speak about the things we are nevertheless able to do.

From Purpose to Risk

The need for science to be human—to be tethered to and work in the service of human integrity—has long been a core concern about biotechnology. This is evident in the more than half a century of explicit debate about the potential of biotechnology to transform (and thereby potentially attenuate) human life. Less explicit but equally important has been the development of approaches to risk and benefit that underpin accounts of the virtues of biotechnological research—of the aspirations, potentials and human futures in whose name it is undertaken.

One of the earliest and most consequential developments in early efforts to govern biotechnology was the advent of a crisp distinction between technical versus social and ethical dimensions of governance. This distinction was codified in the debates surrounding recombinant DNA in the 1970s. The advent of recombinant DNA marked the first moment in which life forms could be reengineered at the genetic level through the incorporation of foreign DNA, including DNA from evolutionarily distant species. The technology raised the specter of radical human control over biological systems whose nature and potential exceeded human understanding.

One of the most immediate concerns about early experiments in recombinant DNA was that they might inadvertently create biohazards such as novel pathogens that would endanger public health or genetically altered organisms that would be ecologically disruptive. However, these worries were nested within a larger range of political concerns about the emerging powers of science and technology and the need for political mechanisms to calibrate them to democratically articulated visions of the good. Thus, the risks that observers of the day associated with recombinant DNA were not limited to the potential for physical harms, but extended to the worry that scientists would charge forward with transformations of the living world that were undertaken without ensuring that scientific projects were genuinely in the public interest (Krimsky 1984). This was as much a worry about how normative judgments about scientific and technological agendas would be made as about which agendas were themselves virtuous or vicious (Hurlbut 2015a, b).

The community of scientists with a stake in recombinant DNA research responded to these concerns by undertaking a set of steps which four decades later continue to be remembered as exemplary of scientific responsibility (Hurlbut 2015b). Leading figures within this community called for a moratorium on research, and in early 1975, an international group of scientists gathered at Asilomar, California to evaluate whether and how recombinant DNA research could proceed. The group of scientists who gathered at Asilomar positioned themselves as the relevant authority to govern this emerging domain of science and technology. They defined what risks warrant worry, what scientific and technological futures are possible, and what questions must be addressed in order to proceed with research. The explicit aim of the meeting was to establish a framework for self-governance in order to forestall and avoid formal regulation and the forms of political deliberation that would accompany it.

One of the most important outputs of this meeting for self-governance was the concept of biological containment: the idea that experimental organisms could be engineered to be self-limiting and thus could do the work of governance themselves, for instance by being dependent on specific culture conditions that (presumably) guaranteed that they simply could not survive were they to escape the space of the laboratory. Biological containment seemed to solve a problem that other forms of governance could not. Whereas expert judgments about risk, protocols for laboratory safety and techniques of physical containment all relied upon human beings acting responsibly and in accordance with rules, biological containment rendered the problem of governance a technological problem that could be solved completely and exclusively through technical means (Hurlbut 2018a, b).

Importantly, in 1975 biological containment was a proposal, not a technology. There was not yet a way to do it. Although a whole range of technologies of have since been developed, it was initially—and remains—first a reformulation of the problem of governance and only secondarily a technological solution to it. Its effect in 1975 (reproduced countless times since then) was to transform a question of "what are the right modes of deliberation and judgment to govern scientific know-how?" to "what forms of scientific know-how can make the problem of governance, and the corollary question of who should decide and in what terms, go away?" In effect, the problem of scientific responsibility could be delegated to an automaton—a biological machinery, whose intrinsically self-limiting character would ostensibly render moot the question of "in the service of what human good are risks being taken?" And it would do so by shifting the question of governance out of "normal expression in speech and thought" into an idiom of know-how that at the same time devalued the significance and authority of political judgment.

Today, biological containment is central to the governance of biotechnology. Synthetic biologists have imagined a menagerie of biological mechanisms for achieving biological containment, particularly for "isolating" an artificial organism from the natural environment even as it is released into the living world (Mandell et al. 2015). This strategy locates guarantees of safety (and, thus, of acceptability) in the technological artifact itself. At the same time, it appears to obviate the need for asking whether that technological artifact is desirable, and, furthermore, what larger aspirations and vision of the good would underwrite the judgment that it is. By relieving synthetic biologists of the need to ask these questions, it allows them to insist that the tools of bioengineering should be made widely available with low barriers of entry. With risk governance "automated" by building mechanisms of containment into biological systems, life can then become an artificial medium for "playing around" with living systems (Endy 2005; Kuiken 2018).

I have argued elsewhere that the concept of biological containment normalized an approach to governing biotechnology by (a) rendering "risks" as technology-specific problems of containment; (b) normalizing and intensifying of an asymmetrical treatment of risks as delimited and manageable and benefits as therefore unmitigated, expansive and socially transformative; and (c) legitimating a politics of agenda setting that defers to scientific declarations of what is possible and plausible as a prerequisite to engaging in normative evaluation and governance (Hurlbut 2018a, b). These three moves, which have become normative in American governance of biotechnology, are also consequential for the ways they re-figure the implications of biotechnology for human life. They refigure it first by treating the space of artifice (the laboratory) and the space of nature as radically separate; and second, by rendering questions of the good, which are intrinsically political, as subsidiary to questions of technically described risk. Put differently, the task of "think[ing] and speak[ing] about the things which nevertheless we are able to do" becomes dependent upon prior scientific authorization of political deliberation, with authorization based upon expert judgments about the (un)reality of risk and thus the (ir)relevance of ethical considerations. Science becomes less human insofar as human concerns about the good—and the modes of political engagement through which they must be confronted—are expelled from the laboratory and marked as occupying a separate sphere.

This is a subtle but significant refiguring of the human. In the last four decades, this now-standard move to delimit the terms and scope of political deliberation by technical judgment has not merely elevated the role of scientific authority in governance. It has also led to refiguring collective political judgment and public reasoning (and the forms of "normal expression in speech and thought" that constitute it) a threat to scientific progress, and thus to humanity as such. Public "misunderstanding," "reaction," and "backlash" are invoked as the primary threat to science and thus to forms of human wellbeing it promises to achieve (Irwin and Wynne 2003).

Here is an example of this discourse in action, a singular instance drawn from a nearly limitless supply of similar examples: Negative assessments that

> cast synthetic biology as a threat to human well-being [...] diminish the fact that the core ethos of synthetic biology [...] is a design process that aims to make human inventions reliable, predictable and safe [...] synthetic organisms might turn out to be the best solution for global health challenges and ecological problems. (Moe-Behrens et al. 2013).

Yet, the authors continue, such assessments themselves point to the important political role of technologies of biological containment by revealing "how the public may react when coordinated efforts towards executing containment and control strategies are not highly visible" (Ibidem).

In this account, technologies of containment are what have been called "experiments of reassurance" (Rabinow and Bennett 2012). They aim to quiet public calls for evaluating the stakes of scientific research by appearing to eliminate risk. In this sense, they are *political* technologies whose role is to dispel ideas of "tampering with nature or 'playing God'" by assigning jurisdictional authority to technical experts to judge whether such morally inflected notions have any relevance (Moe-Behrens et al. 2013; cf. Presidential Commission 2010). The implication is that by constructing a material separation between "contained" engineered organisms and natural environments, nature retains its integrity by definition, and the scientist's action and its consequences are delimited to the sphere of laboratory artifice and thus are in no sense "playing God." Thus, biotechnologists in a single move claim the authority both to govern their own creations and to designate what forms of public sense-making are illegitimate and contradictory to the (political) task of judging what futures are desirable and good.

Thus, refiguring risk as technical problem that is solvable through technological solutions like biological containment has also entailed refiguring moral uncertainty about biotechnology as an expression of public ignorance that undermines good ethical judgment. It is not simply that risk-talk speaks louder than the "normal expression in speech and thought" in the public square, but that such speech and thought have come to be seen at odds with those aspirations to the good that are appropriate to the human condition. Human integrity, wellbeing and progress, in this account, depends upon a politics that defers to the judgment of technical authority in the name of the integrity of that very politics.

The apparent absence of the human from the sterile space of the laboratory is a significant refiguring of the human. This is so because it represents a profound transformation in the modes of sense-making through which human beings constitute political life. Insofar as humanity's relationship to the living world and to itself is both regulated by and constituted through the political work of making sense of and orienting those relationships, these technologies of biological control also represent transmutations of this basic, human, political relationality. These transmutations are made in name of tethering political modes of sense-making to scientific authority, where that authority is exercised to declare both what is real and thus beyond question, and what expressions of moral and political judgment are defective.[1] Thus while laboratory work may, in these contexts, still be undertaken by human hands, it has rejected a crucial dimension of its humanness, namely its embeddedness within and its expressiveness of an imagination of the good.

From Risk to Telos

The politics of judging risk discussed above have been equally consequential for the ways benefits of biotechnology are imagined and debated. As I have already noted, one effect of treating risk as a technical problem is that discussions of the social and moral meanings of scientific research tend to be treated as downstream of judgments about risk. That is, social and moral meanings tend to be treated as "implications", i.e. as consequences of the products science and technology moving out into and impacting the social world, and not as features of those scientific and technological projects themselves. The focus of public ethical debate tends, therefore, to leave aside the projects, processes and animating aspirations of scientific research itself.

In this sense, social and ethical questions tend to be regulated by scientific assessments of what technological advances are plausible and what known risks might accompany them. Generally speaking, a technological scenario must first be deemed plausible and potential risks and "unintended consequences" must be identified before ethical evaluation of "impacts" and "consequences" is seen as appropriate and costs and benefits can be weighed. This stands in remarkable contrast to the ways biotechnology's potential for generating revolutions and "solutions to major societal needs" (National Research Council 2009) tend to be framed as limitless, even if dependent upon science being set free to realize them. For example, the U.S. Bioeconomy Blueprint which was issued by the Obama Administration points to trends and technical advances in the biosciences that include "technologies not yet imagined" that nevertheless "foreshadow major advances" in health, agriculture, energy, and manufacturing (U.S. White House 2014). Risk, safety, and ethical concerns are, by contrast, treated as knowable, manageable, and generally a consequence not of innovation, but of "misuse" (U.S. White House 2014; cf. Doezema and Hurlbut 2017).

This asymmetry between revolutionary benefits and circumscribed risks has become foundational for agenda-setting in the biological sciences, and for a corollary transformation in the professional identity and ethos of a growing community

---

[1] Importantly, the assertion of (and political deference to) scientific judgment is not merely epistemological. That is, it is not merely a matter of deference to (scientific) facts that then inform (political) value judgments. Rather, it is grounded in an imagination of the right ordering of human institutions, where that ordering reflects judgments about what orderings of power and authority are appropriate to human flourishing. In this sense, they are constitutional judgments (Hurlbut 2017)—judgments about what forms of political order are demanded by the nature of human life and world.

within the biological sciences, particularly in synthetic biology (Endy 2005). Synthetic biology grew up less around actual techniques or technologies than around an ethos and approach and an imagination of the eventual usefulness of the field's (future) achievements. Its leading voices proclaimed that synthetic biology would engineer life to achieve forms of human benefit that would otherwise be unattainable.

The project of synthetic biology figures centrally in what has been called the "New Biology" (NRC 2009) a phase of the biological sciences in which they are predicted to "converge" with engineering, computing, and information sciences (Sharp et al. 2011). One of the central, animating visions in synthetic biology has been the promise that life can be rationalized into a repertoire of engineering tools, for instance, standardized parts that, like Lego bricks, can be assembled into novel biological machines that will perform in predictable ways. Just as a programmer need not understand the layers of software and hardware beneath the coding language she is using, the user of these biological parts would not need to know how they function or be able to construct or characterize them.

This is a political vision as well as a technical one. Standardization and black boxing of biological systems would allow for the "democratization" of synthetic biology and the "deskilling" of its practitioners such that there would be a low barrier of entry to participation. This has, for instance, been the animating vision behind the amateur "biohacker" movement. In industrial contexts, synthetic biology-based visions of life as standardizable machinery has underwritten the notion that industrial production of bespoke biotechnological products need not rely on the work of skilled craftsmen, but can be organized into automated workflows that require only relatively unskilled laborers to push the buttons.

What is interesting about this vision of automated bioengineering is that it is central to the public identity that self-styled synthetic biology companies like Amyris seek to project, and yet it is utterly belied in practice. Processes of production that are supposed to be automated and managed by all-seeing software and executed by machines actually require constant intervention by skilled workers who must troubleshoot and tinker to make the biochemistry work because living systems are complex in ways that exceed the powers of machineries and mechanisms that would standardize and discipline them. Both leadership and workers at Amyris know this all too well. Yet, the credibility (and financial value) of the company is tied up with its promise of delivering human-less science—automatic, predictable, replicable, scalable. Thus, when outsiders have been given tours of the Amyris "shop floor," employees have been forewarned to do all tweaking and troubleshooting necessary to make the machines function smoothly *before* the tour group passes through so that the workers' own roles as essential contributors to the work of the lab is rendered invisible. Thus, what the outsider sees on the inside is a false representation of an automated environment, one that is supposed to be proof-in-practice of the industrial controllability of life (Hammang, unpublished manuscript).

Although these frontstage-backstage dynamics are interesting in themselves (Hilgartner 2000), I draw attention to them in this context because they are performances of an elision of labor, and thus of the human beings who do it. This is consequential for three reasons. First, the promise that biotechnological labor can be automated underwrites the epistemological claim that life is information that is rationalizable and manipulable (Bennett 2018). Second, by constructing life as mere information, human life is also recast with the rest of living nature as just one possible version of a malleable (and improvable) program; there is no essence to it beyond the basic code required to make it run. Third, it positions the human qua scientist as occupying an extremely technically powerful but ethically low-stakes role.

The promise of deskilling reduces scientific creation to trivial "playing around" (Endy 2005). The labor of knowing life is reduced to mere tinkering, and one knows enough when one has managed to make life do what one wants. At the same time, it casts scientific work as at once irresponsible and unobjectionable because this sort of labor—and the aspiration to get other things to do your work for you—is foundational and quintessentially human. This characterization of science as play and as fundamentally human marks the sources of contradictory ethical worries—about the integrity of nature and about responsibility of scientific work to shared norms and visions of the good—as contingent, unfounded and less human than this mode of technoscientific creation. In his public talks, Drew Endy, one of the leading figures in (and evangelists for) synthetic biology, sometimes shows an image of playful rabbits with this quote superimposed: "Play is older than culture, for culture, however inadequately defined, always presupposes human society, and animals have not waited for man to teach them their playing." In Endy's account, tinkering with the stuff of life, and thereby making life (including human life) the medium of expression of human artifice, is a natural and appropriate pre-cultural human impulse. It precedes and exceeds culture, society and the limitations on human expression that they impose. It is at once defining of human uniqueness and a superior expression of the continuity between human life and the living world than remaining helplessly subject to what evolution has produced.

In this sense, the promise of automation and the ethic of playfulness figure life in the same way: life is an open-ended, rationalizable information system that is at once infinitely engineer-able and is as-yet imperfect, incomplete and therefore open to innovation and improvement. Remaking life is presented as at once a mundane and yet idealized expression of the basic human impulse to creative artifice—like ancient peoples fashioning tools from stones. Thus, the aspirations behind it, the self-alienation from life–as-given that it entails, and the sorts of freedom from natural constraint that it promises are figured as essentially human because purely playful. Synthetic biologists present this sort of play as beyond reproach because it is an expression of a defining human impulse to creative exploration which is an end in itself.

Arendt wrote: "The human artifice of the world separates human existence from all mere animal environment, but life itself is outside this artificial world, and through life man remains related to all other living organisms" (Arendt 1958, 2). The project of synthetic biology rejects this formulation, turning it on its head. It asserts that life can and will be the very core of the artificial world, and that humanity shall remain related to other living organisms not by being subject to the same nature, but by engaging in the forms of creative play that it shares with other living creatures, and which gave rise via evolution to the diversity of life itself.

The corollary to this construction of science as play is the notion that it is not the ethos of science but only its material effects that should be the focus of ethical judgment. Risk-taking is figured as virtuous: only the specific effects of new technologies should be evaluated and, where risky, guarded against. At the same time, synthetic biology promises that play, even if an end in itself, will be productive: finding ways of improving upon naturally occuring living systems will improve human life—health will be secured, life extended, societies fed, clean energy supplied in abundance, etc. Thus, benefits tend to be presumed (even of "technologies not yet imagined" (White House 2014)), whereas harms have to be demonstrated. Harms are understood as unintended consequences or as "misuses" of the otherwise beneficial free and creative expression of (essentially human) scientific inquiry and innovation.

This asymmetry underwrites a valorization of synthetic biology as able to serve the human in unprecedented ways. Even as synthetic biologists present themselves as simply at play, they simultaneously claim the ability and thus the responsibility to address central, enduring needs of human life—food, health, energy, environmental security, and so forth. The field has constructed itself as able to respond, and thus as the right response, to basic problems of human welfare and security. In the same talk in which he shows the playful rabbits, Endy includes a slide titled "What is our Telos?" The answers he supplies are: "Provide for all of humanity; stabilize and recover natural biodiversity; take infectious & other diseases off the table; Enable a culture of citizenship; understand via building." This is an ambitious and expansive agenda for an enterprise whose ethos is play for its own sake (and whose members nevertheless bristle or roll their eyes at the suggestion that they are somehow "playing God.") Yet it is certainly coherent with a vision of biology in which life is engineer-able from its most basic subunits up, and that treats complexity as a matter of linear scaling up of standardized subunits. (After all, Legoland is nothing more than a multitude of Lego bricks, rationally assembled.) This is not just a scientific vision, but a political one: it is a vision of the right human response to the vulnerabilities and needs of human life, from sustenance to citizenship. In this sense, the human condition is refigured as a design problem, one that both demands and authorizes a scientific enterprise that supplies it with "solutions."

What I want to problematize is not primarily the scope of the technological promises or their credibility, but the way responsibility for the question "what is our telos?" is appropriated, and the way the human is thereby re-figured as in need of, and as rightly subject to this account of purpose and end. The problem of political and normative judgement—of asking what sort of lives we should live, what sort of futures we should aspire to—is positioned as secondary to the question of what beneficent technological futures are possible; and the forms of action and aspiration that are appropriate to human lives and societies are figured as at once dependent upon, and given their fullest expression through, science and technology. Thus, synthetic biology with its (aspirationally) de-humanized machineries of production appropriates the figure of the human as its telos, even as it displaces arenas of speech and action in which the human is crucially given expression.

Figuring the Human

In the previous two sections, I argued that the ways expert judgments of risk and benefit around biotechnology have come to be privileged have entailed not only re-figuring biotechnology and the forms of intervention in the world that it represents, but also re-figuring questions about—and modes of making sense of—what is desirable and good for human lives and societies.

Technological futures are highly significant sites in which human futures are imagined and enacted in contemporary public life, futures which extend beyond material technological changes to notions of the right ways to evaluate and deliberate about the good, in what terms, guided by what norms and priorities, and drawing upon and deferring to what forms of authority. We have seen how scientific experts have sought to draw down or silence certain public ethical concerns—concerns given "normal expression in speech and thought" (Arendt 1958)—by characterizing them as grounded in unrealistic assessments of risk. And we have seen how at the same time those experts demand freedom from constraint by claiming to be able to see the future human benefits of an ever-expanding repertoire of technical know-how. In this section I offer a brief discussion of a different sort of move: where a longstanding ethical limit, apparently grounded in a judgment about human integrity that seems to stand apart from and above the technological conditions of the moment, is declared untenable and unreasonable precisely because it has become technologically feasible to transgress it.

It has been recognized for half a century that genetically engineering future children touches upon fundamental questions of human rights, dignity and integrity that cannot be reduced to questions of technical risk. Partly in recognition of the hard-to-resolve ethical questions associated with human genetic engineering, there has been a longstanding, internationally shared assumption in mainstream scientific and bioethical thinking that human germline genetic engineering is unacceptable. This is reflected in the US Recombinant DNA Advisory Committee's longstanding policy of not entertaining proposals for germline focused experiments, in article 13 of the Oviedo Convention which prohibits heritable human genetic engineering, in declarations of the UNESCO International Bioethics Committee, and in numerous other authoritative declarations (National Institutes of Health 1985; COE 2017; UNESCO 2015).

The advent of genome editing technologies, and in particular CRISPR/Cas9, has reopened the question. CRISPR has made it conceivable that one could make targeted genetic changes to human gametes or embryos, thereby producing heritable genetic alterations in the child that would result. The advent of this technology has led some prominent scientific authorities to declare that the longstanding prohibition was never more than a temporary, an ethical judgment that was really just a scientific judgment that genetic modification was as-yet too technically difficult or risky. And now, because scientific circumstances have changed, the prohibition is obsolete.

Thus, a norm that may have included technical considerations about safety but was by no means limited to these considerations, is being reframed as reflecting no actual consensus beyond technical judgments about the state of a technology. Take, for instance, a statement made by Richard Hynes, co-chair of the U.S. National Academies of Sciences, Engineering and Medicine (NASEM) study committee tasked with evaluating the scientific, ethical and governance issues associated with human genome editing. He made the following statement at the NASEM event marking the release of the committee's report: "There has been a line drawn by many that says…you should refrain. That was mostly because there was no way of considering how to do that at all…so nobody was arguing that it should be done." (NASEM 2017). This major reversal in ethical and policy judgment is figured as a modest and entirely appropriate adjustment made in response to an incremental technological step forward.

I want to point out in particular what this implies for the figure of the human. Whereas in other areas of biotechnology, it has been possible to make the case that the primary ethical concerns are technical and consequentialist (e.g. the potential for negative outcomes like biohazards and similar technically-intelligible risks), in the case of human genetic engineering, there is essentially no disagreement that questions of human rights, integrity and dignity are at stake. Furthermore, these ideas have a long history of robust interpretation, not only in moral and political philosophy, but in the legal and political institutions that have, in many cases, been organized around and codified some version of a concept of—and commitment to—human integrity, rights or dignity. In short, these are not fragile concepts or arrangements, even if their forms vary and are the subject of ongoing interpretation, debate and disagreement.

And yet, where discussions about human genome editing have unfolded within the context of wider discussions about biotechnology, they have tended to slip into the patterns that I pointed to in the previous sections (Jasanoff and Hurlbut 2018). They have privileged discussions of physical risk as the primary ethical concern, they have treated the advent of CRISPR as opening a novel discussion about human integrity rather than treating the technological possibility as one new point in a larger and older constellation of normative issues. And in so doing, they have shifted the idiom of ethical debate into the same kind of reactive consequentialism discussed above: the goods and bads, benefits and harms are understood in terms of what the *consequences* would be of engaging in genetic modification of a person (Hurlbut 2018a, b). The discussions are not about what aspirations lie behind the development of such techniques, nor about what moral imaginations and images of the human should guide those processes. The question of human integrity as such—e.g. the question of whether there is a right not to be engineered—is displaced by the far narrower question of whether a particular genetic intervention might or might not cause a deleterious biological consequence.

This is not merely a transmutation of a categorical prohibition into an exercise in risk-benefit analysis. It is also a refiguring of the human. For the last several decades, the prohibition against human genetic engineering was generally understood as being grounded in a judgment about how humans ought not be treated by virtue of being human. Although the ethical reasoning behind this judgment was elaborated to varying degrees and in different ways in different contexts, there has nevertheless been a categorical commitment to human integrity unqualified by any details about the specific means by which it might be compromised.

Recent statements like the NASEM's, by contrast, call for development of techniques that contravene this longstanding prohibition once risks can be overcome and safety and efficacy can be demonstrated (although what would constitute adequate demonstration is itself an open question, one that occludes—but nevertheless entails—tacit judgments about what lives are worth living, and what sort of people there should be.) At the same time, that report puts forward a list of universal ethical principles to guide and govern human genome editing in the name of ensuring that the integrity of human persons and societies is not violated in the process (NASEM 2017). These principles include, for example, *justice*: the notion that access to these techniques must be equitably distributed lest they lead to the creation of imbalance, disparity or inequality. In this framing, it becomes imperative not only that these techniques are applied to human beings, but that they are applied equally to all human beings. The harms associated with their use are treated as a function not of the technology, but of its unequal distribution and application. In effect, the technology becomes essential to the achievement of (one idea of) the sorts of human lives that ought to be brought into existence (and not) and of right social relations within the human community. It is refigured as a technology for achieving justice and human integrity, even as the meaning of "human integrity" is subtly rewritten in the image of the technology. "Human dignity" is approached in similar terms. The report treats the question of whether human dignity would be compromised as a "speculative consequence" of genetic engineering—the possibility that people might come to think of engineered individuals as somehow less than human, thereby attenuating their dignity. In this sense, "dignity" is understood in terms of the technology and its possible effects, rather than the other way around.

My point is that in these discussions, the human is re-figured as an unstable object of technological intervention and control, whereas the technology and the scientific enterprise, ethos and projects that give rise to it are treated as fixed and given. This is reflected in certain respects even in direct bioethical engagement with human integrity and dignity. In these philosophical discussions, the primary questions tend to include: whether genetic control somehow compromises an essential moral relationship to one's offspring; whether there is an essential human nature or identity that ought not be altered; whether the human genome is a kind of common property that ought not be appropriated and altered by a small minority without the consent of the collective; whether boundaries need to be described that distinguish restoration of normal human health from nature-altering enhancement, and so forth.

The modes of collective moral sense making—that is the work of moral and political custodianship of the human, which is itself a fundamental expression of being human—are also at stake. If collectively imagining futures and making sense of their meanings is a fundamentally human activity, then refiguring the mode in which humans should make sense of their own humanity, and the ways in which repertoires of technical know-how should shape and discipline "normal expression in speech and thought" is itself a profound refiguring of the human. Yet the privileging of technical constructions of risk and benefit, and treating normative questions as attached to the effects, rather than the causes of biotechnology has meant discounting engagement with the forms of meaning which are being deployed "to understand, that is, to think and speak about the things which nevertheless we are able to do" (Arendt 1958). By extracting (scientific) *know-how* from the (political) arena of *knowing-what*, the forms of speech and thought that science valorizes are largely exempted from the question of how we as human beings should speak and think about the human condition. Put differently, the question of how we should understand, and the kind of political engagement that that question demands of us as a human community, is treated as coming after (ostensibly) purely scientific questions about technological capabilities and their potential consequences are already resolved, even though the way those technical questions are framed has a profound effect upon how we understand, and upon the politics of seeking collective understanding.

One important consequence of privileging technical modes of evaluation is that this creates an imperative for "basic" research to evaluate and address questions of safety. In these reports, and in much of the discourse emanating from scientific circles more broadly, the notion that the human germline is inviolable has been eclipsed by the inviolability of basic research. Science that understands itself to be the custodian of the human because uniquely able to supply the knowledge and know-how necessary to imagine, identify and enact good human futures, also understands itself as an exemplar of true human freedom. Science that figures itself as an agent of human emancipation simultaneously figures the essence of human freedom as the unrestricted play of scientific inquiry and artifice. In this sense, science re-figures the human, even as it figures itself as quintessentially human and in the service of the human.

What Science, Which Human?

The question of whether science will remain human depends in profound ways upon the role that scientific authority plays in the way the human is figured. I have argued that the development of biotechnology has been accompanied by the development of ways of speaking and thinking that position science as an essential locus of authority for declaring what the human condition is, and what it ought to be—that is, for offering images of human life that it claims at once to illuminate and emancipate. I have argued that some of the primary moves that have developed over the last four decades to draw down public deliberation from explicit engagement with questions of human integrity, purpose and good, and to refocus it on epistemic questions of risk and benefit have, in fact, played a crucial role in re-figuring the images of the human that guide and govern modes of moral sense-making. These ways of refiguring the human are embedded not only in the way science and technology are talked about in the public sphere, but in the practices, approaches and forms of self-understanding that constitute science itself.

By segregating scientific authority from questions that are marked as "moral and religious," science has also sought to reframe those questions as "merely" political and therefore as contingent, parochial and irresolvable and therefore distinct from the ostensibly definitive, universal, and answerable epistemic questions over which it claims jurisdiction. This is, I have argued, a mode not only of carving up and asserting jurisdiction in relation to other forms of social authority, but a wholesale refiguring of how the figure of the human should be understood, and the forms of politics appropriate to that task. This transformation of the contours of politics is itself a refiguring of the human because it is an account of what ordering of human institutions is in the service of—and represents a commitment to—human integrity.

It is unquestionable that the ways of thinking and speaking that I have described demand philosophical reflection and critique. Yet I want to emphasize that the question of whether science will remain human—and what image of the human science will imagine and invoke to authorize its agendas—is a historical and cultural question as well as a philosophical one. The ideas of the "human" against which the ethos and ambitions of science are measured are not exogenous, at least not in the ways this question is asked in public life. Rather, notions of both "science" and "human" have to be recognized as culturally situated and as regulative of the modes of normative and political sense-making through which the question is asked and answered.

Indeed, one of the primary reasons for attending to the canonical scientific moves that I have explored in this paper is that they have consequentially informed the concepts, terms and loci of debate around human integrity and progress in public life more generally (although it is beyond the scope of this paper to demonstrate this empirically). The figure of the human that informs practices of political and moral sense-making is at once a cause and an effect of those practices. Transformation in the ways questions are asked, the imaginaries that give shape to debate, and the ordering of the forms of institutional authority that weigh in on political deliberation not only shape what normative answers are seen as right and acceptable, but themselves become referents for culturally shared imaginations of rightness—both epistemic and moral—that privilege certain ways of speaking and thinking while displacing others. This is reflected in arrangements of forms of knowledge and inquiry beyond science itself, including, for instance, the kinds of philosophical and ethical inquiry that have grown up in conjunction with biotechnology (e.g. professional bioethics). Forms of ethical inquiry that are preoccupied with the "impacts" and "consequences" of particular technology, or which position themselves as policing the terms of discourse, for instance translating "religious" formulations into secular terms or displacing questions about "the fundamental nature of human life and the dignity and worth of the individual human being" with more narrow questions about the risks and benefits of the biotechnology *du jour* (Presidential Commission 1982; cf. Hurlbut 2018a, b), have played a central role in facilitating the sorts of transformations that this paper has explored.

Science plays an outsized role in shaping how we see we behold the human—*ide ho Anthropos*—in contemporary public life, even as it distances itself, and thus washes its hands of responsibility for the social meanings of its role. It invites (and demands) that we behold the human in the light of particular imaginations of progress, power and virtue, and in ways that take as given particular forms of knowledge, authority and judgement. In this sense, science takes responsibility by denying its own responsibility, claiming the capacity to offer life-transforming know-how and asserting the authority to declare what futures are possible, desirable and good, even as it declares that the ambiguous moral meanings and potentially undesirable consequences associated with these futures are problems for others.

Yet science is a human enterprise. If science is to remain human, it must cultivate ways to recognize itself as central to, and not outside of and above, the all too human arena of political life. For it is within this arena, not in spite of it, that we as a human community must seek to "understand…that which nevertheless we are able to do," and to aspire to those forms of "normal expression in speech and thought" through which we can calibrate our modes of collective speaking and thinking to the good.

References

Arendt, Hannah. 1958. *The Human Condition*. Chicago: University of Chicago Press.

Bennett, Gaymon. 2018. Digital enhancement: Life and the future of gene editing. In *Gene editing, law, and the environment: Life beyond the human*, ed. Irus Braverman, 169–186. New York: Routledge.

Committee on a New Biology for the 21st Century: Ensuring the United States Leads the Coming Biology Revolution; National Research Council. 2009. *A New Biology for the 21st Century*. Washington, DC: The National Academies Press.

Council of Europe. 2017. Statement on Genome Editing Technologies. *Bioethics*. Accessed 17 Feb 2017. http://www.coe.int/en/web/bioethics/news/-/asset_publisher/DcE9CvEiHMnp/content/gene-editing

Doezema, Tess, and J. Benjamin Hurlbut. 2017. Technologies of Governance: Science, state and citizen in visions of the bioeconomy. In *The bioeconomy: Life, technology and capital in the 21st century*, ed. V. Pavone and J. Goven, 49–71. Cham: Palgrave Macmillan.

Endy, Drew. 2005. Foundations for engineering biology. *Nature* 438(7067): 449–453.

Evans, John H. 2002. *Playing god? Human genetic engineering and the rationalization of public bioethical debate*. University Of Chicago Press.

Habermas, Jürgen. 1986. *The Theory of Communicative Action*. Cambridge: Polity Press.

Hammang, Anne. Unpublished Manuscript. Routine Interruption: Standardizing Life and Labor in Synthetic Biology.

Hilgartner, Stephen. 2000. *Science on Stage: Expert Advice as Public Drama*. Redwood City (CA): Stanford University Press.

———. 2015. Vanguards, Visions and the Synthetic Biology Revolution. In *Science and Democracy: Making Knowledge and Making Power in the Biosciences and beyond*, ed. Stephen Hilgartner, Clark Miller, and Rob Hagendijk. New York: Routledge.

Horkheimer, Max, Theodor W. Adorno, and Gunzelin Schmid Noerr. 2007. *Dialectic of Enlightenment*. Trans. Edmund Jephcott. Stanford: Stanford University Press.

Hurlbut, J. Benjamin. 2015a. Remembering the Future: Science, Law, and the Legacy of Asilomar. In *Dreamscapes of Modernity: Sociotechical Imaginaries and the Fabrication of Power*, ed. Sheila Jasanoff and Sang-Hyun Kim, 126–151. Chicago: University Of Chicago Press.

———. 2015b. Limits of Responsibility: Genome Editing, Asilomar, and the Politics of Deliberation. *Hastings Center Report* 45(5): 11–14.

———. 2017. *Experiments in Democracy: Human Embryo Research and the Politics of Bioethics*. New York: Columbia University Press.

———. 2018a. In CRISPR's World: Genome Editing and the Politics of Global Science. In *Cambridge Handbook of Genomics and Society*, ed. Stephen Hilgartner and Barbara Prainsack. Cambridge UK: Cambridge University Press.

———. 2018b. Laws of Containment: Control Without Limits in the New Biology. In *Gene Editing, Law, and the Environment: Life Beyond the Human*, ed. Irus Braverman, 77–94. New York: Routledge.

Huxley, Julian. 1963. In ed. Wolstenholme, Gordon. *Man and His Future*. Boston: Little Brown & Co.

Irwin, Alan, and Brian Wynne. 2003. *Misunderstanding science?: The public reconstruction of science and technology*. Cambridge University Press.

Jasanoff, Sheila. 2004. Ordering Knowledge, Ordering Society. In *States of Knowledge: The Co-Production of Science and Social Order*, ed. S. Jasanoff, 13–45. London: Routledge.

Jasanoff, Sheila, and J. Benjamin Hurlbut. 2018. A Global Observatory for Gene Editing. *Nature* 555 (7697): 435.

Kelly, Kevin. 2010. *What Technology Wants*. New York: Viking.

Krimsky, Sheldon. 1984. *Genetic Alchemy: The Social History of the Recombinant DNA Controversy*. Cambridge (MA): The MIT Press.

Kuiken, Todd. 2018. Vigilante environmentalism: Are gene drives changing how we value and govern ecosystems? In *Gene editing, law, and the environment: Life beyond the human*, ed. Irus Braverman, 95–112. New York: Routledge.

Mandell, Daniel J., Marc J. Lajoie, Michael T. Mee, Ryo Takeuchi, Gleb Kuznetsov, Julie E. Norville, Christopher J. Gregg, Barry L. Stoddard, and George M. Church. 2015. Biocontainment of genetically modified organisms by synthetic protein design. *Nature* 518 (7537): 55–60.

Moe-Behrens, Gerd H.G., Rene Davis, and Karmella A. Haynes. 2013, January 25. Preparing synthetic biology for the world. *Frontiers in Microbiology* 4.

National Academies of Science, Engineering and Medicine (NASEM). *Public Report Release Event for Human Genome Editing: Science, Ethics, and Governance*. February 14, 2017. Washington DC.

National Institutes of Health. 1985. Recombinant DNA research; proposed actions under guidelines. *Federal Register* 50(160): 33464.

National Research Council. 2009. *A new biology for the 21st century*. Washington, D.C.: The National Academies Press.

Rabinow, Paul, and Gaymon Bennett. 2012. *Designing human practices: An experiment with synthetic biology*. Chicago/London: University Of Chicago Press.

President's Commission for the Study of Ethical Problems in Medicine and Biomedical and Behavioral Research. 1982. *Splicing life: A report on the social and ethical issues of genetic engineering with human beings*. Washington, D.C: The Commission.

Presidential Commission on Bioethical Issues in Biomedical Research. 2010. *New directions: The ethics of synthetic biology and emerging technologies*. Washington DC: The Commission.

Sharp, P.A., C.L. Cooney, M.A. Kastner, J. Lees, R. Sasisekharan, M.B. Yaffe, and S.N. Bhatia, et al. 2011. The Third Revolution: The Convergence of the Life Sciences, Physical Sciences, and Engineering. *Massachusetts Institute of Technology*.

Sinsheimer, Robert L. 1969. The Prospect for Designed Genetic Change. *American Scientist* 57 (1): 134–142.

US White House. (2014). National Bioeconomy Blueprint. http://www.whitehouse.gov/administration/eop/ostp/library/bioeconomy.

UNESCO International Bioethics Committee. 2015. Report of the IBC on updating its reflection on the Human Genome and Human Rights(HS/YES/IBC-22/15/2 REV.2), Paris. October 2.

The Dehumanization of Technoscience

Alfredo Marcos

Diagnosis of a Problem: The Dehumanization of Technoscience

Technoscience and Its Semantic Field

Science and *technique* are two different realities from a historical and conceptual viewpoint. Indeed, there is technique since the origin of the human being, while science was born, at the earliest, alongside writing, around the year 3000 BC. There have been civilizations whose technical development was not accompanied by a corresponding development in science, as it was the case in ancient China and late medieval Europe. Inversely, some cultures, like the Greek culture, witnessed major scientific breakthroughs without a parallel technical evolution.

However, at a given moment, at the beginning of the Modern Era, they started to cooperate. Baconian philosophy issued an invitation to explore nature scientifically in order to master it technically. The two philosophical roots of Modernity, empiricism and rationalism, favored the cooperation between science and technique. Moreover, in some sense, modernity is precisely this: an age guided by the cooperation between scientific research and technical development. *Technology* was thus born. Their collaboration increased to such a point that, at the end of the modern era, it became a symbiotic entity that we may call *technoscience*.

In recent decades, technoscience has turned its attention toward living beings, considering them both an object of study and a field of intervention. The discovery of the structure and function of nucleic acids opened up the path toward *biotechnology*. Biotechnology may also have a profound impact on human beings, and

A. Marcos (✉)
Facultad de Filosofía y Letras, Universidad de Valladolid, Valladolid, Spain
e-mail: amarcos@fyl.uva.es

M. Bertolaso, F. Sterpetti (eds.), *A Critical Reflection on Automated Science*, Human Perspectives in Health Sciences and Technology 1,
https://doi.org/10.1007/978-3-030-25001-0_13

therefore in this case we may speak of *anthropotechnics*. It is noteworthy that the application of technics to life forces us to reconsider the traditional distinction between what is natural and artificial and, as Hans Jonas says (1979), engenders new responsibilities.

This terminological clarification once done, I shall from now on use the term technoscience to refer to this ample semantic field (technique, science, technology, technoscience, biotechnology and anthropotechnics).

The Symptoms of the Problem

Now, technoscience is experiencing a twofold dehumanizing process: in the stage of production and in the stage of application.

On the one hand, there is a growing dehumanizing trend with regard to those who are doing technoscience. Technoscientific activities are carried out in a growingly automated context. The goal of technoscience seems to have turned from understanding the world to obtaining and processing big data. 'Omic' sciences provide a recent, clear example of such a trend, although their antecedents can be traced back to far earlier times: the idea of an *ars magna* proposed by Ramón Llull, Leibniz's *characteristica universalis* or the *logical structure of the world* advanced by Carnap were already paving the way to the mechanization of knowledge, a knowledge understood as a automatable combinatorial analysis. More items could be added to this list, such as, for instance, the historical case of eighteenth-century relational chemistry (Serres 1989, 297–319). Traditionally, chemistry was focused on the properties of substances. Relations were just an occasion for the properties of substances to show up. However, in relational chemistry properties are substituted by relations. Thus, comprehensive tables of relations were drawn; the objective was to study *all* possible relations among substances. Traditional chemistry could be content with knowing the most significant reactions, that is, the reactions that unveil the properties of each substance, without the need to undertake a comprehensive compilation. I mention this historical example because we may learn some lessons for our present situation. When there is no prudent—and therefore human—estimate of the significance and meaning of data, all of them have the same relevance and therefore it becomes necessary to perform a comprehensive, combinatorial and automated search.

Something similar is now taking place with regard to omic sciences (genomics, epigenomics, proteomics, metabolomics, connectomics), born as a result of the Human Genome Project. The suffix *-oma* comes from Greek and suggests a notion of totality. Sciences ending in -omics tend to enumerate, with the help of computers, the totality of elements in a given area (all the human genes, all the proteins, all the metabolic pathways…). Curiously, the more traditional suffix applied to sciences comes from the Greek term *logos* (biology, geology…) and it refers to a knowledge full of wisdom and not to a mere encyclopedic knowledge.

In my opinion, there can be no objections against omics, since they may result highly valuable; yet, the attempt to reduce all sciences to a sequencing exercise must be opposed. This understanding of science threatens to marginalize all that is truly human, all that goes beyond the mere registry and combination of elements, that is, anything that has to do with genuine creativity, emotions, meaning, significance and values—even moral and aesthetic values—, reflection, and dialogue. Data collection is undoubtedly a significant step in the progress of science, but a science reduced to gathering and combining data would be an effort as expensive as useless.

On the other hand, the most recent applications of technoscience, as it is reflected in anthropotechnics, become a risk to human nature itself. *Trans-* and *post-*humanist projects point towards the dissolution of what is human through the use of technoscience (Diéguez 2017).

Furthermore, technoscience's applications may transform nature, rendering it hostile to the necessary requirements of a human ecology (Valera 2013). Likewise, in this area it is possible to refer to technoscience as a dehumanizing agent.

To sum up, dehumanization is—obviously—a problem for the human being, since it is expelled from the practice of science and technically manipulated, but it is also a problem for technoscience itself, when its thirst for understanding is replaced by a blind power based on the brute force of computing. Giving in to dehumanizing trends, technoscience would risk becoming both stupid—since it will provide no understanding—and useless—since it will provide no service to anyone.

Possible Causes

Among the possible causes of the detected problem, my intention is to consider two of them: a naturalistic ontology and an erratic anthropology. I am talking about an ontology that reduces all existing things to natural things, and indeed to material things (Soler 2013). Were we to accept the tenets of radical naturalism, it would be quite improbable that technoscience may acknowledge its inner diversity and its outer boundaries. Therefore, it will be unlikely for technoscience to search for legitimate partners to cooperate with in the complex task of knowing. In the eyes of the radical naturalist, the sources of knowledge external to technoscience would be, at the very best, provisional and superficial, if not plainly illegitimate.

For instance, in the field of biology, genetic reductionism was trendy for a while. According to genetic reductionism, every living being would be absolutely knowable and explainable once its genes were explored. Every living being was seen as a mere combination of genes; in fact, somehow as an epiphenomenon of its genes. Consequently, some scholars thought that the enumeration of genes would derive into a profound and complete knowledge of all living beings, would open unlimited possibilities for its technical manipulation and would be a panacea for curing all illnesses. This is the reason behind the origin of the major *genomic* projects that have been the beneficiaries of sizable endowments. However, the development of such projects has allowed us to perceive their limits, since not everything is held in the genes (Marcos 2011). The verification of this fact paved the way for many other *omic* disciplines, a welcome step towards a certain plurality that needs to become even greater.

Understanding living beings calls for a pluralistic research, at very different levels and using various methodologies, including the philosophical method. We can hardly understand what a living being is without concepts such as complexity, function, identity, information, conscience or meaning. Some of us think it is not possible to translate these concepts into naturalistic jargon. When dealing with a phenomenon as complex as life, we must resort, of course, to the brute force of computing, but also to all imagination, creativity, intuition and prudence that the human spirit is able to master.

The second root of dehumanization lies in an anthropology we have labeled as erratic. Nowadays, fashionable intellectual trends go from the existentialist or nihilist denial of human nature to its radical naturalization. Both extreme positions, while seemingly the opposite, have similar practical consequences. The denial of human nature invites to its technical production, whereas the radical naturalization of human nature makes humanity as available to technical manipulation as any other object of nature. In both cases we face a kind of anthropotechnics with no boundaries or criteria, leading to the dehumanization of the human being on account of its unrestricted artificialization. It is possible to say that there are no criteria because both the nihilist and the radical naturalist have renounced a normative vision of human nature (Marcos 2010).

Searching for a Solution

After explaining the main points of the problem, my intention is to offer some philosophical ideas that may contribute in the search of a solution. In the first place, it seems that the assertion of a pluralist ontology may advance the cause. This type of ontology is in line with the promotion of a plurality of scientific methods and with the acceptance of a diversity of legitimate epistemic sources, thus going beyond the limits of technoscience, while at the same time interacting with it. In the second place, dehumanization is fought against by showing the irreplaceable function of persons in the production of technoscience. Technoscience is from its very inception a personal action, as I shall claim below. Finally, in the third place, I shall argue for a guiding concept of human nature, which may become a criterion, and so may protect us from the danger of dehumanization caused by an uncritical use of the anthropotechnics. In sum, anthropotechnics must be used in order to improve human life, to make it *truly* human, not to transcend—and supposedly enhance—what is human.

A Pluralist Ontology and a Systemic Model

The dehumanization of technoscience is fought against from its very ontological roots, assuming a pluralist ontology that provides a shelter for the many various entities that compose the abundance of reality (Feyerabend 1999), including basic forces and material elements, living beings, plants, animals and people especially. People are endowed with a subjective conscience that allows them to do technoscience and that probably lies beyond the limits of technoscience itself (Arana 2015). Moreover, technoscience does not only have limits; it was precisely born out of a methodological process of limitation (Marcos 2016). Galileo laid the methodological foundations of modern science when he delimited the aspects of reality under scrutiny.

Acknowledging the existence of boundaries requires us, at the same time, to accept the existence of other sources of knowledge and of other practices as legitimate as technoscience, although external to it. Moreover, such acknowledgment paves the way for the cooperation between technoscience and other areas of human life that may provide it with further knowledge and meaning. These mutual, collaborative relationships can be conceived in light of a systemic model (Agazzi 1992) that considers technoscience as a system of human actions, an autonomous but necessarily limited system related to other systems. The systemic model highlights the feedback processes in which technoscience is involved. For instance, if technoscience ends up being perceived by the public opinion as a threat to the environment, as a dehumanizing agent, and as an enormous pile of hard-to-grasp and even harder-to-convey data, the support it presently receives (whether in the form of financing, legal protection, new technoscientific vocations, media exposure or academic promotion) would experience a likely decrease. Thus, the systemic model suggests that the dehumanization of technoscience is, among other things, a threat to the functioning of technoscience itself.

Technoscience as a Personal Action

In the second place, I consider necessary to emphasize the irreplaceable function of persons in the production of technoscience. Technoscience's call to understand can only be fulfilled when it is carried out by persons and not merely entrusted to machines. Technoscience is a *personal* action: all the dimensions of a person are, to a greater or lesser degree, involved in doing technoscience (Marcos 2014). The tendency toward dehumanization in technoscience derives from a misunderstanding of the scientific method and of the concept of objectivity. It might seem that objectivity in technoscience is achieved by suppressing or standardizing everything that is subjective and by replacing it with some kind of automatically executable method. Rather, the subjective-personal dimension is a necessary prerequisite in building technoscientific objectivity. Precisely, the personal dimensions of the subject allow for the creativity and the sensible judgment that are indispensable for the production of an intelligent, communicable and useful technoscience.

For some philosophers, particularly in the positivist tradition, science is made of theoretical and observational assertions, the type of assertions typically found in articles, lab reports and books. The formulation of such assertions would be guided by a supposedly scientific method, exclusively based on observation and logical inference. However, not even observational reports and logic inferences are fully automatable; moreover, this vision of science is, to say the least, partial, and nowadays it is considered outdated.

All human faculties—not only logical and observational abilities—take part in the different stages of technoscientific research. It is obvious that the initial stage of research—the identification of a problem—depends upon our capacity to wonder and our curiosity, as well as upon the social context and the traditions in which we are rooted. Certainly, there is no algorithm capable of identifying problems in a purely mechanical manner. The identification of a problems calls for the concurrence of all dimensions of a person.

After identifying a problem, researchers start to formulate hypotheses in order to solve it. The formulation of hypotheses greatly depends on the imaginative abilities of the person, and sometimes the origins of a hypothesis have to be found in distant places from technoscience, such as someone's artistic education or someone's metaphysical or religious convictions.

After the problem is proposed and the hypotheses formulated and filtered, we must consider the stages of empirical testing. In order to test a hypothesis, we must draw its empirical consequences, an impossible task unless we include some presuppositions besides the hypothesis. Therefore, we have to choose the auxiliary assumptions and decide how much trust we place in them. Once again we face an operation that can only be performed if the whole person is involved.

It may be possible to think that, once this point is reached, the empirical stage is absolutely free from any personal involvement. However, we know that is not the case: even the simplest observation is conditioned by our expectations. In many cases it is the action of researchers that originates a given object or phenomenon. Perception, even in the most straightforward cases, is not a passive action; it only emerges as a result of someone's activity.

Once the empirical results, whether by observation or experimentation, are obtained, they may totally or partially agree with expectations or they may differ from them. If there is no total agreement, we must resort to some kind of estimation ability in order to ascertain whether the difference is significant or not. After deciding if the disagreement must be taken into account or not, we find that the empirical testing can either reinforce or weaken the hypothesis. Yet, pure logic does not allow us to determine whether a hypothesis has been definitely proved or refuted. We somehow know that a correct empirical test tends to argue in favor of the hypothesis; nonetheless, the assessment of the empirical test cannot be done by pure logic or probability calculation, but rather by the combination of human abilities that a person can harmonize thanks to his or her common sense. Similarly, an empirical mistake should lead us to reconsider either the hypothesis or some of the auxiliary assumptions; logic by itself cannot tell us which way to follow. It is following our common sense that we decide which hypotheses are to be accepted and how much trust we can place in them.

At the more technological stages of research, we know that automatically applying a theory to the resolution of a practical problem is hardly a possibility; rather, it requires some kind of art, some complex tradition. The resolution of a problem does not derive from a simple, mechanical translation of a theoretical science; it calls for a balanced effort of a large number of human capacities.

Therefore, the success of technoscience, in terms of truth and usefulness, depends on the correct dosage of each one of our personal abilities in each stage of research. And it is prudential reason that helps us find the correct dosage and rhythm. The one doing technoscience is always a human person. Logic reasoning and observation are indeed human abilities, but a person is far more than that. Technoscience derives from the whole person, with all his or her (dis)abilities, attitudes and circumstances. This includes one's emotions, feelings, motivations, affections, interests, attention, intuition, imagination, memory, aesthetic sense, social context (conversation), historical context (traditions) and so on. Someone might say that all these personal—and thereby subjective—dimensions are precisely those that must remain at the door of the lab so that scientific objectivity is not compromised. I do not believe such thing. I rather think that it is not possible to do science unless the whole person is involved. Moreover, the integration, dosage and balance of all human abilities are achieved thanks to the common sense, or, in more philosophical terms, thanks to a prudent attitude.

If we accept the unavoidable personal nature of technoscientific research; what is more, if we acknowledge that the objectivity in technoscience is precisely the result of the sensible actions of people, we shall find a remedy against the dehumanizing trends.

Technoscience at the Service of a (Truly) Human Life

Finally, in the third place, technoscience makes sense when it is placed at the service of human life. However, the *trans- / post*-humanist project stands for an in-depth modification of the human being by using anthropotechnics. In this business of allegedly enhancing humanity, we are dealing with an abstract idea of perfection that someone wishes to impose upon real human beings. When a particular and real human being is compared to an abstract idea of perfection some undesirable effects emerge (Sandel 2009). For a start, it is an unfair comparison that often derives into rejecting or despising human life in general. Moreover, history has taught us that when the senseless dream of utopian perfection has prevailed, a disastrous collateral effect is the disdain for the weakest, most vulnerable and disabled people. The real human being, imperfect as it is, comes off badly when compared to the designed utopia. This leads us to overlook that our human life is already a gift.

We cannot forget that human nature functions as an essential normative indicator. If human nature is not respected and taken as an ineludible reference point, technoscience itself loses its meaning and becomes sterile. It might produce changes in the world, but those changes cannot be considered as an improvement, since they precisely destroy any possibility to reach a shared normative reference.

On the contrary, both the common sense and the Aristotelian tradition argue for a normative understanding of human nature, including its animal, social and spiritual aspects. These three dimensions of human life cannot be reduced to one of them and are not simply juxtaposed. Each one of them completely permeates the other two. Furthermore, we must always keep in mind that humanness is a wholesome, single and indivisible attribute in each person. Human nature, thus understood, provides technoscience with a meaning and a purpose; it must be place at the service of the well-being of persons and their environment; at the service of a just and peaceful social living; at the service of the multiple possibilities opened for a full spiritual development; in sum, at the service of the wholeness of person.

Concluding Summary

We have detected the existence nowadays of some trends toward the dehumanization of technoscience, both at the time of its production and of its application. Such dehumanizing trends are a true problem for human life and for the continuity of technoscience itself.

A false concept of objectivity leads to the necessary exclusion of the personal subject from the process of doing technoscience and to its replacement by automated systems that enable the gathering and processing of big data. With regard to the application of technoscience, the ecological environment that provides a home to humanity is being endangered and there is an attempt at implementing anthropotechnic projects to reinvent human beings—that are viewed as deprived of a nature—as artifacts (posthumanist project) or to transform its own nature—that is seen as a strictly natural reality—at the deepest level (transhumanist project).

Once the symptoms are identified, we have tried to point out the possible causes of such ills. A poor, radically naturalist ontology endorses the idea that a science without subject is the one legitimate source of knowledge. An erratic anthropology, swinging between nihilism and radical naturalism, fosters the idea that human beings must be at the mercy of anthropotechnics. Therefore, these philosophical trends, insofar as they are a current intellectual fashion, pave the way for dehumanization.

In the final pages of our paper we have attempted to point out some possible solutions to the problem of dehumanization. In line with our proposal, we have argued for a *pluralistic ontology* that, in turn, calls for a methodological diversity that include both sciences and humanities. Furthermore, we have suggested that the *systemic model* can shed some light when reflecting about these issues. Thus, technoscience would be seen as an open subsystem, exhibiting its own inner complexity and interacting with many other subsystems proper to human life (educational, political, economic, moral…).

Moreover, we have recommended a revision of scientific methods, emphasizing the essential relevance of personal aspects, even as we try to be objective. If we perceive technoscience as a personal action, it would be harder to give in to the current dehumanizing trends.

Finally, we stand for a concept of human nature that provides technoscience with a purpose. Precisely, it is meaningful when it is placed at the service of *a truly human life*, that is, a life according to the nature of human beings.

References

Agazzi, Evandro. 1992. *Il bene, il male e la scienza*. Rimini: Rusconi.

Arana, Juan. 2015. *La conciencia inexplicada*. Madrid: Biblioteca Nueva.

Diéguez, Antonio. 2017. *Transhumanismo*. Barcelona: Herder.

Feyerabend, Paul. 1999. *Conquest of Abundance*. Chicago: University of Chicago Press.

Jonas, Hans. 1979. *Das Prinzip Verantwortung*. Frankfurt am Main: Insel Verlag.

Marcos, Alfredo. 2010. Filosofía de la naturaleza humana. *Eikasia. Revista de Filosofía* 35: 181–208.

———. 2011. Filosofía post-genómica. *Investigación y Ciencia* 418: 46–47.

———. 2014. El pulso de la ciencia. In *Una ciencia humana*, ed. Alicia Villar and Antonio Sánchez, 169–182. Madrid: Universidad de Comillas.

———. 2016. Rescher and Gadamer: Two Complementary Views of the Limits of Sciences. In *The Limits of Science*, ed. Wenceslao J. González, 167–182. Leiden: Brill.

Sandel, Michael. 2009. *The Case Against Perfection*. Cambridge, MA: The Belknap Press.

Serres, Michel. 1989. *Éléments d'histoire des sciences*. Paris: Bordas.

Soler, Francisco. 2013. *Mitología materialista de la ciencia*. Madrid: Encuentro.

Valera, Luca. 2013. *Ecologia umana. Le sfide etiche del rapporto uomo-ambiente*. Roma: Aracne.

What Is 'Good Science'?

Christopher Tollefsen

Introduction

The authors of the papers collected in this volume were invited to investigate a "renewed understanding of the human character of science."[1] To a considerable, but not, I think, exclusive extent, that understanding is envisaged as a corrective to certain trends in scientific method, trends which seek to reduce as much as possible the stamp of the human on the activity of knowing as such. Scientific practice, in these new trends, "takes place in a context increasingly populated and executed by machines." Complexity is reduced to complication, and the effort to understand nature is replaced with the attempt to aggregate data and information, with little attention to the way in which human creativity is necessary for the understanding of scientific problems, and human insight necessary to grasp to significance of the data. Thus the concern of the volume: will science remain human?

This paper approaches that question, and the question concerning the "human character" of science, from a different, but I think complementary, direction. Various controversies over particular scientific endeavors have led some commentators to assert the impropriety of imposing moral, political, or religious limits on scientific inquiry. A paradigm case of this is to be found in the controversy which I will take as exemplary for the purposes of this paper: the disputes over the moral permissibility of research destructive of human embryos.

Important instances of this trope were found, for example, in the aftermath of the announcement, in November 2007, that scientists had succeeded in reprogramming

[1] Quotations in this paragraph are from the invitation to participate in the March 2018 conference, "Will Science Remain Human? Frontiers of the Incorporation of Technological Innovations in the Bio-Medical Sciences," received by email July 19, 2017.

C. Tollefsen (✉)
Department of Philosophy, University of South Carolina, Columbia, SC, USA

M. Bertolaso, F. Sterpetti (eds.), *A Critical Reflection on Automated Science*, Human Perspectives in Health Sciences and Technology 1,
https://doi.org/10.1007/978-3-030-25001-0_14

skin cells back to a pluripotent state. While some defenders (see, e.g., Krauthammer 2007) of President Bush's policy of August 2001 permitting federally funded research on existing stem cell lines believed that the new discoveries vindicated the President's emphasis on embryonic stem cell alternatives, the New York Times editorial page writers demurred (Editorial Board of the *New York Times* 2007), writing, "President Bush can try to defend his restrictions on stem cell research as good religion or good morality [...]. What he can't do is defend it as good science" (Ibidem).

Similarly, Alta Charo was quoted as saying, "Criminalizing any area of science, as opposed to merely regulating it, would be contrary to the political and constitutional traditions of academic and scientific freedom, as well as the historical spirit of inquiry that characterizes this country" (Chapman 2007). For a final example of this idea, we could turn to President Bush's successor in the White House, President Obama. In his remarks announcing his decision to "Remove Barriers to Responsible Scientific Research," the President said

> This Order is an important step in advancing the cause of science in America. But let's be clear: promoting science isn't just about providing resources – it is also about protecting free and open inquiry. It is about letting scientists like those here today do their jobs, free from manipulation or coercion, and listening to what they tell us, even when it's inconvenient – especially when it's inconvenient. It is about ensuring that scientific data is never distorted or concealed to serve a political agenda – and that we make scientific decisions based on facts, not ideology. (Obama 2009)

Here too, I suggest, do we have an attempt to minimize the "human character" of science, and, in particular, of "good science." But consider the implications of the *Times*' claim. By this logic, Nazi science, so long as it was not needlessly inefficient, say, or headed down unnecessary blind alleys, could have been "good science" – independent of its obvious moral failings. Of course, this might be a defensible view; good science might be entirely independent of good morality. Yet who would really want such science? If Nazi science was good, then good science is simply not good enough.

Similarly, Charo's distinction between regulating and criminalizing is also problematic. It is true that violations of the research regulations of the Common Rule[2] for research ethics are not as such criminal acts; yet those regulations in many cases are intended to rule out certain kinds of actions as unacceptable. Moreover, in some cases, acts which are wrong as contrary to research regulations are *also* criminally wrong. Some instances of failures to obtain informed consent are also instances of assault; some instances of fraudulent research are also criminally fraudulent. And it would be an error to suggest, as perhaps President Obama did, that what was not learned in virtue of complying with such regulations had been thereby "distorted", "concealed", or governed by merely "ideological" concerns.

So even if "good science" has nothing to do with "good morality," there is every reason to think that morally bad "good science" ought to be, and often is, not merely regulated in the sense of being "guided," but regulated in the sense of

[2] The name given to the Code of Federal Regulations (CFR) Title 45 (public welfare), Part 46 (protection of human subjects).

being "ruled out," and even prohibited in criminal law. No one genuinely operates with such an un-nuanced view of the separation of science and morality in real life – only in the pages of partisan disputation for the sake of some particular favored scientific project.

Nevertheless, I want here to look more closely at the relationship between science – "good science" – and morality. This relationship exists, I'll argue, on at least three axes which I shall call the external ethics of science, the social ethics of science, and the internal ethics of science. Of these, *only* the first – the external ethics of science – suggests that good science may not necessarily be good morally. Yet, when looked at in conjunction with the other two axes of ethical science, it becomes clear that the separation of science and morality is not nearly so obviously justified as is sometimes thought. "Good science" turns out to be a more complicated reality than the stereotype frequently presented to us in the scrum of political engagement.

But, as I'll suggest very briefly in my concluding comments, if good science has the characteristics I describe in this paper, then the project of science "populated and exercised by machines" is chimerical. Good science *as science* is indeed a human activity permeated by human concerns and norms.

The External Ethics of Science

What is the truth behind the claim that good science is not itself moral science? It is this: many of the moral norms that govern ethically responsible science bear no necessary relationship to the practice of science itself. That is, these norms may be understood independently of an understanding of the nature of science; they may be applied to areas of human activity other than science; and their justification does not turn on any particular claims about science.

Consider the requirement of informed consent. Contemporary justifications of this requirement in biomedical research refer often to the value of autonomy, which is breached when experimental subjects are duped or coerced into participating in research without their knowledge (see Faden and Beauchamp 1986). Other justifications incorporate concerns for bodily integrity: it is considered, at law, battery to perform a medical intervention on another contrary to their expressed wishes (Jackson 2006). Privacy too is infringed upon when subjects are studied without their knowledge, and hence consent. Finally, and perhaps because of the deep connection between medical research and the practice of medicine, a concern for the good of the experimental subject has to some degree militated against the use of subjects who, even with full consent, could expect no real medical benefit from the research but some risk of harm.[3]

All these justifications, with the possible exception of the last, for the most widely accepted norm of clinical research, depend not upon the nature of science for their sense, but on the nature of persons, and what is valuable for persons, generally.

[3] I have made these points at greater length in Tollefsen 2008, Chapter Two.

All persons, in almost all contexts, are better off when their autonomy, their bodily integrity, and their privacy is respected. And these values may be violated in numerous ways across numerous contexts: it is not only in scientific experimentation that these norms have application. So informed consent is necessary not just in experimental interventions, but also in purely therapeutic interventions. And analogues to the requirement of informed consent may be seen in the areas of journalism, and even criminal investigation, as when criminal suspects are read their Miranda rights.

Similarly, human persons are held to be morally immune from unprovoked violence: it is wrong everywhere and always to intentionally kill an innocent human being. This norm is understood by reference, again, to the nature and value of persons, and its application is of wide scope: no killing in domestic disputes; no killing to get an inheritance; no killing because you just don't like the look of the other fellow. And of course, no killing in science, a norm flagrantly violated by the Nazis, but *nearly* universally accepted today.

But not universally. For this norm is routinely violated in embryo-destructive research, research in which human beings, at their earliest stages of development, are killed as part of the projects of obtaining new knowledge about embryogenesis, obtaining new knowledge for the advance of assisted reproduction, and the obtaining of pluripotent cells for stem cell research.

Now these aspects of embryo destructive research would not be morally problematic were the early embryos in question *not* human beings, or *not* human persons; such research would capitalize upon human biological material in more or less the same way that research on other cell cultures does. And both claims, that the embryo is not a human being, and that the embryo is not a person, have been defended, by philosophers and others, in order to justify the termination of embryonic human lives. But I believe that both claims are false.[4]

As regards the first denial, that human zygotes and human embryos are early stage human beings, members of the same biological species to which you and I belong seems to me to be well-established scientific fact. In standard textbooks of developmental biology, we find passages such as the following, which are part of received scientific wisdom:

> *Human development begins at fertilization* when a male gamete or sperm (spermatozoon) unites with a female gamete or oocyte (ovum) to produce a single cell – a zygote. This highly specialized, totipotent cell marked the beginning of each of us as a unique individual. The zygote, just visible to the unaided eye as a tiny speck, contains chromosomes and genes (units of genetic information) that are derived from the mother and father. The unicellular zygote divides many times and becomes progressively transformed into a multicellular human being through cell division, migration, growth and differentiation. (Moore and Persaud 2003, 16).

If such a view is correct, then the destruction of embryos, whether left over from *in vitro* fertilization (ivf) procedures, or created by ivf or somatic cell nuclear transfer

[4] For the claim that the early human embryo is not a human being, see, e.g., Smith and Brogaard 2003. For the claim that embryos, even if human beings, are nevertheless not persons, see Harris 1999. These essays are merely representative.

for the purpose of scientific research, is the destruction of human beings. And at least on the surface, this would appear to be seriously morally wrong.[5]

Many philosophers, of course, deny this claim, on the grounds that a being, such as the embryo, which is not currently capable of rational thought or self-consciousness, is not a *person*. Such a claim denies that all human beings are worthy of fundamental moral respect, and holds instead that only a subset of human beings is so worthy, the subset which is capable of rational activity, or of self-consciousness, or something similar.

Yet "capable of" and "capacity" are ambiguous terms. No being goes from an inability to think to an ability to think, without changing what it fundamentally is, unless it *already* has a capacity – we could call this a *radical* capacity to indicate that it is there from the beginning of the entity's existence – to develop itself to the point of being able to manifest thought, self-consciousness, and so on. The capacity to think is not instilled in a being from without, but is the realization of a radical capacity already present in beings of that sort; and if the capacity is not present, as it is not in cats and dogs, then it is *not* a being of that sort, and no amount of development will ever bring *that* being, without a change of substantial kind, to the state of rationality.

Human zygotes and embryos thus *do* have a capacity for rationality (and consciousness, self-consciousness, and various other personal attributes); the question is then whether it is the status of possessing those capacities in radical form, or the achievement of having developed those capacities, that qualifies a being for full moral respect, of the sort that renders a being immune from destruction for the sake of another's purposes. I believe this question is a fundamentally moral question: it is the question of whether to treat all beings of the same kind as us – human beings – in the same fundamental ways that we would wish to be treated – or whether it is permissible to introduce radical differences of treatment – killing some, protecting others – on the basis of non-essential differences, such as age or stage of development. Fairness, I believe, requires that one treat all human beings with the same fundamental forms of moral respect. But because persons just are those beings to whom fundamental moral respect is owed, a consequence of this argument (rather than, strictly speaking, a premise of it) is that all human beings are persons (see Tollefsen 2011).

By the argument above, therefore, the killing of embryos for research purposes must be considered a deep moral wrong, a wrong of such gravity that legislation for the protection of these innocent human beings is surely called for. Just legislation in this area would not simply "regulate" in the sense of guide; it would attempt to prohibit, whether by criminal or other penalties, the practice of such research, for precisely the same reasons that other deliberately lethal endeavors are similarly "regulated": the killing of innocent human beings is radically contrary to the common good it is one of the primary purposes of a just state to prevent such injustice from taking place.

---

[5]A further review of such claims, and responses to objections, is to be found in George and Tollefsen 2008. Jason Morris, a biologist, has criticized our claims in Morris 2012; George and I, with Patrick Lee, respond in Lee et al. 2014.

So even the best embryo destructive research is bad: morally deficient. Yet we can see that in an obvious sense, the norm about killing, like the requirement of informed consent, emerges from outside the domain of scientific practice and is "imposed" on science, as it were, externally. If science has its own ethic, it is not clear that this ethic need concern itself as such with killing (or with consent); it is the value of persons, rather than the good of science, that justifies these norms. And so, while the claim that these norms should be ignored or abandoned where science is concerned is clearly false, the claim that good science and good morality are two different things has something to be said for it.

The Social Ethics of Science

We might speak, in a sense that will be familiar to readers of Alasdair MacIntyre's *After Virtue,* of a practice, or, better, practices of science. MacIntyre defines a practice, in a well-known passage, as:

> [...] any coherent and complex form of socially established cooperative human activity through which goods internal to that form of activity are realized in the course of trying to achieve those standards of excellence which are appropriate to, and partially definitive of, that form of activity, with the result that human powers to achieve excellence, and human conceptions of the ends and goods involved, are systematically extended. (MacIntyre 1984, 187)

The description seems aptly applied to science as currently practiced: it is a socially established and cooperative activity whose structure is to some extent defined by the history, or tradition, of that activity. And, more importantly, science has an internal good, in MacIntyre's sense, around which those cooperative activities are structured, namely, the good of truth and generalizable knowledge about nature. There are, further, virtues that are necessary for the achievement of that good in a cooperative setting. As we shall see in the next section, this understanding sets the stage for an internal ethics of science, distinct from the external ethics just discussed.

But science is more than a practice or set of practices: it is a profession. Like many other practices, such as medicine, to which it is closely related, science has an internal good that is of wide and abiding social value. Society is better off, both instrumentally, and intrinsically, when science flourishes. In the former way, the benefits of science are too numerous and obvious to require extensive discussion: computers, medical therapies, the technologies of travel – all are the result of science, and all improve our lives considerably.

But even considered in itself, science benefits a society, which is better off just for having scientific awareness of the way the world is and goes than it would be if it remained in ignorance. Unlike the relatively enclosed, albeit still valuable, practice of chess, say, the goods of science are wide and deep: they can be pursued in a multitude of ways, and systematic pursuit along any one of the axes of scientific inquiry tends to open up new forms of inquiry previously unimagined. Successful (and even, to some extent, unsuccessful) pursuit of scientific inquiry thus expands the horizons of human well-being in radical ways; so science should be considered a social good of considerable significance.

For this reason, the practice of science becomes professionalized: scientific practice becomes embedded in an institutional context in order that its social benefits may be more easily achieved, promoted, regulated, and distributed. The institutions of science provide the external goods – money, fame, encouragement – that science needs to progress, while monitoring the conduct of scientists, distributing the benefits of science to the rest of the community, and initiating new members into the practice. Moreover, because these goods, and their regulation and distribution are central to the common good, and because it can be difficult to coordinate and fund efforts distributed across a wide variety of agents, the state rightly can play a role, I believe, in providing assistance to the profession, in the form, negatively, of laws, of course, but positively in the form of grants, institutes, universities, symbolic expressions of support, and the like. Professionalization without the helping hand of the state would likely be less adequate, possibly to the considerable detriment of the common good, and so the helping hand is reasonably offered.

Even before professionalization, much less the involvement of the state, science is intrinsically social; it must, to be done at all well, be done cooperatively. But in virtue of this professionalization and the new ties of science to the state, professional science becomes even more a matter of public concern in at least four ways.

First, as we have already seen, science is public in its benefits. But professionalization enables science to achieve these beneficial goods more easily, and allows for a more efficient distribution of those goods to society at large. Science carried out only by the wealthy, in a self-funded and insular way, would be relatively private; contemporary professionalized science is massively public.

Second, science is methodologically public. Scientists rely on a method of inquiry, largely, although not exclusively, modeled around hypothesis, prediction, and experiment, that is deeply social. Scientists work together to design and carry out experiments, promulgate the results of their research, and then rely on the work of others both to test their findings, and to pick up where they left off in the advancement of understanding. Scientists share their findings, pass them on from generation to generation, and together constitute a tribunal before which each new block in the architecture of understanding must come under scrutiny before being accepted. Professionalization of science facilitates this by making the ideal of a scientific community more of a reality.

Third, by a society's choice of what practices to professionalize, the society determines and expresses its moral identity: a society that cares not a whit for science and scientific knowledge, and only, say, for knowledge of its sacred scriptures, determines and expresses itself in a way different from the shape taken by Western society over the past several centuries. Similarly, by its choice of projects and methods, professionalized society expresses and shapes the nature of the wider society. A scientific community that observes but does not cure poor black men of syphilis both reflects and creates racism in the larger social realm.[6]

[6]The history of this particular form of racism in science is told in Jones 1992.

Finally, by its reliance on the state, positively as well as negatively, the practice and profession of science is a matter of public concern as regards questions such as: the state funding of scientific research; the just distribution and availability of scientific results to a state's citizenry; and broadly speaking, the question of legitimacy, for a state's citizens have a stake in desiring that the state's role in the promotion of science not go beyond that to which citizens would consent, or beyond the state's just authority.

Professionalization is thus like the signing of a contract between the practitioners of science and the wider society – and state – within which science is practiced, a contract for mutual aid and benefit. Without professionalization, science remains marginal and moribund; and society fails to receive on a large scale the many benefits that science has to offer. This surely brings with it obligations on the part of society: to aid in the provision of resources to science, to promote scientific understanding amongst its members, and to protect those liberties necessary for scientists to make advances in the pursuit of truth.

But professionalization must likewise bring with it obligations on the part of science. The public nature of science – in its benefits, methods, expressive power, and state support – means that science is extraordinarily situated not only to benefit society, but also to harm it, precisely as a society. Here I do not have in mind simply the sorts of obvious consequential considerations usually raised, about viruses that can escape or weapons of mass destruction. Rather, science and scientific practice have the potentiality to harm society *as society*, in ways that correspond to its fourfold social nature.

Consider a society widely shaped by benefits it would be wrong to accept: the benefit of massively extended lifespans at the cost of future generations, or widespread asexual reproductive technologies such as cloning, or crime control techniques that operated by subversion of a criminal's ability to make free choices (the sort of technologies envisaged in an extreme way in *A Clockwork Orange*, but perhaps possible even now on a more modest scale by the use of pharmacological interventions). Or consider a society in which the methods of science were tainted by unethical approaches: racism, violations of autonomy and privacy, dishonesty, and so on. Third, consider a society whose science was deliberately set against the values of that society, whose scientific culture was, for example, deeply atheistic, and dedicated to conflict, on every possible front, with any consideration that seemed even remotely "religious." Finally, and relatedly, consider a political society promoting, with tax dollars, forms of research considered by many citizens to be grievously immoral.

In all such cases, the overall effect of science on society, culture, and politics would and could only be corrupting. We could expect three possible outcomes from the ensuing crisis: reform, rebellion, or capitulation.

Working backward, a society would capitulate by falling sway to the appeal of illicit benefits, by cooperating in immoral methods, by gradually ceding questions about religion to the scientific "experts," and by accepting the state's role in promoting wicked science. Remade in the image of science, such a society would have, effectively, lost its conscience.

Perhaps somewhat better off – but only perhaps – is the society that rebels against science so conceived, a society that itself wars against professional science and turns elsewhere for all its sources of knowledge and power. Such a move might seem barely an option today in the West, yet its motivating appeal is recognizable in some aspects of the curricular battles and retreats from science evident in the current culture wars, wars, incidentally, in which both sides seek to appropriate the various powers of the state. And the *image* of a society, and some of its leaders, as in open, and unreasonable, rebellion against science is one that defenders of embryo destructive research and other disputed theories and practices have relied much upon as a rhetorical trope.[7]

Reform of science, by contrast with both capitulation and rebellion, operates with an image of the relationship between science and society similar to that detailed in my description of the professionalization of science: science neither divorces itself from the moral ideals and aspirations of the society to which it belongs nor rebels against them; nor does that society deny the worth or legitimacy of scientific knowledge, nor the proper province of its method, nor the appropriate role of the state in making science possible. "Good science" in such a society is not clearly distinguishable from that society's moral good, for it is shaped by society's moral sense of itself, and its needs, and its ethical boundaries. As Benjamin Hurlbut and Jason Scott Robert write,

> Good science is an achievement of a good society, and not the reverse. The ways we reason together as a society, and the policy resolutions we settle on, should reflect this. Facile references to politics "standing in the way of good science" – too common in the hESC controversy – denigrate the foundations of science and democracy alike. Science depends on a space in which it can observe its internal norms and exercise its specialized skills, but in a way that coheres with, rather than contradicts, the norms and aspirations of the democratic societies in which science is embedded. For science itself is but a piece of the larger project of civilization. (Hurlbut and Robert 2012, 712)

We have here moved away from the stereotypes set in the pages of the *Times*' editorials and the texts of the lobbyists for unconstrained inquiry. And from this standpoint, given not just the nature of embryo-destructive research, but given also the nature of the United States and its people, President's Bush's policy looks increasingly like *good* science.

We should note here that whichever route is taken, or is being taken – capitulation, rebellion, or reform, in some sense embryo-destructive research remains at the heart of the current crisis of science that prompts a need for some response. Many scientists and politicians have waged a steady campaign against the idea that opposition to embryo destructive research is principled and humane, suggesting that it is motivated instead by a religious hostility to the advance of knowledge, and by "blind" faith, and

[7] For example, Owen-Smith et al. write: "Though history has taught us the benefits of thoughtful, rationally driven science policy (the response by the scientific community to recombinant DNA research offers a historical case in point), some types of research are chilled by thoughtless, irrationally driven science policy. We take the response by the political conservatives to climate change and human embryonic stem cell science to be exemplary of this negative case" (Owen-Smith et al. 2012, 727).

that those who hold up progress in embryo research are "responsible" for the deaths that could otherwise have been prevented by the development of timely cures. The overall effect is to make it increasingly difficult to be a "hold-out" against such research; capitulation inches closer. Yet at the same time, the most extreme responses have been of a piece with rebellions against other perceived domains of scientific hegemony, such as Darwinian evolution and global warming. In such cases, not only the *moral* legitimacy of what are clearly non-scientific claims – "there is no God," "we must have recourse to massive government intervention to stave off inevitable catastrophe," "embryonic stem cell research is the only *humane* and *responsible* option" – but also the scientific legitimacy of descriptive claims – "evolution of species proceeds by random mutation and natural selection," "there is significant and possibly harmful global warming," "research on human embryos offers certain technical advantages over research on alternatives such as adult stem cells" – is questioned or rejected out of a recoil at the overreaching of science.

The demands of the social ethics of science are thus not negligible, and they extend to the need for a reform of the relationship between science and society, a reform that recognizes the legitimate authority of science in some domains, but recognizes as well that this authority is limited, and situated within a broader set of obligations and expectations to which science, and scientists, are rightly subject.

The Internal Ethics of Science

We can go further still, however. It is clear, even from what was said earlier about the internal goods of science, that there must be an internal ethics of science corresponding to those goods. For the goods of a practice require virtues, excellences which make possible the successful pursuit of these goods, individually and with others, and the creative opening up of new avenues of achievement.

Few would doubt that there are indeed excellences shared by scientists that aid them in pursuit of their ends. Scientists are clever, imaginative, foresighted, and capable in many cases of difficult feats of ratiocination and abstraction. They have technical skills with their equipment and a specialized knowledge not shared by most. Yet these excellences are not *moral* excellences. Thus, the suspicion might exist that moral excellences can do no more than keep scientists from falling afoul of extra-scientific considerations. Science itself, the thought might go, is not as such benefited from any moral excellences.

Such an image of science could only be appropriate if science itself were to exist only within the domain of the technical, of what is manufactured by human beings for the pursuit of ends whose normative assessment lies outside the consideration of the artifact itself. The practice of science might be thought, on this picture, to be itself a kind of instrument or technique. It would, accordingly, have no internal ethics.

Yet science is centered around the pursuit of a genuine human good that is not identical to any goal that might be pursued in service of that good, and scientists can be successful only if they commit themselves to that pursuit in something like a vocational fashion: they must give that good a kind of organizational priority in shaping their lives and make many of their more limited choices with a view to how those choices cohere with the commitment to scientific truth itself. And, just as the commitment to a marriage, or to a religious or medical vocation demands a certain reshaping of an agent's moral character in line with the goods being pursued, so does the commitment to science.

Consider, for example, the importance of honesty and intellectual integrity for the scientist. Honesty is a virtue for everyone; but for the scientist, it must take on a special importance as *the* virtue of truth. Moreover, as the scientific community is bound together in that pursuit, and around that good, failures of honesty not only corrupt the scientific inquirer's vocational commitment; these failures eat away at the moral bonds holding the community together. Plagiarism, fraud, and lying in the dissemination of research thus spill over the boundaries of honesty and become failures of justice towards fellow scientists.

Similarly, intellectual courage is no mere technical skill, but reflects a spirited response to the various dangers to truth presented by society, self, and even the scientific establishment. Everyone *wants* the truth to be of a certain sort; it takes courage to see and accept the truth for what it is. Moderation establishes the desire for truth as greater than those desires whose satisfaction – in intemperate living – are incompatible with the life of contemplation, and those desires – for fame and fortune – whose satisfaction threatens to undermine the desire for truth itself. And discipline, named *studiositas* by Aquinas, steels the soul to the labor that is necessary if truth is to be obtained despite many obstacles. These are all moral virtues, possession of which is critical to good science.

Yet it might still seem that possession of all these virtues is compatible with the a-moralism of single-minded devotion to human embryo research, or even the depredations of Nazi science. Could not the Nazi scientist have possessed these virtues, and thus been *fully* "good scientists", yet still have been immoral in an extra-scientific sense? Could not research on human embryos continue by scientists genuinely possessed of all the virtues proper to science as such, with these scientists going wrong morally only as regards external and social considerations? The picture would look something like this: the Nazi scientist is, in a scientific sense, courageous, moderate, disciplined, and honest, even though, from a wider, fully moral, standpoint, he is a bad human being.

It is this question that I wish to address in the remaining discussion. For it seems to me that the answer to this question can't be an unqualified affirmative: "good science" is never so distant from the fully good, the truly virtuous, and the genuinely right as this picture of virtuous but not morally good science suggests.

First, consider again the relationship between science and society. Good science shapes and expresses the values of the society within which it lives, while at the same time being shaped and guided by those values. To move away from this paradigm is to move away from science as a profession, which is never good for science. But this means that in a society with a firm sense of the moral good and right, and in which science exists as a profession in a normatively appropriate way, *good* science will increasingly be morally upright science. Indeed, there is some reason to think that in a society with a morally and *religiously* upright character, science would exist as well in an amicable relationship to religious worship and truth, a relationship unlike, no doubt, that which exists between much contemporary science and religion.

Second, and even more importantly, recall the foundational desire of the scientist, the desire on which the vocational commitment to science was founded. This is the desire for *truth*. Was Plato not correct in claiming that one who really loves something loves the whole of it (*Republic*, 474c)? And that therefore one who loves the truth loves the whole of it, and entirely shuns falsehood?

But truth itself is a whole. There are many truths, but all, in the end, hang together in an ordered fashion. These truths, which together comprise the truth that is the overall object of the inquirer's passion, include practical truths, the truths of ethics. One cannot be a genuine lover of truth but be indifferent to the truths of ethics.

Thus the Nazi scientists' love of truth was, considered as such, deficient; they may have been lovers of truths (and here we should keep in mind Plato's distinction, at *Republic*, 475d, between the philosopher and the one who is merely a lover of sights and sounds), but they were not unconditional lovers of the truth. Indeed, we may see this in an assessment of how they stood in regard to some of the virtues of inquiry discussed earlier. The Nazi scientists were not honest: their work was carried out for the sake of ideological ends, and some of it was shaped accordingly; and while the Nazis achieved some genuine scientific insights, they also indulged in pseudo-scientific inquiry, often motivated by their racist presuppositions. Moreover, Nazi scientists can rightly be described as intellectual cowards, afraid to acknowledge plain truths about the human persons they subjected to their brutal experiments; and some were cowards in more straightforward senses, doing research because they were afraid of what would happen if they were to refuse. Or again, their lack of the virtues of inquiry shows forth in their methodology, which was often wasteful and inefficient by any reasonable scientific standards even apart from its obvious moral deficiencies.[8]

Concern for truth thus should radically shape the internal ethics of science, and indeed all inquiry. But if so, this raises, I believe, troubling questions about the extent to which scientific inquiry that involves human embryos has been adequately shaped by that concern. For the area of embryo research is full of the plain truths, both theoretical and practical, which I discussed earlier. The theoretical truth about embryos is that they are human beings, just as you and I are. And the practical truth is that, as human beings, those embryos are entitled to the moral respect due you or me; any other claim is arbitrary and unfair to those who are like us in every essential way. Yet even in its infancy, the field of embryo research has seen fraud, in the work of Hwang Woo Suk; dishonesty, as in the invention and marketing of the term "pre-embryo"; and partisanship presented as disinterested scientific "analysis." Scientists have participated in the political charade that cures are right around the corner, and they have willingly gone along with, and often encouraged, the myth that scientific expertise is

---

[8] For a helpful discussion of Nazi science which reveals many of these vices, see Evans 2009, Chapter 6. Nazi *medical* science was, of course, especially egregious.

as such sufficient to provide good policy guidance on issues that are intrinsically moral. None of this can be considered good science. It is not a question here of values being imposed upon science from outside, but of scientists failing in their vocation *as scientists*, as human beings vocationally committed to truth in all its forms.[9]

Conclusion

The implications of all this for the central question of this volume should be clear. Were science merely a technical pursuit, existing entirely within the realm of instrumental rationality, then perhaps there could be a science carried out by machines. But even the technical aspects of science must be governed externally by human moral norms.[10] And if my accounts of the social and internal ethics of science are correct, then science is not, and cannot be, a merely technical activity governed solely by considerations of instrumental rationality. It is, rather, fully human action, to be governed by sound practical reasoning, both individually and communally; so understood, its "human character" is essential to its continued existence and a professionalized set of practices.

References

Chapman, Dennis. 2007. Reprogramming the Debate: Stem-Cell Finding Alters Ethical Controversy. *University of Wisconsin-Madison News*, November 20. http://www.news.wisc.edu/14475. Accessed 5 July 2018.

Editorial Board of the *New York Times*. 2007. A Stem Cell Break-Through – No Thanks to Mr. Bush. December 6. http://theboard.blogs.nytimes.com/2007/12/06/a-stem-cell-breakthrough-no-thanks-to-mr-bush/. Accessed 5 July 2018.

Evans, Richard J. 2009. *The Third Reich at War*. New York: The Penguin Group.

Faden, Ruth, and Tom Beauchamp. 1986. *A History and Theory of Informed Consent*. New York: Oxford University Press.

George, Robert P., and Christopher Tollefsen. 2008. *Embryo: A Defense of Human Life*. New York: Doubleday.

Harris, John. 1999. The Concept of the Person and the Value of Life. *Journal of the Kennedy Institute of Ethics* 9: 293–308.

Hurlbut, Benjamin J. 2017. *Experiments in Democracy: Human Embryo Research and the Politics of Bioethics*. New York: Columbia University Press.

---

[9] This bleak verdict seems to me to be substantially confirmed by the work of J. Benjamin Hurlbut in his study of the history of human embryo research; see Hurlbut 2017. I should stress that while the failures of Nazi science are used to illustrate a point about the internal ethics of science, and while I think that contemporary human embryo research also involves moral failings, I do not intend to suggest any moral parity between Nazi researchers and contemporary scientists.

[10] Nor, of course, do I assume that even epistemically science is an entirely technical activity. But that question, though central to this volume, is beyond the scope of this paper.

Hurlbut, Benjamin J., and Jason Scott Robert. 2012. Stem Cells, Science, and Public Reasoning. *Journal of Policy Analysis and Management* 31: 707–714.

Jackson, Jennifer. 2006. *Ethics in Medicine*. Cambridge, UK: Polity.

Jones, James H. 1992. *Bad Blood: The Tuskegee Syphilis Experiment*, New and Expanded ed. New York: Free Press.

Krauthammer, Charles. 2007. Stem Cell Vindication. *Washington Post*, November 30. http://www.washingtonpost.com/wp-dyn/content/article/2007/11/29/AR2007112901878.html. Accessed 5 July 2018.

Lee, Patrick, Christopher Tollefsen, and Robert George. 2014. The Ontological Status of Embryos: A Response to Jason Morris. *Journal of Medicine and Philosophy* 39: 483–504.

MacIntyre, Alasdair. 1984. *After Virtue*, 2nd ed. Notre Dame: University of Notre Dame Press.

Moore, Keith L., and Persaud, T.V.N. 2003. *The Developing Human*, 7th ed. New York: W.B. Saunders.

Morris, Jason. 2012. Substance Ontology Cannot Determine the Moral Status of Embryos. *Journal of Medicine and Philosophy* 37: 331–350.

Obama, Barack. 2009. Remarks of President Barack Obama on Signing an Executive Order Removing Barriers to Responsible Scientific Research Involving Human Stem Cells and a Memorandum on Scientific Integrity, Washington, DC, March 9. http://www.presidency.ucsb.edu/ws/index.php?pid=85833. Accessed 5 July 2018.

Owen-Smith, Jason, Christopher Thomas Scott, and Jennifer B. McCormick. 2012. Democracy Is Working. *Journal of Policy Analysis and Management* 31: 726–728.

Smith, Barry, and Berit Brogaard. 2003. Sixteen Days. *Journal of Medicine and Philosophy* 28: 45–78.

Tollefsen, Christopher. 2008. *Biomedical Research and Beyond: Expanding the Ethics of Inquiry*. New York: Routledge.

———. 2011. Fetal Interests, Fetal Persons, and Human Goods. In *A Critical Analysis of Pro-Choice Arguments*, ed. Stephen Napier, 163–184. Dordrecht: Springer.

Cultivating Humanity in Bio- and Artificial Sciences

Mariachiara Tallacchini

The Humanity of Technoscience in Biotechnology

The question asked in this book, *Will Science Remain Human?*, focuses on the fact that contemporary science is primarily performed through technological means. This unprecedented technological apparatus is not only reducing some human roles, but seems to have the potential for altering some fundamental structures of science itself. The supply of data from all knowledge domains as well as the availability of artificial forms of reasoning, judging, and deciding is radically transforming how science has been working lately, and even more how it will operate in the future. From this perspective "a renewed understanding of the human character of science" (Bertolaso and Sterpetti, Introduction, Chap. 1, section "The theme of the volume") seems necessary as rationality tends to be constructed more and more as a sequence of machine-readable codes and a series of ultrafast processes for elaborating information. Moreover, the risk exists for scientific knowledge to be represented in a simplified way, hiding human responsibility, freedom, creativity and choice (Bertolaso and Sterpetti, Introduction, Chap. 1).

The question, however, also invites other perspectives on the current and future relations between the most valued forms of human cognition and practice—science or, more properly, technoscience, technology as inseparable from the scientific enterprise—and the increasing autonomous development of scientific and technological tools in the face of potentially new ways of defining humanness. Massive technological innovations in biomedical sciences have been challenging and redefining several aspects of life, including human life, for several decades; and further forms of biotechnology, gene editing, and synthetic biology will introduce even deeper transformative outcomes. Not only life sciences remain a crucial site for

M. Tallacchini (✉)
Dipartimento di Scienze Giuridiche, Università Cattolica del Sacro Cuore, Piacenza, Italy
e-mail: mariachiara.tallacchini@unicatt.it

M. Bertolaso, F. Sterpetti (eds.), *A Critical Reflection on Automated Science*, Human Perspectives in Health Sciences and Technology 1,
https://doi.org/10.1007/978-3-030-25001-0_15

reflection, as the engineering of life has led to a conception and application of technoscience seemingly disregarding its own deeply human nature. But also, relevant connections and similar trends exist, both at the theoretical and practical level, between the biological and artificial domains.

The following pages offer some reflections on the chapters in this book that are primarily considering the human character of science from the perspective of biological and biomedical sciences. More specifically, Benjamin Hurlbut, Alfredo Marcos, and Christopher Tollefsen's contributions have focused on this different aspect of the human decline in science. The Authors propose three different approaches to the human characters of science and what it means: the pervasive and multifaceted presence of ethics in science and its inextricability from the notion of good science—(Tollefsen, Chap. 14); a systemic perspective where science cannot be seen as separate from other social systems and where only a plurality of languages can legitimately contribute to figuring out the world we live in (Marcos, Chap. 13); science and politics as sites for co-production of knowledge and values (Hurlbut, Chap. 12).

Although the proposed philosophical (Tollefsen and Marcos, Chaps. 13 and 14) and sociological (Hurlbut, Chap. 12) accounts of the potential threats to the humanness of science follow different trajectories—and definitely diverge in their commitments to an ontological or a constructivist vision of science and humanity—, similar or complementary arguments run across their narratives: revealing that critical voices toward a potential loss of humanity through technology come from multiple and pluralistic perspectives—often under-represented in the public debate.

In these pages I will focus on some shared concerns and conclusions. First, the Authors converge in arguing that the technologies of life represent a major place where the humanity of science has been and is under threat. Second, they similarly highlight some assumptions that inform, connect, and circulate in, both life sciences and artificial sciences. Third, directly or indirectly they all argue that the question about "what kind of humanity" we should nurture and reinvigorate should be placed at the core of the human cognitive and political enterprise.

Technologies of Life and the Separation of Facts and Values

Even though the sciences of artificial life and artificial thinking have become the source and the main place for both epistemic and ethical concern, biotechnology has been the domain where scientific knowledge started revealing its potential loss of humanity and the first major expression of technoscience directly challenging humanness. Its evolution toward synthetic biology and gene editing confirmed and perhaps exacerbated this trend. The widely popularized expression "playing God" aimed at highlighting the dangerous underestimation of the uncertainties involved in intervening in complex living systems through discrete, surgical-like insertions, combination, and deletions of biological and genetic components. More specifically, the potential threat of altering specific, if not ontological, characters of human life, as well as of re-defining the beginning of what should count as human life.

The ideological use of the separation between facts and values in the practice of biological sciences has been critical to this end. Indeed, the logical distinction of descriptive and prescriptive statements, identified a few centuries ago as the impossibility to derive ethical guidance from natural events, has become the metaphoric signature of modernity. However, as Sheila Jasanoff has pointed out in looking at the relations between knowledge and normativity in biosciences regulation and policy, the "is and ought" of life and their practical separation in scientific work is much more recent and is strongly connected to the development (especially the economic development) of biotechnology (Jasanoff 2011).

The "ideological separation" of the "is-ought" distinction is seen as crucially connected to the process of dehumanization of science. According to Tollefsen, in biological applications (especially in relation to embryos) the argument of "the impropriety of imposing moral, political, or religious limits on scientific inquiry" (Tollefsen, Chap. 14, section "Introduction") has been widely evoked and has been used as a way to exclude values from discussions on the use of science. At the same time, the Author argues, morality is widely recognized as pervading science in a variety of forms–namely externally, socially, and internally— and, indeed, "morally bad" "good science" has been ruled out in situations such as the unethical experimentation in Nazi concentration camps. The presence of ethics in science, and its inextricability from it, shows that science is a human activity permeated by human concerns and norms; however, according to Tollefsen, the separation of ethics from science has been instrumental to opening the path to the use (and abuse) of human life in specific cases—such as embryos, whose humanity is denied. And science is dehumanized when it is turned against human beings.

But the path connecting the divorce between facts and values and a dehumanized science can be also framed within other visions. Marcos identifies the causes of the dehumanization of technoscience in an oversimplified ontology and in an erratic anthropology, swinging between nihilism and radical naturalism. Dehumanization is the consequence of denying the existence of a normative "human nature" through a reductionist and mistakenly objectivist vision of science. Instead, according to the Author, ethics is an expression of humanity and an inseparable part of being human; this is why it is structurally embedded in scientific thinking and practice.

Hurlbut's constructivist approach and coproductionist language introduces a further point of view, providing a sociological and science policy account of how the separation and purification of the language of facts from the language of values has taken place in the policy and regulatory process of biotechnology.

The meaning and practice of coproduction implies that scientific and regulatory processes are not separated in society, but they reciprocate in influencing and legitimizing themselves to allow new technological entities being stabilized in society. Coproduction offers a complex and dynamic vision of how things become what they are as it follows their development through the linguistic negotiations and the practical agreements made by scientists and policy makers. By looking at the relations, boundaries, and interfaces between knowledge and values in science and/as politics, this approach shows how facts and values become established among scientific communities, institutions, and citizens.

Hurlbut examines three different modes of the current discourses and practices in biotechnology and their role in reconfiguring the human. A first narrative concerns the normalization of risks through the idea that they can be contained and confined in controlled environments. The second consists of the re-description of biological entities—and their potentially unknown emergent properties— in engineering terms (e.g. the "brick" metaphor in synthetic biology). The third explores how the arguments about "serving versus violating" the threats to human integrity in human gene editing have been framed and used.

In all these cases, that have specifically characterized the US policy debate—but that can also be traced in slightly different terms in the European Union regulatory context (Jasanoff 2005)—, biotechnological development has been portrayed in neutral and reassuring terms, where risks are redefined as merely technical and technically governable issues, and the relevance of public debate and values has been managed to remain epistemically marginal.

In synthetic biology this reduction and normalization of risks has been performed by suggesting that the mechanistic language of engineering represents the most adequate description for new cells and organisms. Also, as a consequence of treating biotechnological unknowns just as technical issues, the authority to set both the scientific and political agenda has been widely deferred to scientists. The separation of the lab from nature and the reduction of questions of the "good" to questions of "technically manageable risks" have proved successful in creating the rhetoric of scientific authorities as the legitimate actors for social deliberation. At the same time, public discussion has been depicted as a threat to full human development. Human integrity, wellbeing and progress, Hurlbut argues, depend on a politics that legitimize technocracy in the name of the integrity of scientists.

Tollefsen and Marcos's philosophical and ontological account of a dehumanized science that has been deprived of its "very human nature" is thus enriched by Hurlbut's approach of coproduction, and by the narrative of how scientific discourse has succeeded in capturing and substituting the political and public debate. "The problem of political and normative judgement—of asking what sort of lives we should live, what sort of futures we should aspire to—is positioned as secondary to the question of what beneficent technological futures are possible; and the forms of action and aspiration that are appropriate to human lives and societies are figured as at once dependent upon, and given their fullest expression through science and technology" (Hurlbut, Chap. 12, section "From risk to telos"). Conversely, "a longstanding ethical limit, apparently grounded in a judgment about human integrity that seems to stand apart from and above the technological conditions of the moment, is declared untenable and unreasonable precisely because it has become technologically feasible to transgress it" (Hurlbut, Chap. 12, section "Figuring the human").

A common thread seems to emerge here through different theoretical routes. The externalization and the denial of a place for human values in "good"—namely valid or sound— science, the reduction of science from human activity to mere technique, and the marginalization of values reframed as controllable technical issues: all these conceptual maneuvers illustrate how facts and values have been instrumentally separated to make technoscience appear as a neutral endeavor.

Reductionist Assumptions in Life Sciences and Artificial Sciences

Biotechnology and its evolution are not the single place where science as a fully human activity has been challenged. Until recently, life sciences have been considered more ethically controversial than artificial sciences. While life sciences were directly intervening on the main components of nonhuman and human life, digital tools and artefacts have kept appearing much less problematic, with implications still far from becoming a threat to human life. Now, while life sciences continue to generate concerns, the artificial domain has clearly displayed its challenges.

Since its beginning biotechnology has been problematic from the regulatory point of view, due to its complex effects. This is why ethics has been called in quite early as a strategy to cope with potentially unpredictable impacts and the need to give a voice to broad public concerns (Jasanoff 2005). This has been definitely the case for the European Union, where four main directives were enacted, dedicated ethical institutions set up, and procedures for consulting the public introduced for the first time (Tallacchini 2015).

A completely different regulatory approach has been applied to the domain of ICT and artificial sciences, where privacy and data protection have been early identified as the main (and even single) values at risk. For a long time and quite uncontroversially the legal protection of personal data seemed to have taken care of all information and digital issues. No further need for reflecting on other ethical values was envisaged (Sabelli and Tallacchini 2018). This landscape has dramatically and rapidly changed in the past few years, when the emergence of other values (other fundamental rights and ideals of distributive justice) and the need for ethics in the digital domain have become suddenly urgent—even more due to the previous lack of awareness and reflection about the potential threats.

If biotechnology raised concerns about humans (and humanness) as an "object" of technology and control in altering and editing the genome, artificial sciences challenge humans also as the traditional "subjects" of knowledge. They ended the assumptions about the uniqueness of humans as rational beings: the most performative and efficient forms of computation are no longer human and the meaning of knowledge is now largely identified with the power of generating and elaborating data.

The two phenomena, however, are not unrelated. Not only is biotechnology widely and practically affected by artificial sciences, but similar theoretical assumptions underlie the vision of knowledge in both domains. Biological and artificial sciences are both informed by some specific elements in the vision of knowledge, such as a scientistic approach to reality, the turn from "knowing what" to "knowing how," and a computational understanding of reasoning.

Scientism can be defined as the vision that identifies science as the "language of reality" (Jasanoff 2016). This vision has several implications and shades of meaning. It may imply that all languages other than "science"—still portrayed as a single all-encompassing term— should be reduced to scientific language in order to be relevant, both in epistemic terms and when used for social deliberations. Also, scientism is the ideology according to which everything that is scientifically (and technologically) do-able should be done.

In legal and policy contexts, scientism informs the argument according to which only evidence-based policy and legislation can be considered transparent, accountable, and legitimate. Purifying science policy from disagreements based on values and depicting science as the solution to social problems represents a straightforward strategy to annihilate public discussion and deliberation.

The attempt to make law and politics more reliable, namely less subjective and arbitrary, by reference to the certainty and objectivity of science and scientific method has characterized a significant part of philosophy of law since the origin of the modern State and the construction of law as a scientific domain (Ezrahi 1990). Reference to sciences and their presumed evidence as a strategy to make normative statements more objectively founded, and therefore acceptable, has increasingly dominated the vision of what can be arguably considered as rational in legislation and politics.

This tendency, cutting across all technological domains, is currently taking the shape of what is called "datafication," namely the reduction of knowledge to data. The social effects of this form of reductionism—triggered by the digital domain, but that has equally colonized biology—, was already clearly identified in 1996 by legal scholar James Boyle in his essay *Shamans, Software and Spleens: Law and the Construction of the Information Society*, where the inevitable links between the biological and digital worlds were already unveiled. "The tendency is toward the economic and conceptual separation of the informational message from the medium—cells, diskettes, telephone directories, or whatever—and of the progressive devaluation (literally, the diminishing marginal cost) of the medium as compared with the message" (Boyle 1996, p. 7).[1] In the legal and economic domains and in discussing the allocation of intellectual property rights, Boyle showed how courts' decisions in favor of "patenting life" were increasingly constructing and legitimizing information as more relevant than "raw materials," both inorganic and organic.

According to Marcos, datafication of knowledge and life is leading to an impoverishment of science and is depriving science and technology of a fully human, meaningful purpose. "This understanding of science threatens to marginalize all that is truly human, all that goes beyond the mere registry and combination of elements; and therefore, all genuine creativity, emotions, meaning, significance and values —even moral and aesthetic values—, reflection, and dialogue. Data collection is undoubtedly a significant step in the progress of science, but a science reduced to gathering and combining data would be an effort as expensive as useless" (Marcos, Chap. 13, section "The symptoms of the problem"). In the Author's perspective, isolating data from the systemic contexts where technoscience operates is also a distortion of a sound vision of science: in fact, several applications of technoscience may transform nature in ways that do not meet the necessary requirements of a human ecology.

[1] The title refers to the landmark judicial case of John Moore, whose spleen gave rise to a cell-line that was patented as an intellectual invention owned by the researcher who used Moore's biological materials (see *Moore v. Regents of University of California*, 51 Cal. 3d (1990)).

This scientistic vision of knowledge as data is also influencing what kind of questions can be legitimately asked. The broad implications of science and technology—namely what kind of society we would like to set up, what kind of humanity we are nurturing, what kind of institutions we should commit to—is substituted with statements about the "know how," namely how to allow technoscience to proceed by removing the obstacles to its luminous progress. This passage from "knowing" to "knowing how" is another expression of a narrow epistemic perspective.

A further major consequence of this reduction of knowledge to data concerns the interpretation and simplification of reasoning and rationality as mechanistic computational processes. Computation appears as an objective way of looking at how the mind works. Algorithms and algorithmic forms of reasoning and learning seem to be able to provide a more objective foundation also for social decision-making.

Indeed, the tendency toward considering thinking and computing as synonyms is spreading from science to the regulatory domain. The concept of "smart regulation" has been transnationally adopted as the current mode to reframe and replace the traditional value of the "rationality of the law" and of the legislator (EC 2010; Drahos 2017). Law-making is "smart" when it operates as a mechanistic step-by-step program that imitates machine functioning. Smart regulation will prospectively take place by substituting human regulators with automatic execution. As a complement to the automatization of the legislator, the "smart" legal program is also converting citizens into "nudgeable" objects of norms as determined by behavioral sciences.

For Science to Remain Human: Normatively Defining Human Nature or Cultivating Human Skills?

Will science remain human? Tollefsen sees the potential dehumanization of science mostly as an abstract exercise: science is not a technical endeavor dominated by instrumental rationality, but a "fully human action, to be governed by sound practical reasoning, both individually and communally" (Tollefsen, Chap. 14, section "Conclusion"). "(I)f good science has the characteristics I describe in this paper, then the project of science 'populated and exercised by machines' is chimerical. Good science as science is indeed a human activity permeated by human concerns and norms" (Tollefsen, Chap. 14, section "Introduction").

The threat for science to be dehumanized appears quite realistic to Marcos, both in epistemic and ethical terms. In fact, the reductionist lines along which knowledge is interpreted and shaped provide the premises for how technoscience is produced and practiced. Against a humanly impoverished science Marcos suggests three integrated strategies: the adoption of a pluralist ontology, the recognition of science as a fully human activity, and a vision of technoscience serving a meaningful human life. A "pluralist ontology," namely "the promotion of a plurality of scientific methods" and "the acceptance of a diversity of legitimate epistemic sources" interacting with scientific knowledge would avoid the narrow vision of science. Instead, the mechanization and automatization of scientific procedures inevitably favor an unreflective attitude towards knowledge as mere computation rather than human dialogue.

The roles of humans and humanness cannot be taken for granted as intangible and ontological characters of science, but need to be actively maintained if science has to remain fully human.

The current vision of technoscience is forgetting that this is an activity defined by specific limits—in terms of methods and meanings— and therefore that there should be room for other discourses. One of the most valued aspects of scientific knowledge, objectivity, is radically misconceived when it is represented as the suppression of all personal and subjective dimensions, and when it is equalized to automatically executable methods. On the contrary, the human ability to formulate scientific theories and hypotheses may depend on forms of personal knowledge not related to science, and even less to data. Indeed, art and philosophy can nurture these skills. This is because the whole person is involved in doing science. And only a wholesome conception of humanness provides technoscience with a meaning and a purpose.

To this end Marcos argues in favor of a guiding concept of human nature, a normative vision of the human that can act as a protecting criterion against the danger of an uncritical use of anthropotechnics. In the light of this anthropology, technoscience would be bound to keeping human life truly human without attempting to transcend or enhance what is human (Marcos, Chap. 13). However, a normative vision of human nature can be problematic if only authoritatively asserted.

In Hurlbut's STS and coproductionist account of science and humanness the answer about maintaining science human is related to "our" ability as human beings to integrate and legitimize a multiplicity of discourses and languages respecting the varieties of human voices and values in shared, public arena.

Even though in pluralistic and complex societies evoking a "we" is not free from ambiguities, the attempt to open up possible new paths for dialogue, confrontation, and decisions—instead of closing them down through a single, pre-identified solution—proceeds along the more acceptable lines of democratic societies. "The question of whether science will remain human depends in profound ways upon the role that scientific authority plays in the way the human is figured" (Hurlbut, Chap. 12, section "What science, which human?"). The language of science is currently politically endowed with the power to dismiss how the "we as human beings should speak and think about the human condition" (Hurlbut, Chap. 12, section "Figuring the human").

Even from a very mundane perspective and in the absence of strong ontological commitments, some fundamental human capabilities and skills can and need to be actively practiced and maintained. The potential for unlearning specific forms of knowledge and actions not mediated by digital means and forms of discourse other than science, as well as the possibility to forget and loose specific ethical abilities such as making moral choices or bearing responsibilities, are very real threats and are already happening.

In particular, self-reflexivity—as the capability to step back from our cognitive-and-value assumptions to reconsider them with a fresh look—is an invaluable human character that, at the individual and collective level, should be trained and cultivated. In this respect the use of multiple "technologies of humility" to connect science, society, and institutions (Jasanoff 2003) may favor humble self-reflexivity. Technology of humility is the expression introduced by Sheila Jasanoff to refer to public discursive practices around technoscientific options and social decisions. Technologies of humility may be relevant here in creating conditions (time and space) for debate; in performing public assessments of framing, vulnerability, distribution, and learning; in envisaging forms of transparency, openness, and participatory design for social algorithms; in discussing trusted scientific models and technologies; in introducing extended peer-review; in acquiring all relevant knowledge dispersed in society for decision-making; and, in general, in democratizing science as a way to increase awareness (Liberatore and Funtowicz 2003; Hilgartner et al. 2015).

This approach is even more relevant in relation to artificial sciences and machine learning algorithms, where, faced with the complexities and autonomous processes of data acquisition and elaboration, also experts cannot be able to fully understand the knowledge outcomes even by performing reverse engineering. Automated thinking has already introduced forms of reasoning very distant from human thought—with the currents attempts to re-orient human reasoning towards more machine-like processes—and involves the challenge of moving towards our inability to understand and interpret machine-functioning according to human-like meanings, categories, contexts.

Maintaining the full humanity of science and politics implies that social decisions about technoscience should be discussed in the wider political arena, where the human community decides through "normal expression in speech and thought" (Arendt 1958). Hannah Arendt's quote from *The Human Condition*—repeatedly evoked in Hurlbut's contribution—should remind us that our shared humanity has to remain at the core of our politics.

Can "we" humans converge and agree about the sort of humanity we need to cultivate and nurture? And, will democratic societies develop the ability to preserve and encourage a real plurality of discourses and languages for public dialogues on technoscientific directions?

Several recent declarations and recommendations on AI, big data, and algorithms have similarly called for these normal expressions of speech and thought. These documents ask for "value alignment," "widely shared ethical ideals," "responsibility" and compatibility "with ideals of human dignity, rights, freedoms, and cultural diversity" (Future of Life Institute 2017); or make reference to the "fundamental ethical principles, based on the values laid down in the EU Treaties and the EU Charter of Fundamental Rights" (EGE 2018); or invoke the concept of "human flourishing" as the overarching principle that should guide future research and technoscientific development (British Academy and the Royal Society 2017).

The principle of human flourishing is defined as including "wellbeing and the need for individuals and communities to thrive, but it is deliberately broad." This is because "(a)t moments of contention, the principle should serve to reflect the fundamental tenet that society does not serve data but that data should be used to serve human communities" (British Academy and the Royal Society 2017, p. 9).

In the light of this definition, proposed not as an answer, but as a direction, for the question asked in this workshop, human flourishing may be interpreted as the need for a self-reflexive attitude, always ready to revisit and discuss which human characters should be maintained as human as possible.

References

Arendt, Hannah. 1958. *The Human Condition*. Chicago, IL: University of Chicago Press.

Boyle, James. 1996. *Shamans, Software and Spleens: Law and the Construction of the Information Society*. Cambridge MA: Harvard University Press.

British Academy and the Royal Society. 2017. *Data Management and Use: Governance in the 21st Century a Joint Report by the British Academy and the Royal Society*. https://royalsociety.org/~/media/policy/projects/data-governance/data-management-governance.pdf. Accessed 18 July 2018.

Drahos, Peter, ed. 2017. *Regulatory Theory. Foundations and Applications*. Acton: ANU Press.

EC (European Commission). 2010. *Communication from the Commission to the European Parliament, the Council, the European Economic and Social Committee and the Committee of the Regions, Smart Regulation in the European Union*. Brussels, COM (2010) 543 final.

EGE (European Group on Ethics in Science and New Technologies). 2018. *Statement on Artificial Intelligence, Robotics and 'Autonomous' Systems*. Brussels: European Commission.

Ezrahi, Yaron. 1990. *The Descent of Icarus*. Cambridge, MA: Harvard University Press.

Future of Life Institute. 2017. Asilomar AI Principles, *Asilomar*, January 2–5. https://futureoflife.org/ai-principles. Accessed 18 July 2018.

Hilgartner, Stephen, Clark Miller, and Rob Hagendijk, eds. 2015. *Science and Democracy. Making Knowledge and Making Power in the Biosciences and Beyond*. London/New York: Routledge.

Jasanoff, Sheila. 2003. Technologies of Humility: Citizen Participation in Governing Science. *Minerva* 41: 223–244.

———. 2005. *Designs on Nature. Science and Democracy in Europe and the United States*. Princeton, NJ: Princeton University Press.

———, ed. 2011. *Reframing Rights. Bioconstitutionalism in the Genetic Age*. Cambridge, MA: MIT Press.

———. 2016. *The Ethics of Invention. Technology and the Human Future*. New York/London: W.W. Norton & Company.

Liberatore, Angela, and Silvio Funtowicz. 2003. 'Democratising' Expertise, 'Expertising' Democracy: What Does This Mean, and Why Bother? *Science and Public Policy* 30 (3): 146–150.

Sabelli, Chiara, and Mariachiara Tallacchini. 2018. From Privacy to Algorithms' Fairness. In *Privacy and Identity Management. The Smart Revolution*, ed. Marit Hansen, Eleni Kosta, Igor Nai-Fovino, and Simone Fischer-Hübner, 86–110. Cham: Springer.

Tallacchini, Mariachiara. 2015. From Biobanks to Genetic Digital Networks: Why official pre-identified values may not work. In *Science, Philosophy and Sustainability. The End of the Cartesian Dream*, ed. Ângela Guimaraes Pereira and Silvio Funtowicz, 98–111. London/New York: Routledge.

The manufacturer's authorised representative in the EU is Springer Nature Customer Service Centre GmbH, Europaplatz 3, 69115 Heidelberg, Germany. If you have any concerns regarding our products, please contact ProductSafety@springernature.com

Printed and bound by CPI Group (UK) Ltd, Croydon, CR0 4YY
30/09/2026
02225237-0007